Industrial Maintenance Management

INDUSTRIAL MAINTENANCE MANAGEMENT
Ninth Edition

Dedicated to effective industrial maintenance management

KENDALL/HUNT PUBLISHING COMPANY
4050 Westmark Drive Dubuque, Iowa 52002

Graphics © Corel Corporation/Image Club Graphics

Copyright © 1995 by Paul D. Tomlingson

ISBN 0-7872-0202-9

All rights reserved. No part of this publication may be reproduced, stored in a retrieval system, or transmitted, in any form or by any means, electronic, mechanical, photocopying, recording, or otherwise, without the prior written permission of the copyright owner.

Printed in the United States of America.
10 9 8 7 6 5 4 3 2 1

Contents

Preface	iii
Introduction	iv
About the Author	vii

Section I - The Foundation of Effective Maintenance

1 The Production Strategy Guides the Maintenance Program	1
2 A Maintenance Objective to Support the Strategy	5
3 Establishing Policies for the Conduct of Maintenance	7
4 Adhering to the Principles of Maintenance Management	11
5 Developing an Effective Organization	19
6 Securing Management, Operations and Staff Support	27
7 Defining the Maintenance Program	29
8 Educating Personnel on the Maintenance Program	33
9 Specifying the Maintenance Work Load	37
10 Establishing Maintenance Terminology	41
11 Achieving Productivity and Performance	43
12 Evaluating to Improve Performance	49

Section II - Maintenance Organization in Transition

13 Examining Organizational Options	59
14 Implementing Successful Team Organizations	75
15 Making Effective Organizational Changes	83
16 Downsizing the Maintenance Organization	91
17 Essential Duties of Key Maintenance Personnel	97
18 Carrying Out Craft Training	105
19 Conducting Supervisory and Staff Training	115

Section III - Tuning up the Maintenance Program

20 How to Define the Maintenance Program	123
21 Examining Preventive and Predictive Maintenance Essentials	141
22 Conducting Effective Planning	155
23 Forecasting and the Component Replacement Program	163
24 Performing Effective Scheduling and Work Control	169
25 Conducting Essential Maintenance Engineering	177
26 Developing and Utilizing Standards	183
27 Conducting Shop Operations and Support Services	193

Contents

28	Achieving Effective Material Control	197
29	Conducting Non maintenance Project Work	205
30	Establishing Information to Support Maintenance	209
31	Conducting Mobile Equipment Maintenance	217

Section IV - Maintenance Management Information Systems

32	Conceptualizing Information Needs	227
33	Developing the Information System	241
34	Implementing the Information System	247
35	Work Order Systems	259
36	Utilizing the Information System Effectively	279

Section V - Supporting Maintenance

37	How Management Can Help	323
38	The Operations Role in Successful Maintenance	327
39	Staff Support for Maintenance	333

Section VI - Achieving Productivity and Performance

40	Measuring Maintenance Productivity	335
41	Establishing the Essential Performance Indices	339

Section VII - Evaluating to Sustain Effectiveness

42	Conducting Effective Evaluations	345
43	Converting Evaluation Results into Improvement Actions	371

Appendices:

A	The Maintenance Objective	375
B	The Maintenance Work load	377
C	Maintenance Terminology	379
D	Maintenance Policies	387
E	Duties of Key Maintenance Personnel	391
F	Predictive Maintenance Techniques	393
G	Maintenance Performance Evaluation	409

List of Illustrations	443
Index	449

Preface

Industrial Maintenance Management is a constructive guide for developing, implementing and improving the maintenance organization and its program in the industrial environment.

From plant manager to operator and craftsman, *Industrial Maintenance Management* shows how to enhance the maintenance organization and its performance.

Combining savvy and practical ideas of maintenance programs, organization, people and control, *Industrial Maintenance Management* tells how to ensure plant maintenance will contribute successfully to profitability.

Introduction

There are some facts about industrial maintenance that must be taken into account concerning its management.

- Maintenance is the largest controllable cost in most plants. It often represents over 35 % of operating costs. If not controlled, a plant cannot expect to be profitable.
- Maintenance labor exists to install materials. Therefore, the efficiency with which maintenance personnel install materials is a primary factor in their ability to control costs. Thus, the maintenance organization and its program must ensure productive utilization of labor.
- Unless planning is first rate, maintenance won't be able to install materials effectively and reduce or control costs. Therefore, maintenance must ensure effective planning.
- Planning is only possible if there is a quality PM program. When PM inspections and testing are inadequate, deficiencies are not uncovered in time to plan the resulting work. Instead, emergency repairs predominate. Therefore, a technically competent PM program must exist and its services carried out properly if planning is to be done effectively.
- Many observers of the industry have concluded that the overall cost of plant maintenance will not go down unless it is done with fewer people and done less often. Thus, smaller, more efficient maintenance organizations become requirements for survival in today's competitive industrial world.

Fewer people means downsizing. Doing maintenance less often means improving the maintenance program.

Thus, to reduce costs and sustain profitability, many plants will be faced with the need to downsize the maintenance organization and improve their maintenance programs. They must emerge from the downsizing experience with a smaller, more efficient organization with a better program. In the process, they must also observe issues of job security, alternate use of redundant personnel and the intelligent use of attrition strategies.

Introduction

Thus, organizational change, new maintenance organization arrangements like Total Productive Maintenance (TPM), the use of teams and better information must guarantee more effective maintenance.

Unfortunately, the adoption of new organizations for industrial maintenance, brought about by the need to downsize, will not be easy because of long standing behavior that must be changed. In most maintenance organizations, for example, maintenance supervisors are selected solely on the basis of their technical skills. As a result, they focus on fixing things rather than on effective labor control. Then, all manpower is assigned to the supervisor resulting in few controls as to how they utilize their crews. It follows that, if such a supervisor becomes superintendent, as many will, his instincts may detract further from the plant's aspirations for more effective control of work. For these reasons, the education of all supervisors is a key factor in any successful improvement effort.

Teams are a consistent organizational objective for many plants as they seek to downsize, retain effectiveness and trim costs. However, moving from a traditional craft organization to a team will be difficult for most plant maintenance organizations. There is often no concept of how the team should operate. Everything is new to the craftsman, supervisor and planner as they emerge from the craft organization and attempt a team approach. Some supervisors, because the team may have no appointed leader, may even see a team as a treat to their job security and resist its implementation.

Logically, there must be an interim step between today's craft organizations and the team organization of tomorrow. Some plants are using area organizations in which one supervisor is responsible for all maintenance within a reasonably sized geographical area. He is given the necessary craft resources to accomplish the work. Thus, the combining of crafts in the same crew to improve productivity may spell the end of traditional craft organization. Next, the supervisor will realize that, with total responsibility for an area, he can no longer say "its the electricians' fault". In an area organization, for example, the supervisor's total activity can be budgeted and controlled. Therefore, the work order system he used to ignore is now suddenly useful and necessary. Soon, he realizes that leadership must displace his full attention to only the technical aspects of jobs.

Some plants have pushed area maintenance a step further by adding a craft pool. Day to day maintenance remains the responsibility of the area supervisor but, manpower for shutdowns is centrally controlled.

Introduction

Thus, labor is allocated against work priorities and able to be used more productively. It is no longer exclusively in the hands of the supervisors who may lack an appreciation of good labor control and productivity.

Steps like this will start the behavioral changes that help to make the team organization more successful. Considerations like this must be built into the organizational development process.

People Considerations

Assuming that the basic organization was agreed on, a new program developed and an improved information system were being implemented, the biggest, most difficult problem remaining would be dealing with resistance to change. People must be convinced that the changes are beneficial, first to them as individuals, then to their immediate work groups (crews) and finally to the plant. The most effective ways to overcome resistance to change are:

- Education - Let people know what the changes are and how they will be affected.
- Participation - Give personnel an opportunity to think through the changes and make alternate proposals that will get the job done.
- Time - Give people an opportunity to absorb the changes.
- Security - Change will not be successful unless both salaried and hourly personnel are assured that their jobs, thus, their economic security will remain intact.

Most maintenance personnel are aware of some pending change. However, if they are not informed on the nature of the changes, they will be uncertain how they are affected. Thus, resistance rather than help may be their response. Their resistance will be in proportion to their degree of knowledge about the changes. Therefore, the sooner the education process begins, the greater are the chances of overcoming resistance and getting on with beneficial changes. It will be important, in this era that dictates change, for the plant to establish how the changes will be implemented. Therefore, getting people involved and securing their informed participation and support will yield positive, beneficial change.

About the Author

Paul D. Tomlingson is a management consultant specializing in the design, implementation, and evaluation of maintenance management programs for industry. He is a veteran of 24 years of world-wide maintenance management consulting, the author of three textbooks and over 100 published trade journal articles on maintenance management. Tomlingson is a 1953 graduate of West Point and in addition to a BS in Engineering, he holds an MA in Government and an MBA, both from the University of New Hampshire. In addition to the presentation of his own public and on site seminars, he has appeared in seminars at the University of Wisconsin, Hofstra University, the University of Denver, National and Regional Plant Engineering Shows, and the Society of Mining Engineers Conferences. Mr. Tomlingson has been listed in *Who's Who in the West*.

I

The Foundation of Effective Maintenance

1

The Production Strategy Guides the Maintenance Program

The Production Strategy

The production strategy is the plan for attaining the profitability of the plant. It provides a mission for the total organization and assigns specific objectives to operating, maintenance and staff departments. The strategy also provides policies or guidelines for day to day department operations. Based on this guidance, individual departments develop procedures for carrying out their services and advising other departments how to obtain or utilize the services they provide.

The plant manager is the developer of the production strategy since he is ultimately responsible for assuring the plant's profitability.

While the primary intent of the production strategy is to assure effective daily operation of the plant and its profitability, the manager also addresses activities such as:

- The expenditure of capital funds
- The use of contract services

As the plant operates, he utilizes information to judge the performance of individual or collective departments in meeting the overall mission of the plant. In addition to monitoring how well the plant is operating by observing cost and performance, the manager visits field operations and staff activities frequently to obtain a first hand picture.

Maintenance and the Plant Manager

Because maintenance is a service, they depend on operations, for example, to make equipment available so that required services or repairs can be carried out on time. Similarly, maintenance depends on warehousing or purchasing to provide dependable service so that materials will be available when needed.

Not surprisingly, maintenance cannot compel operations to follow its program, nor can they demand services from material control departments. Therefore, there must be sufficient clarity in the industrial manager's guidelines to impress all departments that the success of maintenance lies in their understanding and cooperation. Through his actions, the manager creates an environment in which the maintenance program can be successful.

The Plant Manager's Role - The plant manager ensures that the maintenance program will be successful by:

1. Assigning maintenance a clear objective to guide their actions within the production strategy.
2. Providing policies or guidelines, applicable to all departments, explaining how maintenance is to be performed and how other departments are to support or cooperate with them.
3. Requiring that maintenance develop a program which, when properly executed, will ensure the reliability and maintainability of production equipment and facilities. As necessary, he assigns maintenance responsibilities for buildings or outside areas. He reviews and approves the program.
4. Ensuring that maintenance documents its program properly and takes the initiative in educating operations, management and staff departments on its provisions.
5. Requiring that maintenance organize themselves to yield a well led, responsive work force capable of executing the maintenance program effectively.

Monitoring Maintenance Performance - A portion of the plant manager's task is his continuous, critical assessment of maintenance performance. His assessments of are not limited to formal computer generated information. They include both the objective and the subjective.

The Production Strategy Guides the Maintenance Program

Typical indices of maintenance performance:

- Is compliance with the PM program at least 85% ?
- If not, did operations fail to make equipment available or did maintenance fail to carry out the prescribed services on time ?
- What is the labor cost to install each dollar of material ?
- Has maintenance measured labor productivity ?
- When was the last evaluation of maintenance performance ?
- What has been the progress in developing standard bills of materials for all major component replacements ?
- When will the equipment data for the new information system be completed ?
- What is the progress in testing the first team organization in the concentrator ?

Interest in Maintenance Performance - Each of these elements of information is related to the plant's production strategy and the manager must be aware of the contribution of maintenance toward meeting the strategy. Maintenance is the plant's largest controllable cost. Therefore, its performance and costs must be monitored. But, often plant managers do not take sufficient interest in maintenance performance. There are several explanations:

- Many have limited maintenance backgrounds. They often come from operations or plant engineering. Thus, their knowledge of maintenance may be insufficient to question their performance.
- Other activities may command their time and attention causing them to leave maintenance to their own devices.

This should be corrected through more involvement in maintenance matters such as attending the weekly scheduling meeting or visiting a shutdown area.

Helping Maintenance Leadership - Some plant managers assume that the maintenance leadership needs no help because they have been doing it for so long. This can be a mistake. The majority of maintenance superintendents have always been in maintenance and inbreeding is common. Many superintendents have risen from craftsmen to supervisors to planners and general supervisors on the way to becoming superintendent. While these are good, practical people, they may not have had training in managing large organizations with complex programs. Therefore, a plant manager is well advised to examine the managerial backgrounds of key maintenance personnel to determine whether his guidance is needed.

Operations Support - In similar fashion, the manager's production strategy must also assess operation's support of maintenance. Are there are any key operating personnel who see meeting production targets as their primary reason for existence ? Unless, operations can assure the plant manager that equipment is operated properly and that they understand their responsibility for the effective use of maintenance services, operations may thwart the maintenance effort unintentionally. Consider reversing this by:

- Making operations responsible for the cost of maintenance
- Having operations explain why equipment missed PM services

Staff Support - Likewise, the plant manager must assess how well staff departments support maintenance. To illustrate:

- How many major jobs were not done because purchasing failed to deliver materials on time ?
- What good is a cost report that is two weeks late ?

Summary- The plant manager uses the production strategy to express the mission of the plant. He then assigns specific, mutually supporting objectives to each plant department to guide them in the support of that mission. Then, through policies or guidelines, he creates the basis for day to day procedures.

For maintenance specifically, he:

- Verifies the effectiveness of the maintenance organization
- Approves the maintenance program
- Ensures operations is responsible for using maintenance effectively
- Ensures staff departments support maintenance with good service

The overall effect is to create a favorable working environment in which maintenance, as a service, can successfully carry out its program.

2

A Maintenance Objective to Support the Strategy

The Maintenance Objective

Within his production strategy, the plant manager develops an overall objective. Departments then link the specific tasks implied by the their objectives with the plant's objective. Department objectives are mutually supporting and consistent with the plant's production strategy.

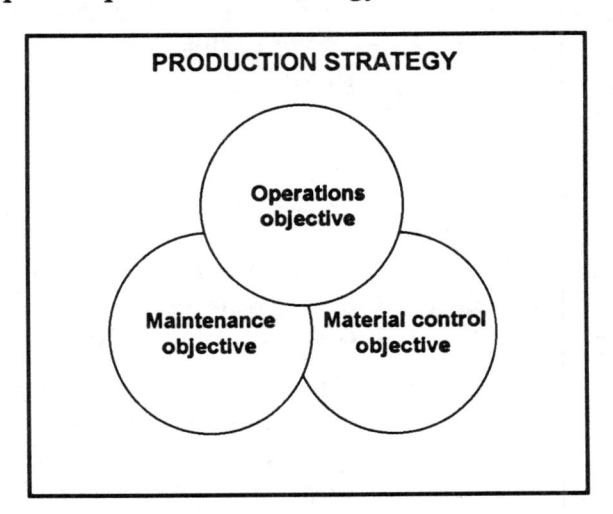

```
PRODUCTION STRATEGY

        Operations
        objective

  Maintenance    Material control
  objective      objective
```

Typically, the objective assigned to maintenance includes the following language:

Objective - The primary objective of maintenance is the repair and upkeep of production equipment to keep it in a safe, effective, as designed, operating condition so that production targets can be met on time and at least cost.

Often, there are supplemental objectives. For example, a secondary objective of maintenance is to perform approved, properly engineered and correctly funded non maintenance work (such as construction and equipment installation) to the extent that such work does not reduce the capability for carrying out the maintenance program. In addition, maintenance will operate support facilities (such as hoist operation). Maintenance will also monitor the satisfactory performance of contractor services.

Interpreting the Objective - There are several key phrases used in the illustrated objective. A primary as well as a secondary objective are provided to establish precedence. Production equipment maintenance is first while, project work follows, resources permitting. The use of phrases like 'as designed' means that equipment modification is excluded from the primary maintenance task (it isn't maintenance). The plant manager would expect that all non maintenance work be approved, properly engineered and correctly funded. He also intends that such work only be done if it does not reduce the capability of maintenance to meet its primary objective of carrying out the maintenance program. The objective further states that, if maintenance is to perform operating functions, such as hoist operation, they will be properly staffed to do so. With an objective such as that illustrated, there is little question about priorities and limits as to what and how much maintenance can do. Maintenance knows its responsibilities and organizes to carry them out. Operations is also aware of maintenance limitations and requests support accordingly. The objective ensures that the maintenance program will be carried out effectively.

The absence of a clear objective can yield unfortunate consequences. Typical:

In a large manufacturing plant, the 210 man maintenance work force was having difficulty carrying out the maintenance program. During shutdowns it was necessary for them to be augmented by contractors to catch up. An investigation showed that 32% of work requests by production were not maintenance. Rather, they were equipment modifications or installations. As a result, 21% of maintenance manpower was used on this work, detracting from manpower available for basic maintenance. The plant manager clarified the maintenance objective and educated personnel in the proper use of the maintenance work force, correcting the problem.

Summary - The assignment of an objective provides a clear target for each department and clarification among departments as to their common goal.

3

Establishing Policies for the Conduct of Maintenance

Maintenance Policies

Maintenance policies are ground rules established by plant managers to preclude misunderstanding of the roles and responsibilities of key departments. Adherence to the policies assures that each department contributes to profitability within the plant's production strategy. In turn, policies are the basis of day to day procedures for each department as they provide or obtain services. Typical policies are suggested below as they relate to specific activities such as the maintenance program or the control of labor.

On department relationships:

Operations will be responsible for the effective utilization of maintenance services.

Maintenance will be responsible for developing a pertinent maintenance program, educating personnel on its elements and carrying it out diligently. They will also make effective use of resources and ensure that quality work is performed.

Each department manager will ensure compliance with the policies covering the conduct of maintenance.

Each department will develop and publish procedures by which other departments may obtain its services.

On the maintenance program:

Maintenance will publish work load definitions and appropriate terminology to ensure their understanding and proper utilization.

Maintenance will conduct a detection oriented preventive maintenance (PM) program. The program will include equipment inspection, condition monitoring and testing to help avoid premature failure. The PM program will also provide lubrication, servicing, cleaning, adjusting and minor component replacement to help extend equipment life.

PM will take precedence over every aspect of maintenance except bona fide emergency work.

No major repairs will be initiated until the PM program has established the exact condition of the equipment and elements of the repair have been correctly prioritized. In all instances, major repairs will be subjected to planning procedures, unless an emergency repair is indicated.

The work order system will be used to request and control work.

Planning and scheduling will be applied to comprehensive jobs (e.g. overhauls, major component replacements, etc.,) to ensure that work is well organized in advance, properly scheduled and completed productively and expeditiously.

Maintenance will publish a priority setting procedure which allows other departments to communicate the seriousness of work and maintenance to effectively allocate its resources. The procedure will facilitate the assignment of the relative importance of jobs and the time within which the jobs should be completed.

Maintenance will develop and use information concerning the utilization of labor, the status of work, backlog, cost and repair history to ensure effective control of its activities and related economic decisions such as equipment replacement. Minimum necessary administrative information will be developed and used. Performance indices will be used to evaluate short term accomplishments and long term trends.

Maintenance will return unused stock materials to the warehouse. Maintenance will not attempt to store such materials.

Establishing Policies for the Conduct of Maintenance

On the control of labor:

The maintenance workload will be measured on a regular basis to help determine the proper size and craft composition of the work force.

The productivity of maintenance will be measured on a regular, continuing basis to monitor progress in improving the control of labor.

Every effort will be made to implement and utilize organizational and management techniques like teams or Total Productive Maintenance (TPM) to ensure the most productive use of maintenance personnel.

On maintenance engineering:

Maintenance engineering will be emphasized to ensure the maintainability and reliability of equipment.

Current technology will be utilized to facilitate effective maintenance.

No equipment will be modified without the concurrence of maintenance engineering.

All new equipment installations will be reviewed by maintenance engineering to ensure their subsequent maintainability.

The overall preventive maintenance program will be assessed annually by maintenance engineering. They will ensure that it covers all equipment requiring services and that the most appropriate types of services are applied at the correct intervals. The performance of the PM program in reducing equipment failures and extending equipment life will be verified.

On material control:

Procedures for obtaining stock and purchased materials or services will be strictly adhered to.

Parts will not be removed from any unit of equipment and used to restore another unit to operating condition without explicit authorization from the maintenance superintendent.

On non maintenance work:

Engineering, operations and maintenance are jointly responsible for ensuring that all non maintenance projects (construction, modification, equipment installation etc.,) are necessary, feasible, properly engineered and correctly funded before work commences.

Maintenance is authorized to perform project work such as: construction, modification, equipment installation and relocation only when the maintenance workload permits. Otherwise, contractor support will be obtained subject to the current labor agreement.

Equipment modifications will be reviewed to determine their necessity, feasibility and correct funding prior to the work being assigned to maintenance. All such work will be reviewed by maintenance engineering before work commences.

Summary - Policies are intended to cause certain interactions and behaviors by personnel that directly support the production strategy. However, policies or guidelines often go through several development stages. Initial policies might be derived from an evaluation to establish the exact nature of a problem requiring correction and the precise language of the policy. For example, in the policy stating that:

The overall Preventive Maintenance program will be assessed annually by maintenance engineering. They will ensure that it covers all equipment requiring services and that the most appropriate types of services are applied at the correct intervals. The performance of the PM program in reducing equipment failures and extending equipment life will be verified.

The plant manager wished to give the PM program strong emphasis and felt he should establish a minimum requirement through a policy. Subsequently, if an evaluation indicated that the problem was successful corrected, he might elect to change or omit the policy.

See also Appendix D - Maintenance Policies.

4

Adhering to the Principles of Maintenance Management

The Principles of Maintenance Management

Maintenance is a complex discipline. Not only is it a service that must support a specific production strategy but, it must be responsive, often to the unexpected. Its personnel are dependent on the cooperation of production and the support of staff departments in order for their program to be successful. In addition, there must be sanction and support from plant management.

In the previous chapters, the role of the plant manager in providing a clear objective and policies for the conduct of maintenance was discussed.

Complex activities are built on principles so that those carrying out the activities will have explicit guidelines to follow. Adherence to principles ensures that there is consistency as the activity is carried out and all personnel involved understand what is to be done, how any why.

Therefore, as organizational changes are considered or the maintenance program is changed, for example, the specific changes should be weighed against the principles of maintenance management.

Maintenance is supported by 20 maintenance management principles. They apply equally to maintenance personnel as they do to operations, staff departments and plant management.

Each of these 20 principles of maintenance management is discussed to provide an appreciation of their importance to successful maintenance.

01 - Involve plant management in the utilization of maintenance as a integral part of the plant's production strategy.

Maintenance is a service. Its primary task is to ensure the safe, effective operation of equipment so that production targets can be met on time and at least cost. Maintenance cannot compel production to follow its program nor can it require staff departments to support them. Caught in the middle, they are dependent on plant management to take the steps necessary to ensure the cooperation of operations and a high level of service from staff departments like purchasing or warehousing. Accordingly, maintenance should verify that plant management is sufficiently involved to cause this support and cooperation to take place.

02 - Clarify the objective of maintenance within the plant's production strategy.

As an extension of its production strategy, plant management assigns mutually supporting, consistent objectives to its operating, maintenance and staff departments. Maintenance must examine its assigned objective in the light of the overall strategy and ensure that its organization and their program are fully capable of supporting the objective.

03 - Ensure policy guidelines are supportive of the assigned objective.

Plant management amplifies the maintenance objective by providing policies to ensure important aspects of the program are well understood and carried out consistently throughout the plant. These policy guidelines, in turn, guide maintenance in the development of their day to day procedures.

04 - Define the maintenance program consistently within the assigned maintenance objective and policy guidelines.

Maintenance must spell out for production, staff departments and its own personnel how maintenance services are requested, planned, scheduled, assigned and controlled.

An overview must then be amplified with procedures which are coordinated with those of other departments. For example, the work order system procedures must be coordinated with inventory control procedures to cause material data to flow into the maintenance information system. Interaction with operations might be illustrated by depicting the weekly operations - maintenance scheduling meeting and the flow of information elements to individual decision makers should be shown.

Adhering to the Principles of Maintenance Management

05 - Define and publish terminology used in maintenance.

Confusion arises when people do not understand one another. If maintenance wishes to avoid this confusion they must define and publish the terms used in everyday communication with production, staff departments and within maintenance.

06 - Establish a responsive organization.

Different environmental circumstances or plant configurations require corresponding maintenance organization alignments. Methods of communication differ. Crew size and craft composition are influenced by the kind of equipment maintained and its tolerance for delay. Amounts of routine versus controlled work and preventive maintenance requirements differ. Every effort must be made to implement and utilize organization arrangements that maximize maintenance performance and productivity.

07 - Establish an effective work order system.

The work order system is the communications network of maintenance. It is the means by which all types of work are requested and, thereafter, planned, scheduled, assigned and controlled. The work order system is also the link with the accounting system by which field data on resource use (labor and material) are brought into the information system. The work order system must have the capability of handling all types of work from routine PM services (inspections etc.,), to unscheduled, emergency or planned and scheduled work, as well as non maintenance work (like construction).

08 - Ensure the information system supports essential information needs.

The information system is divided into elements of decision making information such as: labor utilization, cost and repair history, as well as administrative information like absentee reports. These information elements must be carefully identified. Sources of field data and the means to convert them into information must be checked for completeness, accuracy and timeliness. Personnel who use the information must be knowledgeable in its use.

09 - Conduct effective preventive maintenance services.

The heart of a PM program is equipment inspection, condition monitoring and non destructive testing. This detection orientation uncovers serious deficiencies in time to avert emergencies. It also discovers major repair needs sufficiently far in advance to allow the required work to be planned. As a result, each major, planned job is carried out with less manpower and in less elapsed downtime.

PM also includes lubrication, cleaning, adjusting and minor component replacement (belts, filters etc.,) to extend equipment life. Repairs are not conducted while PM services are in progress as there is danger that the services may not be completed. All PM services are routine, repetitive activities. That is, the same service is repeated at a fixed interval (1 week, 2 weeks) or at a variable interval (500 operating hours, 3000 miles) etc. All personnel should understand the impact of PM on cost savings and reduction of downtime.

10 - Plan major jobs to ensure better resource use and reduced downtime.

Two comparable major jobs, one planned the other unplanned result in dramatically different performance data. The planned job is invariably completed with 12 - 15% less manpower and in about 6% less elapsed downtime. The equipment returns to a productive status faster. Considering that downtime is about 3 times the cost of maintenance that could have avoided it, maximum planned work is a desirable objective.

11 - Apply standards in planning.

Standards are goals for selected major jobs. There are quality standards which prescribe the desired end result or quality level of the work. There are also quantity standards which prescribe man-hours by craft, job duration and cost. Some standards are carefully engineered for unique jobs. Other standards are developed historically as certain jobs are repeated. Standards also include task lists, bills of materials or tool lists that are applied with each repetition of selected major jobs. Thus, planning can be expedited and done more consistently. The information system, in combination with the work order system, should allow standards to be set and subsequently, actual performance compared with them.

12 - Schedule major jobs and services to ensure best use of maintenance resources and least interruption of operations.

Scheduling should be a joint maintenance - operations activity. It can never be an unilateral action. In the case of major jobs like component replacements or overhauls, operations must make the equipment available and maintenance must make resources available to do the work. For routine services like PM inspections, production must agree in advance to the scope of services, their cycle (or frequency) and abide by the results of the inspection (I.E., a serious safety deficiency found during a PM service may require immediate equipment shutdown and repair.).

Adhering to the Principles of Maintenance Management

When maintenance and operations carefully observe each others needs (for equipment by production) and constraints (availability of repair resources from maintenance), the scheduling procedure results in the successful achievement of its objective: least interruption of operations and the best use of maintenance resources.

13 - Require first line maintenance supervisors to develop and follow a daily work plan.

As the individual totally responsible for getting the actual maintenance work done, the maintenance supervisor can use all of the help he can get. However, he also desires to have some control over the work to be done. He must exercise judgment and provide latitude to his crew members to accomplish their tasks within established guidelines. He should prepare a daily work plan drawn from: the approved weekly schedule; the PM schedule etc. He must also incorporate the latitude in his plan to be responsive to the unexpected. The maintenance supervisor objects to having his work plan spelled out by a planner or clerk who is remote from the field situation. The effective supervisor knows the daily work plan and its success are his responsibility. If a team organization is adopted by maintenance, the need for a daily work plan is even more urgent, as all members must be totally informed of the actions and objectives for each shift.

14 - Measure the work load to determine the work force size and composition accurately.

Guesswork often prevails in work load measurement: matching the work load (essential work required of maintenance) with the work force (the right number of personnel of the proper crafts to get the work done). Work load measurement starts by dividing essential work into identifiable, measurable elements such as PM services, routine services (custodial work), planned and scheduled maintenance (overhauls or component replacements) and making reasonable allowances for emergency and unscheduled repairs. Man-hours required for all of this work must be assessed against current productivity levels. (Is each hour of work paid for equal to an hour of productive effort ?) The result is a preliminary work force of the approximate size and craft composition to carry out the estimated work load.

At this point, elements of the information system feed back the necessary information to fine tune the work force and make it more responsive to work load changes. Specifically, a labor utilization report showing by craft, the manpower used on each type of work would confirm initial estimates made. Next, the backlog, showing the rise and fall of residual man-hours by craft as the work load changes, would allow adjustments in the work force size and craft composition to be made.

Information would then be used to confirm that more PM or planned and scheduled maintenance has displaced emergency and unscheduled work, producing better productivity and reducing the backlog.

15 - Establish limits for non maintenance work (like construction or equipment installation) and apply planning and scheduling procedures to it.

Many maintenance departments are required to perform non maintenance work (like construction, modification, equipment installation or relocation). In these instances, they must ensure that this additional work does not preclude carrying out the basic maintenance work load. Maintenance should, after determining its work load, establish limits for non maintenance work it can accept.

Thereafter, to ensure effective control of non maintenance work, each job should be subjected to planning and scheduling. Suitable management information must be available to ensure control of this type of work whether it is carried out by maintenance or contractors. Also, the work order system must provide an engineering work order to handle the approval and funding aspects of non maintenance jobs.

16 - Control labor.

Once maintenance work is underway, the only direct means maintenance has to control cost is the efficiency with which they use labor to install materials. The cost of maintenance will not go down unless it can be carried out with fewer people or done less often. This means that the maintenance work force might be downsized. But, before downsizing can take place, labor productivity must be improved. In turn, labor productivity is unlikely to improve if the maintenance program is in disarray. Thus, there is a need to upgrade the maintenance program in conjunction with any effort to improve productivity.

Neither Total Productive Maintenance (TPM) nor the formation of teams in themselves will improve productivity. Further, if the current habits of maintenance personnel yield poor productivity, it is unlikely that the formation of a team or the adoption of TPM alone will improve productivity. There must be a fundamental change in the way maintenance operates. At the base of that change must be effective control of labor. These facts make the control of labor an important maintenance function.

17 - Ensure quality material support.

Maintenance is the plant's largest consumer of materials. It receives material support from the warehouse, purchasing, shops and commercial sources.

The complexity of the material control function and the requirement for quality administration and accountability require it be well managed. The assignment of inventory control to maintenance can split the material control function and should be avoided. Maintenance must ensure its personnel follow procedures for obtaining materials to uphold accountability principles. Material control personnel must be familiar with the maintenance program as maintenance relies on them to provide materials on time and in proper quantities.

18 - Train supervisors, staff personnel and craftsmen.

Help from everyone is needed for a successful maintenance program. Yet, never was the axiom that: "if you expect someone to help, you must first tell them how", more needed than it is in maintenance. More potentially successful programs have failed as a result of a failure to educate personnel.

19 - Educate operations, staff departments and management on the total maintenance program.

Operations must cooperate with the maintenance program and staff departments must support it if it is to be successful. While the appreciation of operations and staff departments of the maintenance program is helpful, it is enhanced dramatically by their understanding of the program. Typical is the ability of these personnel to use information.

For example, once operations finds out that repair history reveals certain failure patterns only they can control (mal operation, equipment abuse etc.), they will scan the repair history as eagerly as maintenance does. Similarly, when material control personnel learn that the linkage between the work order system and inventory control system can produce standard parts lists for repetitive jobs, they will eagerly support the development of this information. Likewise, when the plant manager learns that indices can correctly identify costly units, the value of information will increase. Thus, the education of production, staff departments and plant management on the maintenance program can pay worthwhile dividends.

20 - Measure performance on a regular, continuous basis.

The phrase: set goals, observe feedback, make corrections, improve performance, is especially meaningful because performance is linked with productivity and profitability. The maintenance department which does not measure its performance on a regular, continuous basis stagnates and fails. It makes no contribution to profitability.

By contrast, the maintenance department that regularly, continuously measures its performance shows constant improvement and makes a significant contribution to plant profitability through gains in productivity and performance.

Summary - We have now observed that a sequence of events has transpired. First, the plant manager developed and published his production strategy. Next, as a part of that strategy, he identified a plant objective. To ensure that all key departments understood their roles in supporting the production strategy, he assigned each department an objective and amplified the objective with policies. In turn, each department developed procedures based on the policies. Now, as maintenance organizes, develops its program and interacts with other departments, they should observe the principles of maintenance management.

5

Developing an Effective Organization

An Effective Maintenance Organization

The principles of organization have been surprisingly consistent over the years. In fact, if they are compared between 1916, 1964 and today only the words differ, the message is the same.

Kootz & O'Donnell (1964) [1]

Division of work
Parity of authority and Responsibility
Unity of command
Unity of direction
Unity of objective
Balance
Scalar chain
Functional definition
Efficiency
Span of management
Exception principle
Responsibility
Feasibility
Continuity
Leadership facilitation

Henri Fayol (1916) [2]

Division of work
Parity of authority and responsibility
Discipline
Unity of command
Unity of direction
Subordination of individual to group
Renumeration of personnel
Degree of centralization
Scalar chain
Order
Equity
Stability of tenure
Initiative
Esprit de corps

Today different words are used. Among them:

- Employee empowerment (initiative, responsibility and leadership facilitation)
- Teams (subordination of individual to group, order, equity and esprit de corps)
- Work control (division of work, balance and exception principle)
- Mission (unity of direction)
- Productivity and performance (efficiency and order)

The application of the principles of organization should bring about a high level of organizational efficiency. However, as is the problem with every organization, the principles are not applied effectively. Maintenance organizations are no exception. Organizations themselves with departments and hierarchies often contribute to the unhappiness of individuals who make up the structure of the organization. In every organization, individuals eventually come to the point where, because of personal limits, organizational constraints or some other factor, upward progress stops or slows down sharply. If the persons affected do not make personal adjustments, frustration, anger and resentment appear. These facts account for today's focus on individual employees and efforts to implement an organization that neutralizes these circumstances.

Yet, there are other individual factors. For example, every adult must realize that certain jobs are closed to him because of deficiencies in training, ambition, luck or sheer ability. Unfortunately it takes some time before a person realizes he has reached his upper limit. On the other hand, if individual levels of aspiration conflict with the limits of advancement attainable within the organization structure, there is discontent. In school, for example, the completion of average work results in promotion at the end of the year. Such advancement does not occur in the work environment. The school environment suggests that all have equal opportunity for advancement. In the work environment, this is not true. Unequal men often get the promotions. The organization itself must take into account that it is a structure of people. If that organization is to be effective it must satisfactorily blend together the objectives of the organization with the aspirations of those who constitute the organization. In short, every successful organization has been built upon the proper consideration of the human element that makes it up. Thus, teams, employee empowerment or Total Productive Maintenance (TPM), all of which focus on the individuals contribution to the organizational objectives, are popular goals today. Therefore, the principles of organization are again confirmed and the plant that aspires to successful organizational change must adhere to them.

Developing an Effective Organization

The Maintenance Organization Shell Game

In most maintenance organizations there are three organizational configurations:

- The official organizational chart appears in all the company's literature and is usually displayed prominently. Unfortunately, this is the chart that no one pays any attention to.
- The next organizational chart is the one in the leader's desk drawer. It is marked with red lines showing realistic changes that should be made to get the organization to function well (or at all). The leader is trying to overcome all of the forces that will allow him to install this organization.
- The most important organizational chart is the unpublished power structure. These are the people who make the organization work. Chances are they aren't on any official organizational chart. They usually are on the red lined proposal. They are invariable the people that maintenance goes to when they want positive results.

The real trick of organization is to somehow bring individual and organization goals into close harmony. Resolving the technical and human problems encountered in the process of organizing the maintenance function, requires a deliberate approach. Consider the following:

Objective - Determine the goals, objectives, programs and plans that best meet the overall objective of maintenance. There is a distinct objective of the maintenance function in the production environment: keeping production equipment in a safe, effective operating condition. The objective must be clearly stated by plant management, placing maintenance in the correct perspective relative to the overall production strategy. From this objective, maintenance can develop meaningful plans, programs and procedures. [3]

Identify Essential Work - Determine what maintenance work must be done in order to achieve the results expected within the objective. Maintenance must give priorities to its work based on its importance. For example, PM must be on the top of the list. Planned and scheduled work must be emphasized. Emergency work must be coped with and unscheduled work must be fitted in as time permits. If project work is necessary, it should be prioritized along with planned and scheduled maintenance.

Determine the Division of Work - Divide, group and relate the essential work in a simple, logical, understandable but comprehensive organizational structure. Certain types of work should be delegated to shops, field crews, utility crews, etc., based on their ability to perform the work.

Establish the Assignment of Work - Assign essential work clearly and definitely to various departments and groups in the maintenance organization. Specific assignments of work should be made to clarify responsibilities. Are component replacements to be done in the shop or in the field ? Should lube personnel be on area crews or should they be centrally controlled ?

Determine Personnel Qualifications - Determine the requirements and qualifications of the personnel needed. Once work has been delegated to a specific organizational element, determine what qualifications are required by those personnel who will direct or staff that part of the organization.

Staff the Organization - Staff the maintenance organization with the best qualified people available. Get the best personnel in key organizational spots.

Establish Policies - Establish policies and procedures designed to help achieve the organization's goals. Develop and announce policies that make it possible for subordinate maintenance supervisors to build realistic procedures to comply with these policies.

Using this approach, individual versus organizational goals can be reconciled to ensure the organizational change gets off to a good start.

Examining the Traditional Organization

Within the traditional maintenance organization there are three levels of management:

- First line supervision - These are the supervisors who directly supervise the craft level.
- Mid level supervision - These are the general supervisors who supervise several first line maintenance supervisors.
- Top level supervision - These are the maintenance managers or superintendents who control the entire maintenance function.

Developing an Effective Organization

Collectively, these supervisory levels perform the traditional managerial functions:

- Planning - The activity which shapes the future direction of the organization by developing actions and guidelines in the present
- Organizing - Identifying work to be done and assigning the work in a logical, orderly way
- Staffing - Identifying the organization's personnel needs, obtaining the personnel to meet those needs and placing them within the organization to best meet individual and organizational needs
- Leading - The act of motivating personnel to reach organizational objectives and attain sufficient morale levels to be able to work both for themselves and the organization
- Controlling - The activity that checks actual progress against planned progress

In performing these managerial functions, the first-line maintenance supervisors must pay primary attention to human relations with less attention to applying technical skills.

Checking and inspection of completed work increases, as does time spent on obtaining tools, determining job methods or training crew members.

In general, supervisors are moving away from doing work themselves and are spending more time supervising the work being done by their crews.

In examining first line supervisory activities, 66 % are related to humans: [4]

Meeting Schedules	29
Human Relations	14
Quality Control	11
Cost Control	9
Methods	8
Personnel Functions	7
Tool, equipment	6
Materials and Supplies	6
Working Conditions	5
Other Activities	5
	100%

At the general supervisor level, there is a shift from doing to controlling. At this middle supervision level there is concern for matters such as evaluating, counseling, training, motivating subordinates, monitoring wage and salary systems, holding coordinating meetings and performing personnel oriented functions. Middle level supervisors are essentially watchers. They are the observers who broadly direct the maintenance operations that get the job done.

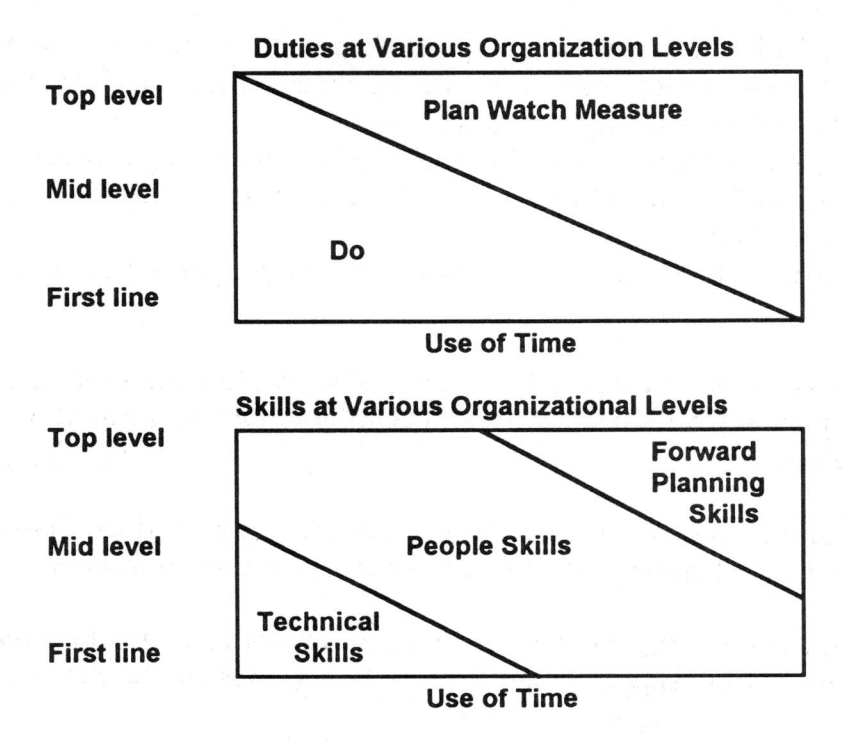

Duties at Various Organization Levels

```
Top level    |  \        Plan Watch Measure
             |    \
Mid level    |      \
             |   Do   \
First line   |          \
             +------------
              Use of Time
```

Skills at Various Organizational Levels

```
Top level    |                    Forward
             |                    Planning
             |                    Skills
Mid level    |        People Skills
             |  Technical
First line   |  Skills
             +------------
              Use of Time
```

At the manager or superintendent level of maintenance, duties are more forward looking. Policies are made. Concepts are developed. Fewer technical skills are needed. There is generally less emphasis on working with people to carry out specific plans or activities. There is more emphasis on controlling resources through subordinates. Tomorrow is emphasized rather than today. There is concern for performance, long range planning and achieving the organization's objectives.

Each level of supervision has a specific time frame within which their principal activities occur. The supervisor should be concerned primarily with today and tomorrow. General supervisors should be looking at next week and beyond. Superintendents must be concerned with the current month. Too often, we see supervisors doing what craftsmen do better. General supervisors sometimes overlap with planners and devote excessive time to procuring materials rather than managing the overall maintenance activity.

Developing an Effective Organization

Therefore, in addition to the general use of time by the various organizational levels, the time frame within which the incumbents principal activities are carried must be examined as well.

ORGANIZATIONAL LEVEL	FOCUS OF ACTIVITY VERSUS TIME
CRAFTSMAN	THIS SHIFT
FOREMAN	TODAY
PLANNER	TOMORROW
GENERAL FM	NEXT WEEK
MTCE ENGINEER	THIS MONTH
SUPERINTENDENT	NEXT MONTH / YEAR TO DATE
MANAGER	NEXT YEAR

TIME →

To achieve better productivity and performance, the maintenance organization may require significant changes in structure and functions. The traditional maintenance organization may be a 'dead horse' in the light of the potential for improving productivity and performance while reducing cost as demonstrated by more responsive types of organizations.

Summary - A careful application of the principles of organization must be made at the outset of any organizational change. This establishes the parameters of the new organization and verifies whether it satisfies the basic principles. The human versus technical needs of the organization must then be brought into harmony to satisfy the diverse needs and talents of individuals. A supervisor, for example, may visualize the future team as a threat to his job security. Therefore, he must know what to expect and what his future job will be.

Similarly, a long time maintenance superintendent may need encouragement to rethink his outlook in applying his talents conscientiously when asked, for example, to control both maintenance and operations.

A look at the tradition managerial functions serves as an additional check. Then, the duties, skills and the focus of time for each organizational level should be examined. Because, the very people who will form the new organization exist in the old one, they must be brought along step by step. If this can be done, they are more likely to be supportive. The challenge of a new organization will require everyone's help.

1. H. Koontz and C. O'Donnell, **Principles of Management**, 3rd Edition, New York: McGraw-Hill, 1964.
2. H. Fayol, **General and Industrial Management**, New York: Pitman 1949 (1916).
3 Ralph Cordinier, **New Frontiers for Professional Managers**, New York: McGraw-Hill, 1956.
4. P. Rigors and C. Myers, **Personnel Administration**, 5th edition, New York: McGraw-Hill 1965.

6

Securing Management, Operations and Staff Support

Maintenance - A Service Organization

As a service organization, maintenance supports operations by applying its program to keep production equipment in an effective operating condition. But, they are dependent on operations to make equipment available so the work can be carried out.

Similarly, maintenance receives services from staff departments like purchasing, warehousing or accounting and again, their effectiveness depends on the quality of service they receive.

Maintenance is in the middle. They are dependent on the cooperation of operations while in need of the support of staff departments. Unless provided, maintenance can have difficulty meeting their objective.

In many industries, difficult working conditions are stressed with the sense of urgency to meet production targets. Often, a plant is attempting to meet cost targets that assure profitability in an intensely competitive world market.

Maintenance is often at the center of the cost reduction effort because it is the plant's largest controllable cost.

Thus, the existence of an orderly environment in which maintenance is performed exerts a tremendous influence on their effectiveness.

Summary - Plant managers can create an environment in which the maintenance program can be successful. The formulation of a production strategy, the assignment of clear objectives to each key department and the provision of clear 'ground rules' assure the success of maintenance and encourage cooperation from operations and support from staff departments. When these conditions exist, maintenance can work more effectively and count on the solid support they know will yield a successful maintenance program. However, there is a two way street in that maintenance must ensure that plant management, operations and staff departments are well informed on their organization, their limits and constraints and the details of their program. [1]

Maintenance needs all the help it can get.

Yet, they bear significant responsibility for defining their program and educating plant personnel about it to obtain the cooperation and support they require.

---

1. Images copyright, New Vision Technologies, Inc.

7
Defining the Maintenance Program

The Maintenance Program

A well defined maintenance program results in efficiency within maintenance because everyone has a solid concept of how maintenance should be carried out. In addition, operations, management and staff departments know how they can best support maintenance. As a result, the desired cooperation is attained and maintenance is able to be more effective. However, what is more important is that the groundwork for an effective plant wide team has been established through the 'road map' or plan of action established by defining the maintenance program. Defining the maintenance program accrues advantages. Among them:

- The existing program is confirmed
- A new program can be developed
- Maintenance personnel are better educated
- Other plant personnel can be educated uniformly
- Personnel participate
- A well informed work force is assured
- Performance improves
- Gains in productivity result
- Product quality improves
- Less downtime results
- Better resource use takes place

Defining the Maintenance Program

A definition of the maintenance program spells out how maintenance work is requested and thereafter, planned, scheduled, assigned, controlled and measured. The program definition is an explanation of how maintenance does its business on a day to day basis.

Program definition includes all of the programs carried out by maintenance:

- Preventive maintenance
- Component replacement
- Planning and scheduling
- Using the work order system
- Utilizing information
- Non maintenance projects, if applicable

In addition, the duties of key maintenance personnel as well as craftsmen are explained as they go about:

- Planning
- Assigning work
- Scheduling
- Controlling and performing work
- Measuring performance

The interaction with operations and staff personnel should be portrayed as they:

- Request maintenance help
- Issue stock materials
- Obtain purchased materials
- Participate in scheduling and coordination meetings

Linkages with other systems should describe how:

- Field data are converted into information
- Stock materials are issued
- Purchasing carries out its functions
- Component rebuilding is managed
- Production statistics are obtained and used
- Payroll uses labor data to generate labor information
- The accounting system functions

Defining the Maintenance Program

Presentation Techniques - The most effective technique for defining the maintenance program is a schematic diagram with an accompanying legend. See Figure 7-1.

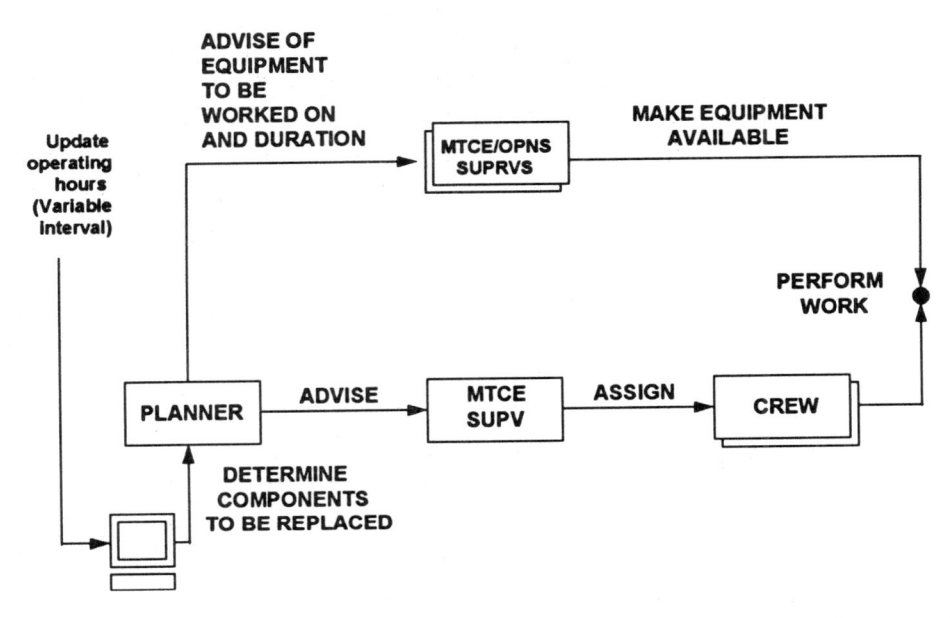

Figure 7-1. In this schematic of the procedure for the replacement of major components, there is a description of the steps required and delineation of the personnel involved and what they do.

Avoid an overly detailed program definition at first, since it must be presented to many personnel in a short time. They will offer comments to change or improve it. Thus, a simple schematic diagram with an accompanying legend works best.

The legend may be on the same schematic diagram, as a border around the diagram or as a separate sheet of paper. The amount of detail contained in the diagram will dictate which technique is best. By using the schematic diagram technique you can:

- Consolidate program details
- Facilitate the education of all personnel
- Keep program definition simple
- Encourage participation during the education phase
- Assure that all personnel can grasp the details
- Modify the diagram easily
- Encourage suggestions
- Have a 'program in everyone's hip pocket'

If the question: What is the maintenance program, gets a look like this . . .

There is a need to better define the maintenance program. [2]

Summary - The definition of the maintenance program assures that all maintenance personnel understand it and can carry out their duties within its framework competently. Similarly, operations personnel will understand the services maintenance wishes to perform in their behalf and they cooperate more effectively. In the same way, various staff departments will better appreciate how their support can be best provided to help ensure a successful plant maintenance program. Plant managers will be gratified with a well defined maintenance program, as it is a reflection of the guidance they have provided through the production strategy, objectives and policies.

Chapter 20 - How to Define the Maintenance Program, is a complete definition of a typical plant maintenance program.

1.- 2. Images copyright, New Vision Technologies, Inc.

8

Educating Personnel on the Maintenance Program

The Education Process

At the conclusion of a comprehensive evaluation of maintenance at a large chemical plant, there were several noteworthy reactions:

- Plant management did not realize the extent to which their actions could influence the success of the maintenance program.
- Operations had not appreciated how much maintenance efforts are hampered when equipment was not made available as scheduled for services offered and work required.
- Staff departments saw how their support was critical to the ability of maintenance to carry out its program efficiently.
- Maintenance determined that they should take another hard look at their program.

Many effective maintenance evaluations include an assessment of what personnel do not know and the extent to which that missing knowledge hampers their performance. Often evaluations reveal wide differences between what maintenance believes they do well and what operations thinks. Invariably, the truth is not so much a poor maintenance performance but, rather a need to inform operations of the nature and extent of the activity.

In the previous chapter, the purpose of defining the maintenance program was explained. In this chapter, the education that must follow is described.

Using the Chain of Command

Explaining the Program Definition - Once the maintenance program has been defined, it should be explained first within the maintenance organization. The key personnel who drafted the program probably include several planners, a general supervisor and the superintendent. Once the initial draft has been prepared, it should be examined for completeness and clarity. Don't be concerned that certain aspects may have been left out or that some aspects might not be entirely accurate. The first draft is merely an effort to get everyone's collective thoughts on paper. The important part is the education process itself. Starting with a reasonably good and complete diagram, the maintenance superintendent should launch the educational process by explaining the diagram to his immediate subordinates, perhaps the general supervisors and the maintenance engineer. [1]

A Draft Format is Best - The diagram should be deliberately kept in draft form, even pencil at this stage. Comments should be added and modifications made as the explanation goes forward. Working down through the chain of command, each general supervisor would review the diagram with several supervisors. Similarly, the maintenance engineer might discuss the diagram with several planners. In turn, each supervisor would discuss the maintenance program definition with crew members a few at a time. It is important during this process to keep the discussions informal and to welcome comments. All comments with merit should be incorporated into the diagram. Thus, as those offering comments see that their suggestions are now part of the maintenance program, they will gain a sense of ownership and help to sell the program to others.

Educating Personnel on the Maintenance Program

Because the diagram can be a single large sheet of paper, it can easily be reproduced. Take advantage of this and give each person a copy for reference during the explanation and discussion. Afterwards, leave the copies with them and ask that they add comments or modifications if they should think of any additional ones.

Monitor Presentations - Be sure to monitor the presentations, again, using the chain of command. Be especially aware of the presentations being made by supervisors with their crews. Some supervisors are intimidated by their crew and may need some reinforcement.

Other supervisors may not have a positive attitude toward the program addressing their crews with indifferent comments like, "I'm required to talk to you about this". However, the best way to find out if someone understands the program is to listen to him explain it to others.

Talk With Individuals - Crew members, especially in a tough union, can be highly skeptical. Therefore, the supervisor should not attempt a general crew meeting to discuss the program diagram. Rather, he should meet informally with two or three crew members at a time so he can better solicit their comments without the pressure of the skeptics that are in some crews. Once all of the meetings have taken place and comments secured, the master diagram should be updated to show all relevant, helpful recommendations. This should be published.

Tell Operations - Next, operations personnel should be educated. However, their explanation should be abbreviated to include only the essentials. If a plant team has been formed with operations, more detail should be provided.

Don't Overlook Staff Department Personnel - Staff departments follow. Their education should focus on procedures that are carried out jointly with maintenance, such as:

- Warehouse issue and return procedures
- The component replacement program and corresponding purchasing actions
- Explanation of the use of accounting reports

Focus on Management - Finally, management should get an opportunity to review the program. This phase works best when all of the superintendents and department managers can be present to explain their roles in supporting the maintenance program.

Summary - It will do little good to define the maintenance program unless the program definition is accompanied by an educational effort to ensure all plant personnel understand the program and know how they can support it. Defining the program and educating plant personnel can be a very easy step with significant benefits. Perhaps the most important benefit is the potential for productive utilization of all maintenance personnel and confidence that they will perform effectively.

If a team is to be formed, one of the essential requirements for its success is the confirmation of the maintenance program. Thus, program definition and education become essential to successful team implementation.

1. - 2. Images copyright, New Vision Technologies, Inc.

9

Specifying the Maintenance Work Load

The Maintenance Work load

The maintenance work load is the amount of essential work performed by maintenance and the identification of the correct number of personnel by craft, working at a reasonable level of productivity, to carry it out.

The determination of the right number of personnel to perform maintenance work has never been a task that maintenance is anxious to do. In a classic conversation with a veteran maintenance superintendent several years ago his response to measuring the work load was, "to keep adding people until the overtime goes down". While this technique is simply not possible these days, it is illustrative of the lack of interest and concern by maintenance that must be changed toward this necessary task.

Even today, when it has been well established that the cost of maintenance will not go down unless it is done with fewer people or done less often, there are still indications that the task of work load measurement is not taken seriously.

Few maintenance organizations, for example, measure productivity, the best way to determine the quality of labor control. In other instances, we find all manpower assigned to the first line supervisor whose strength is in getting things running again and not the control of labor. He is more likely to throw manpower at the emergencies he created by not performing PM services on time or failing to provide his crew with well defined job assignments. Obviously, the supervisor is not solely responsible for this attitude. However, these observations may indicate a general indifference by maintenance toward the improvement of labor control and, through it, the reduction of maintenance cost.

Measuring the Work Load

Work load measurement begins with the identification of the work required of maintenance. It divides itself into two broad categories: Maintenance and non-maintenance. Consider:

Maintenance Work - The repair and upkeep of existing equipment, facilities, buildings or areas in accordance with current design specifications to keep them in a safe, effective condition while meeting their intended purposes. Maintenance work is expensed as an operating cost.

Non-maintenance Work (project work) - Construction, installation, relocation or modification of equipment, buildings, facilities or utilities. Project work is usually capitalized.

Individual categories of the work load must be further defined to allow them to be broken down into elements that can be measured. See Appendix B - The Maintenance Work Load.

The definition of categories of work within the work load provide:

Clarity - All personnel can identify an emergency repair or distinguish the services that make up the preventive maintenance program.

Common Understanding - Personnel in operations and maintenance are no longer squabbling about what is or what is not an emergency repair. See Figure 9-1.

Figure 9-1. On the left, a routine emergency as seem by operations. On the right a major crisis as seen by maintenance.

Specifying the Maintenance Work Load

Control of Work - Each category of work is controlled differently. For example, preventive maintenance services are done routinely (same check list) and repetitively (every four weeks or every 750 operating hours). Each planned and scheduled job is carefully prepared for and scheduled in consultation with operations. Emergency repairs are reacted to, often as a surprise. These differences are then traced to the work order system where a different work order element is used to control each category of work. A standing work order controls PM services; for example. A formal work order is used for planned work and emergency repairs are often initiated with verbal orders.

Measuring the Work Load - In turn, the definition of work load elements facilitates the determination of the work load. For example, if a mechanic were to carry out a three hour PM service every month, the work load would be 36 mechanical man hours annually for that service. In a similar fashion, the entire PM work load could be computed. Most planned work can be estimated by craft as well. In many industries, for example, major component replacements are repeated for similar units of equipment such as replacing conveyor idler rollers. Each repetition of these actions has a known manpower requirement and the number of repetitions is able to be forecasted (See Chapter 23 - Forecasting and the Component Replacement Program). Other major jobs that are unique are planned. Part of the planning task is to estimate the manpower required, by craft. The end result is a reasonably accurate estimate of the manpower required for planned work. Manpower for routine maintenance like shop clean up or training is allocated. Emergency and unscheduled repairs can be limited by effective PM. Management, through carefully thought out policies should specify limits for the amount of non maintenance work to be done. See Figure 9-2.

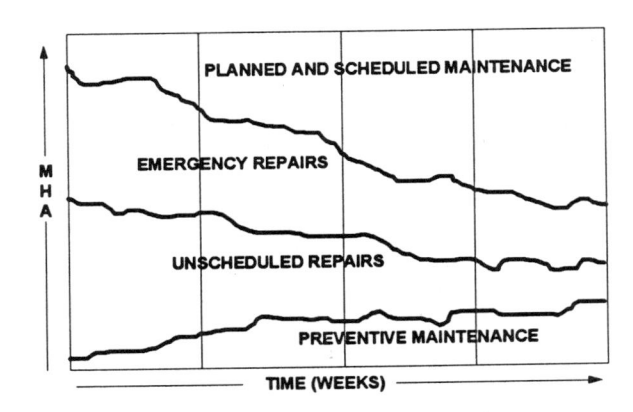

Figure 9-2. More PM reduces manpower used on emergency and unscheduled repairs while increasing manpower for planned and scheduled work.

Summary - Measurement of the work load is essential to the effective use of manpower. It is not possible to improve productivity and reduce labor costs when maintenance is indifferent to how much labor it really needs. There is no magic in the adoption of a team organization or the implementation of Total Productive Maintenance (TPM). An organization does not catch productivity like one catches the flu. Rather, there are concrete steps that must be taken so that productivity can be improved. Work load measurement is such a step. Unless organizational changes like adopting teams are preceded by work load measurement, they are likely to be unsuccessful. Work load measurement is critical to the effective control of labor.

10

Establishing Maintenance Terminology

Maintenance Terminology

Terminology is the language of maintenance. It includes definitions by which plant operations, maintenance and staff personnel can communicate clearly.

Despite the daily conduct of the very activities that should be well defined, they often are not.

Surprisingly, the most common elements are misunderstood and often used incorrectly. Typical:

- Within the same maintenance department, for example, there are typically numerous different, often incorrect definitions of preventive maintenance.
- It is not unusual to hear a plant manager insist that everything maintenance does is 'PM'.
- Emergency repair is often ill defined. This causes the greatest number of unnecessary squabbles between operations and maintenance.
- Especially troublesome is the difference between maintenance and non maintenance. Since one is expensed (maintenance) and the other capitalized, this misunderstanding often has serious consequences that compound cost control efforts.

Within the maintenance department, details that should be clarified are often not. For example:

- An overhaul (total unit) and a rebuild (component) are not the same thing. Yet, they are used interchangeably.
- Modification (non maintenance) and corrective maintenance (done to facilitate repairs) are quite different but, this is seldom acknowledged.

When a maintenance evaluation is carried out, it should include an assessment of how well terminology is defined, publicized and used. In 90 % of circumstances, if the maintenance department has not defined preventive maintenance, for example, more than likely their PM program is faulty and incomplete. Put another way, if maintenance is not certain what something is, they can hardly develop a program to carry it out effectively.

Summary - While this attention to something as basic as talking the right language may seem trivial, it nevertheless is the foundation of good communication and ultimately effective control of maintenance work.

You should ensure that terminology:

- Has been defined
- Has been communicated to all plant personnel
- Is complete
- Is used correctly

See Appendix C - Maintenance Terminology.

11

Achieving Productivity and Performance

Productivity

What is Productivity - Productivity is the measure of the effectiveness of labor control. It is the percentage of time that a maintenance crew or individual craftsmen are at the work site, with tools, performing productive work.

Why is Productivity Not Measured - Regrettably, few maintenance departments measure productivity. Largely, this reaction is defensive. They often feel that another criticism charging them with not working hard enough, is unnecessary.

Ignorance and suspicion also account for the lack of interest in measuring productivity. Generally, these circumstances begin with the hourly work force, who fear:

- Lay offs
- Work force reductions
- Forced changes in work practices
- Tighter work control

First line supervisors, who are on the receiving end of this dialog, are seldom anxious to pursue inquiries into how well they are using their crew members. In turn, senior supervisors and managers visualize that insistence on measuring productivity will cause more problems than cures.

As a consequence, productivity is not measured. The status quo prevails, and poor productivity, that could be easily improved, remains that way.

Many maintenance departments know they must measure productivity but, previous bad experiences may deter a repetition. Thus, the technique of measuring productivity must be effective in correcting the ignorance and dispelling suspicion. It must also be applied in a fair, constructive way. See Chapter 40 - Measuring Maintenance Productivity, for a more complete description of the techniques for measuring productivity.

Causes of Poor Productivity - Among the most prominent causes of poor maintenance productivity are:

- There may not be enough supervision
- Supervision may be carried out inadequately
- Work assignments may be missing or confusing
- Work may be poorly planned
- Scheduling may not be carried out properly
- Job instructions may be incomplete
- Coordination may be inadequate
- Operations may delay making equipment available
- Operations may ignore the work schedule
- Correct labor may not be available
- Materials may be missing or not able to be located
- Tools may be missing or broken
- Support equipment may not be available when needed

A Poor Program Often Means Poor Productivity - Of these, few are the result of any omission by craftsmen. Rather, they are, more often, the result of a poor program, badly executed. See Figure 11-1

Figure 11-1. The craftsman is seldom the cause of poor productivity.

Achieving Productivity and Performance

Thus, a well defined, properly executed maintenance program will help eliminate many of the causes of poor productivity while creating the circumstances that encourage maintenance to check the effectiveness of labor utilization.

The Impact of Productivity - At the management level, the realization of the impact of poor productivity on manpower and cost is sobering. See Figures 11-1 and 11-2.

Productivity versus Manpower Savings

30 Men @ 30 % productivity = 16200 MH/Yr
[30(1800 MH/Man/Yr) (.30)]

30 Men @ 40 % productivity = 21600 MH/Yr
[30(1800 MH/Man/Yr) (.40)]

Added manpower by increasing productivity
21600 - 16200/.40 (1800) = 5400/720 = 7.5 Men

Figure 11-1. Increasing productivity from 30 to 40 % is the equivalent to having 7.5 men added to the work force at 40 % productivity.

Cost of Performing Work

Cost of one hour of Work = Labor + Materials/Hours

Assume: 72 men @ $ 14.00/hr and $ 5100.00 in benefits
Cost per man = $ 34,200.00
Labor cost = 72 x 34200 = 2,463,840
Labor = 42 % of Total cost
Total - Labor = Material = 3,402,446

At 33 % productivity
2463840/(1800 x .33)72 + 3402446/(1800)72 = $ 83.86/hour

At 42 % productivity
2463840/(1800 x .42)72 + 3402446/(1800)72 = $ 66.70/hour

Figure 11-2. Improving the productivity of a 72 man work force from 33 to 42 % results in getting work done for $ 67.00 per hour versus $ 84.00.

Considering the pressure to reduce the cost of maintenance, the measurement of maintenance productivity should be required. It is too vital an element to be overlooked. It is an easy opportunity to significantly improve performance. But, honesty with the work force is a must.

If a plant is to survive in difficult economic times, then all must pull together to achieve the level of profitability that guarantees survival. All must help. The path to profitability begins by ensuring maintenance is performing productively.

Performance

Productivity, performance and profitability are linked. However, maintenance performance cannot be determined with a single index. Rather, there must be a combination of indices that provide an overall performance picture.

Therefore, performance indices must be related to the maintenance organization, the efficiency with which they carry out their program and the quality of interaction with other departments.

Typically, the maintenance organization performance is assessed on:

- Productivity
- Time supervisors spend on control
- Quality of work assignments
- The effective use of labor
- Control of absenteeism
- Control of overtime
- Reduction of accidents
- Compliance with policies and procedures

Execution of the program is judged by observing indices, such as:

- Compliance with the PM schedule
- Completion of the weekly schedule
- The percent of work planned and scheduled
- Reduction of emergency repairs
- Control of the backlog
- Timely completion of major jobs
- Minimizing down time
- Allocation of labor against priorities
- Proper use of the work order system

The quality of interaction with other departments can be judged against such indices as:

- Prompt release of equipment on work completion
- Compliance with stock withdrawal procedures
- Adherence to lead times for purchasing materials
- Quality of labor reporting

Information Needs:

Differing Information Requirements - Within this range of performance indices, different levels of the maintenance organization focus on different time periods. For example:

- The supervisor is most interested in what must be done during his shift. He is less concerned about next week.
- The general supervisor needs information that will produce a successful week for his department.
- The superintendent must look further ahead and divide his attention between a successful overall maintenance program as well as capital work projects.

Also, the focus of interest differs as well. Typical:

Material Control:

- Superintendent -- Is material control effective ?
- General supervisor -- Are specific critical parts available ?
- Supervisor - Can parts be obtained quickly ?

Overtime:

- Superintendent -- Is premium labor under budget ?
- General supervisor -- Is overtime used well ?
- Supervisor --Who is assigned overtime ? Will the union complain ?

Union:

- Superintendent -- Is collective bargaining successful ?
- General supervisor -- Is the plant in compliance with the union contract ?
- Supervisor -- How can various crafts be best used to avoid grievances ?

Generally, the superintendent needs summary and trend information that tells him he is safe, or headed in a safe direction. Consequently, his needs for reports and summaries is greatest.

The general supervisor needs details of labor use, overtime, absenteeism, status of major repairs, and costs. Trends and summaries are helpful, but they are not his prime need. He requires details to attack specific problems. The supervisor wants immediate information. He must know what happened on the preceding shift that may cause problems. He has little interest in trends, reports, and summaries.

Summary - If you want better labor control you must measure productivity. Therefore, policies prescribing that productivity checks will be carried out regularly and improvement targets established are necessary. You must ensure that:

- The hourly work force understands productivity.
- Maintenance supervision has explained it to reduce fear of layoffs.
- Productivity is measured using logical indices.
- Non threatening techniques are used to measure productivity.

Performance must relate to the organization, its program and the quality of interaction with other departments. Performance information will be different depending on the level and interest of the user. But, the information must successfully improve productivity.

12

Evaluating to Improve Performance

Evaluation is the Start of the Improvement Process

It was Will Rogers who, with consummate wisdom, unwittingly commented on the essence of evaluations when he said: "We don't know what they don't know!" In so doing, he helped explain why managers have an obligation to verify that their personnel really know what is expected of them. He further stated that, "What we know, may not be so!" Thus, suggesting that managers must establish the correctness of facts on which they make decisions concerning their organizations.

Yet, there are some managers who may not have considered this basic wisdom. These are perhaps the ones who shun a solid, fact finding evaluation trusting rather to guesswork as a basis for attempting improvement. Thereby, without realizing it, they may be making incorrect decisions harmful to the very organization they are trying to improve.

Plant managers seeking to achieve and sustain profits realize that success is directly linked with their ability to improve maintenance performance. When maintenance performance is poor, costs will be excessive, and unnecessary downtime will cripple efforts to meet production targets and stay within operating budgets.

Unfortunately, desired improvements are not realized because the evaluation of maintenance, the first step of improvement, is not done adequately. Too often, it is not done at all.

Maintenance is Not Evaluated Adequately

Plant managers evaluate production activities frequently, in detail and very effectively. However, many of the same managers make only cursory checks of maintenance performance. By not evaluating maintenance adequately, they deny themselves a starting point for maintenance performance improvements that should be made. This omission cancels any potential contribution better maintenance performance could have made to improving profits. Many plant managers are uncertain about what maintenance should be doing and how their performance should be measured. Compounding this uncertainty is the ineffective maintenance department's resistance. Typical is the proclaimed unpleasantness of the union's reaction to being evaluated. Against such circumstances managers back away from evaluations and necessary improvements aren't identified.

Maintenance Shuns Evaluations

Being evaluated is equated by some maintenance departments to 'a self inflicted wound'. They often view evaluations as misguided 'witch hunts' and, consequently, they resist them. Surprisingly, even first rate maintenance organizations sometimes resist because there are too many factors over which they have no control that influence their performance. Typical:

- A poorly run warehouse managed by a remote accounting department can cripple maintenance performance.
- When operations fail to make equipment available for essential maintenance in an effort to meet unrealistic production targets, maintenance is the ultimate victim.

Why Maintenance Leaders Avoid Evaluations - Maintenance leaders seldom come from outside of maintenance. Invariably they come from maintenance and carry their management styles from the supervisor to superintendent level unchanged. Many are graduates of the 'school of hard knocks'. University degrees, while not a requirement, are rare. Few have had significant management training. They often assume their management responsibilities with little preparation for the challenges they will face. Some may even be unwilling supervisors and even uncomfortable in the role. The majority of maintenance supervisors are ex craft personnel. While they are competent in a single skill, they are often unfamiliar with the overall management principles and techniques necessary to meet their obligations as supervisors.

Evaluating to Improve Performance

Further compounding the situation is the fact that a large number, as ex craftsmen, are former union employees who may harbor an 'us versus management' viewpoint. As a result, many are suspicious of evaluations no matter how logical or necessary they may be.

How to Create a Favorable Evaluation Environment

The best evaluations are carried out in an atmosphere where management uses them constructively to suggest and support improvement while maintenance views them as a way to initiate improvement. Here are some of the ways this environment can be realized:

- Require Evaluations - Management should require that maintenance evaluations be carried out regularly to assure their contribution to profitability. This blunts maintenance resistance and redirects that energy into preparing for evaluations instead.
- Explain Why - The evaluation should be publicized as a checklist describing what maintenance should be doing. Its should describe how they did and serve to guide improvements. It states what is done well, not what is done poorly.
- Identify What - A complete, valid evaluation must include a look at the maintenance organization, its program and the environment in which maintenance operates. Examination of the environment answers questions on the quality of management support or cooperation by operations. As maintenance realizes that the factors over which they have no control will be evaluated fairly, their resistance can be changed to acceptance.
- Schedule When - If management requires that evaluations be conducted on a regular, continuing basis, resistance can be reduced. Maintenance will accept them and try to improve previous results.

A Strategy for Successful Maintenance Evaluations

An 10 - point strategy is necessary to organize and conduct an evaluation and successfully convert the results into improvements.

Here is what managers must do:

1. Develop an Evaluation Policy - There must be a management policy requiring that maintenance be evaluated on a regular, continuing basis. This precludes resistance and redirects the energy of resistance into preparation.

2. Provide Advance Notification - Advise personnel of the content, purpose and use of the results. Avoid surprises.
3. Educate Personnel about the Purpose - The evaluation must emphasize its purpose of obtaining improvements. Misconceptions must be changed through education.
4. Publicize Content - The content of the evaluation must be defined in advance to allow preparation.
5. Schedule the Evaluation - Schedule the evaluation to avoid conflicts like shutdowns or the absence of key people.
6. Commit to the Method of Handling of Results - The plant manager must use the results constructively and convert them into improvements. If the evaluations are corporate sponsored, don't compare mines and create discomfort. Let each plant act on its needs but, don't hesitate to offer support if asked for.
7. Publicize the Results - Publicize evaluation results and offer concrete evidence of management support for needed corrective actions.
8. Take Immediate Action on Results - Organize the improvement effort immediately. Set up action groups and get started.
9. Announce Specific Gains - When gains attributable to the evaluation can be identified, announce them and give credit to the personnel involved.
10. Specify Dates of Next Evaluation - Specify the dates of the next evaluation and identify additional activities for future evaluations. Set new goals and restate the policy of regular evaluations.

Selecting the Best Evaluation Technique

The plant situation influences selection of the best evaluation technique. Acknowledge that some mines may require a thorough evaluation while others may only need a progress check. Three techniques are commonly used:

Physical Audit - This technique includes observation of work, examination of key activities like PM or planning, review of costs, measurement of productivity etc. It relies on interviews, direct observation of activities and examination of procedures, records, costs etc. The technique produces objective and reliable information. It is usually conducted by a team of consultants or company personnel (or both).

Questionnaire - The questionnaire technique targets a cross section of plant personnel typically asking them to compare maintenance performance against well defined performance standards. The results are the views of plant personnel who have consciously identified necessary improvements and desire to achieve them.

Evaluating to Improve Performance

This means potential support for improvement but, only if the results are shared and discussed. Questionnaires must be administered in a business like environment. The questionnaire produces reliable results quickly while minimizing disruption to the operation. Although results are subjective, the participation itself is an education allowing personnel to make an informed commitment to supporting improvement. 1.

Physical Audit with Questionnaire - The combination of a physical audit and a questionnaire provides the most complete coverage. This technique combines the objectivity of the audit with the educational value of the questionnaire.

Recommendations - A physical audit, to ensure objectivity, combined with a questionnaire, to gain early commitment, should be used to ensure the most reliable findings. This technique is the best way of preparing the plant for the improvement effort that must follow. The questionnaire, by itself, is the best choice when a quick, non disruptive evaluation can serve as a reasonable guide in developing a plan of improvement action.

The physical audit alone should be used when management feels that personnel could not be frank, objective and constructive in completing a questionnaire.

Getting Ready for Evaluations

Invariably, the evaluation must include an assessment of the maintenance organization, its program and the environment in which the program must be carried out:

- Organization - Examination of the organization should assess organizational structure and responsiveness as well as supervision, work load measurement, labor utilization, productivity, motivation, craft training and supervisor training.
- Program - An assessment of the maintenance program should include checking the adequacy of its definition, use of terminology, preventive maintenance, the work order system, content and use of information, planning, use of standards, scheduling, work control procedures, maintenance engineering, use of technology and control of the backlog.

- Environment - An examination of the environment should include a review of the objective, policies, support by management, production and staff departments. In addition, material control, safety and housekeeping should be checked. Theses are the factors that influence performance over which maintenance has little control.

The plant should be advised of the evaluation content in advance so they can prepare. If the evaluation is the first one, a full range of activities warrants examination. However, if the evaluation is one of a continuing series, determine what really needs to be looked at based on previous results.

In selecting the evaluation dates, the plant should be aware of potential conflicts such as shutdowns, personnel changes or the absenteeism of key personnel. Rarely does advance notice result in any dramatic shift in performance, so let people know. Then, once the evaluation content is agreed upon, schedule the evaluation carefully, particularly if it will involve a physical audit lasting several weeks.

Often there is genuine fear of audits and evaluations. Some is based on previous bad experiences while some has no basis. Many simply don't want anyone 'looking over their shoulders'.

The danger is that the fear can turn into resistance. Personnel must be educated on the purpose of the evaluation. They must understand how the evaluation identifies problems and starts the improvement process.

Results and How to Deal With Them

The main objective of the evaluation is to convert the results into a plan of improvement action and organize plant personnel to carry out necessary improvements. Several typical case studies illustrate how evaluations were able to uncover significant problems and apply successful solutions to achieve improvements.

Supervision - At a manufacturing complex, maintenance supervisors were criticized for their lack of supervisory capability. Few had received any supervisory training. Most worked directly with their crews rather than leading them. As a result, work control and productivity were poor.

Evaluating to Improve Performance

Solution - Rather than blame the supervisors, the maintenance manager shared the evaluation results with them and asked them to try to explain why they were rated poorly. The discussion revealed they had received no training as supervisors nor were their duties prescribed. They simply carried their craft skills blindly into the unfamiliar environment of the supervisor. As a result of this feedback, training was carried out and duties clarified. Areas in need of improvement were able to be corrected. The willingness of the manager to share results encouraged discussion leading to corrective actions.

Poorly Defined Maintenance Program - At a large paper mill, an evaluation revealed a poorly defined program and, as a result, confusion and poor performance.

Solution - Using the details of the evaluation as a guide, an advisory group was formed to define the program. The group devised a schematic concept drawing, affording a simple, but effective explanation of how maintenance services would be carried out. Training was conducted for key personnel using the concept as a guide. Within a short time, confusion was eliminated and performance dramatically improved. Better communications had made the difference.

Objective - In a large foundry, an evaluation revealed that maintenance was using over 30% of its manpower on non maintenance activities like construction. As a result, it was unable to carry out the basic maintenance program consistently.

Solution - By identifying the cause, a clear maintenance objective was established and procedures set up to allocate manpower equitably between engineering projects and maintenance work.

Maintenance Policies - An evaluation conducted at a chemical processing plant established that most operations personnel had the attitude that once a work request was made their role in supporting the maintenance program was complete. As a result, there was little cooperation in making equipment available or getting operators to perform helpful PM related tasks.

Solution - Realizing that the intended 'team' effort was not present, the plant manager made operations responsible for the 'fact' of maintenance. Operations soon exhibited interest in costs and job performance and challenged existing procedures in a constructive way.

In turn, maintenance reviewed its program more critically. The new level of cooperation led to improvement in performance.

Productivity - At a distribution warehouse, an evaluation revealed poor worker productivity. It was seldom measured nor was the poor productivity ever explained. Misunderstandings by workers showed they were fearful and equated measurements with layoffs.

Solution - Using evaluation results as a guide, an educational program was launched. The outcome was a willingness of workers to conduct measurements to uncover 'factors inhibiting work'. Discussion of results led to beneficial changes that improved both productivity and overall performance.

Preventive Maintenance (PM) - An evaluation in a food processing plant showed that PM inspections were carried out ineffectively resulting in excessive emergency repairs. Evaluation details further showed that few maintenance personnel understood what PM really was.

Solution - As a result of this finding, personnel were educated and the perspective of PM inspections restored. Within a short time the quality and timeliness of inspections improved, accompanied by a steady reduction in emergency repairs and downtime.

Planning - The evaluation of a 6-man planning department in a large sawmill revealed that planners did everything but plan. Evaluation details showed that there were no criteria prescribing what work should be planned. Each planner had a different concept of his job.

Some tried to plan every job. Others acted as material coordinators. The planning department was, as a result, ineffective.

Solution - A discussion of evaluation results with planners and supervisors resulted in the creation of criteria for determining which type of jobs should be planned. The criteria provided a clear division of which jobs supervisors were responsible for and which would be planned. With a clearer mandate, the planning department improved its effectiveness and supervisors established a better working relationship with planners.

Work Order Systems - A locomotive repair shop was having difficulty controlling work and obtaining cost and repair history information. Evaluation details showed there was no work order system.

Rather, there was one piece of paper, inadequate for detailed planning and overly complex for a simple request. No provision was made for the inevitable verbal orders and efforts to suppress them resulted in no work order and no information about the job. Standing work orders were a 'burial ground' for costs and repair history that should have been isolated unit by unit.

Solution - A joint review of the situation by operations, maintenance and accounting led to the development of a workable system that quickly corrected the problem.

Material Control - During an evaluation at a gas generation plant, craftsmen complained of difficulty in identifying materials. Later, warehouse supervisors demonstrated a simple procedure for identifying materials. As a result of these conflicting findings it was learned that craftsmen had not been trained on the procedure.

Solution - Subsequently, training was provided with a significant reduction in craftsmen complaints.

Information Systems - At a foundry complex an evaluation established that, while everyone could easily look up parts, few could manage maintenance with the existing information system. Maintenance had purchased a 'stand alone' package only to find that they needed an integrated system. The program failed to link with field labor and material data thereby blocking access to vital management information.

Solution - Subsequently, the accounting department provided the link to close the information gap based on a review of the evaluation results with maintenance.

Summary - An evaluation establishes the current maintenance performance level by identifying those activities needing improvement as well as those being performed effectively. It not only confirms performance but, is the starting point for necessary improvements. Yet, the most important by products of a well conceived and effectively conducted evaluation are the education of plant personnel and their commitment to provide genuine support for improving maintenance. An evaluation that successfully points the way to improvement makes it more attainable. However, participation of the people who can create the improvements also assures they are ready to help in a cause they believe in.

Notes:

For a more comprehensive discussion of evaluation techniques, additional cases studies and a look at successful conversion of evaluation results into improvement actions, see Chapter 42 - Conducting Effective Evaluations and Chapter 43 - Converting Evaluation Results into Improvement Actions

Appendix G is the Maintenance Performance Evaluation, a self evaluation technique.

1. Images copyright New Vision Technologies, Inc.

II

Maintenance Organization in Transition

13

Examining Organizational Options

The Reorganization Process

Plant maintenance is in transition and many plants that are still poised in traditional organizational patterns are looking hard at:

- Total Productive Maintenance (TPM)
- Self directed Work Teams
- Teams with Coordinators
- Equipment Management Programs

Phrases like World Class Maintenance, employee empowerment and just in time scheduling tempt and intrigue them. Brochures arrive daily proclaiming that:

"A team organization has the potential for improving productivity and performance rapidly. Many long standing problems will be solved by better communications." (Consultant/teacher selling people awareness seminars)

Not all are enthusiastic about the future prospects of the organizational changes being proposed.

Skepticism abounds, typical:

- "Employee empowerment is a stop watch in sheep's clothing." (UAW official)
- "What happens to me after 21 years?" (Foreman on being advised of mandatory early retirement when teams are formed)

Organization Questions - Some realistic questions must be answered about the contemplated changes, among them:

- Will a new team suddenly catch productivity like one catches the flu ?
- Will PM suddenly be successful the day TPM is declared installed ?
- Will a policy on employee empowerment cause the union steward to start 'cheer leading' for more work ?
- If few could use the work order system adequately before, will the adoption of an equipment management program cause a dramatic change ?

The answers to these questions and the solution to the dilemma, lies in appreciating that plant maintenance is different from the manufacturing environment.

Transition to TPM - Many manufacturing plants have made a transition into teams or adopted TPM because their operating circumstances are different:

- Fewer unions
- Less severe working conditions
- Less massive, often simpler equipment to maintain

Facts About Maintenance - There are some facts about plant maintenance that must be taken into account in considering the future organization:

- Maintenance costs often exceed 35 % of operating costs.
- Maintenance is the largest controllable cost in the plant.
- Traditional organizations are difficult to dislodge.
- Unions resist change, often violently.
- All personnel are assigned to foremen who may not understand productivity.
- Productivity is seldom measured.
- Operations may not make equipment available readily.
- Programs are often incomplete.
- PM is sometimes inadequate.
- Information systems are sometimes ignored due to urgency of operation.
- Evaluations are rare.

This is not the end of the bad news. A stark economic fact must be confronted, consider:

The cost of maintenance rarely goes down unless it is done with fewer people and done less often.

Examining Organizational Options

As a result, there are serious considerations about moving from a traditional organization to a team organization. If so:

- The change must start from a successful traditional organization.
- The maintenance program should already be proven successful.
- Everyone must know what, why and how the change will be made.
- Everyone should have a satisfactory answer to "What's in it for me ?"
- Progress and performance must be measured during the change.

Examining the Traditional Types of Maintenance Organizations

Traditional Organization - The traditional organization is a formal, structured hierarchy of line supervisors and staff personnel. Line supervisors are responsible for assigning and controlling work. They prescribe work rules and spend most of their time on work control activities. Staff personnel, like planners, perform administrative tasks like organizing major jobs before they are scheduled. The traditional organization may limit the initiatives of personnel by too rigidly making assignments and prescribing methods of accomplishing work. Compared with a self directed team, the traditional organizations success depends on control whereas, team success depends on the initiatives of individuals who have aligned their personal aspirations with those of the group. See Figure 13-1.

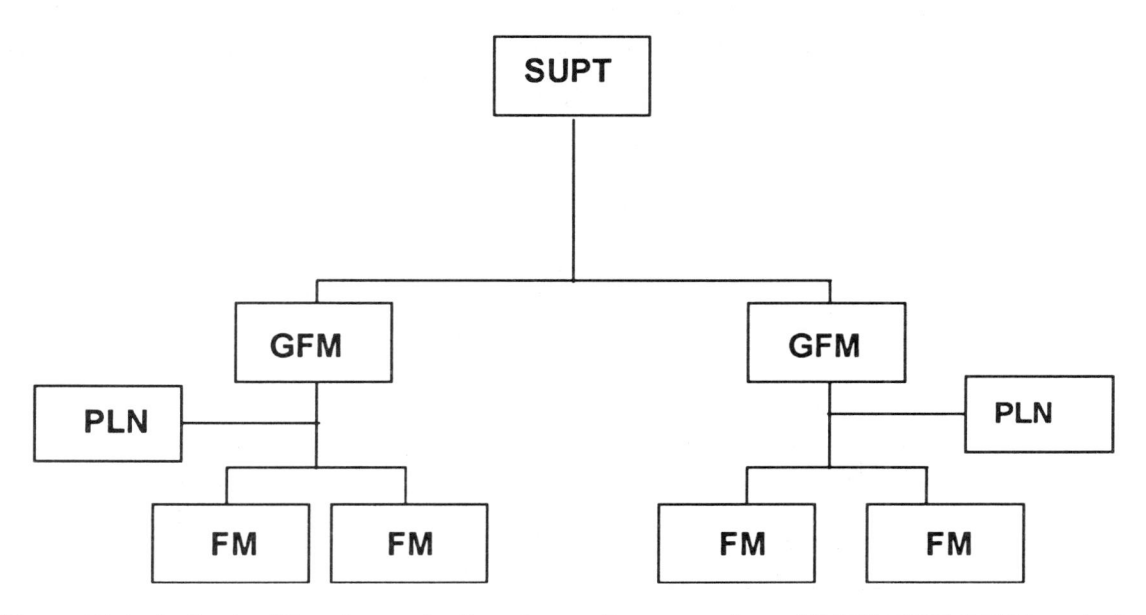

Figure 13-1. In the traditional organization shown, line supervisors (SUPT, GFM & FM) control work while staff personnel (PLN) support them.

Line Staff Organization - There are three types of traditional organizations that are used in plant maintenance: the line staff organization, the craft organization and the area organization. They are illustrated below. See Figure 13-2.

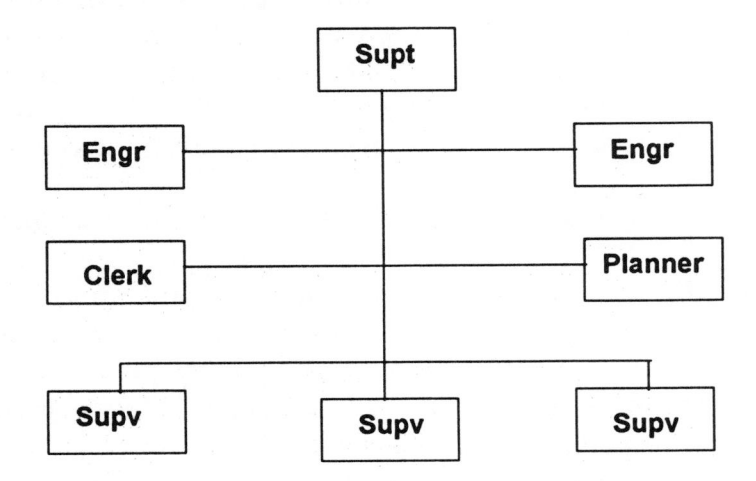

```
                          ┌──────┐
                          │ Supt │
                          └──────┘
                              │
        ┌──────┐                            ┌──────┐
        │ Engr │────────────────────────────│ Engr │
        └──────┘                            └──────┘

        ┌──────┐                            ┌────────┐
        │ Clerk│────────────────────────────│ Planner│
        └──────┘                            └────────┘
                              │
        ┌──────┐           ┌──────┐           ┌──────┐
        │ Supv │           │ Supv │           │ Supv │
        └──────┘           └──────┘           └──────┘
```

Figure 13-2. The line staff organization tends to center control with the superintendent and is often characterized by some inflexibility and, at times, slow response.

Craft Organization - A craft organization is arranged so that one supervisor controls all personnel of the same craft. However, since most work will require more than one craft, many jobs cannot be done without arranging to obtain other crafts from other crews. Supervision of the job is divided as will be the responsibility for work completion and quality. The craft organization tries to relate to work quality but, results do not always agree. See Figure 13-3.

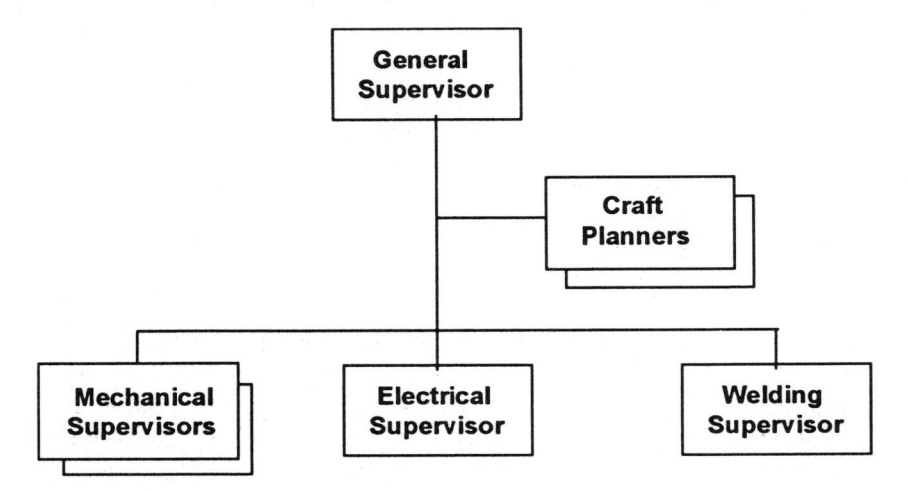

```
                    ┌─────────────┐
                    │   General   │
                    │ Supervisor  │
                    └─────────────┘
                           │
                           ├──────────────┐
                           │              │
                           │       ┌─────────────┐
                           │       │    Craft    │
                           │       │   Planners  │
                           │       └─────────────┘
                           │
        ┌──────────────┬───┴────┬──────────────┐
        │              │        │              │
 ┌─────────────┐ ┌─────────────┐ ┌─────────────┐
 │  Mechanical │ │  Electrical │ │   Welding   │
 │ Supervisors │ │  Supervisor │ │  Supervisor │
 └─────────────┘ └─────────────┘ └─────────────┘
```

Figure 13-3. Many plant maintenance organizations utilize the craft organization. Unfortunately, it presents numerous problems for the organization aspiring to move to a team organization.

Examining Organizational Options

The craft organization creates accountability problems because no one supervisor is responsible for the total maintenance of any unit of equipment or area. Mechanical supervisors blame electrical supervisors and vice versa.

Electrical craftsmen, used to receiving assignments from electrical supervisors, may not feel that mechanical supervisors are competent to supervise them. Similarly, mechanical supervisors with no electrical background may not want to supervise electricians. Control of several crafts may also be a problem for each supervisor.

Most craft supervisors 'grew up' in the crafts they now supervise and continue to think in terms of 'how to make something run again' rather than how to improve work control. Few are interested in improving productivity and tend to feel that "if a crew of nine will do the job, fourteen are better". Poor productivity is further assured by assigning all personnel to a supervisor. As a result, a planner, for example, needing craftsmen for scheduled work must 'horse trade' with supervisors to get the personnel needed.

Craft organizations create enclaves of specialists with limitations in applying skills. Typical, the electrician cannot loosen the bolts to move an electric motor. He must call for a plant repairman to do it. Specializations like 'liner repairmen' abound with no special skills but, rather the job security that guarantees other personnel cannot intrude into their domain. Unions love these arrangements and personnel with seniority move into them while guarding the secret that the job often only takes 25 hours a week of 'work'.

In general, the plant maintenance organization aspiring to a team organization must first escape from the 'mind set' of the craft organization.

Area Organizations - Many large manufacturing plants, especially in heavy industry, have successfully made the transition into area organizations. The area organization helps to establish many of the accountability and control features that team organizations promise. In an area organization, each area supervisor is responsible for all maintenance carried out in his area. He has all of the crafts required to perform day to day maintenance. That is, if he is a mechanical supervisor, he would be supervising other crafts assigned to him including, if needed, electricians. The area organization forces the supervisor to concentrate on work control rather than the technical aspects of each job. As in a team organization, he must rely on the professionalism of each craftsman. Thus, team like circumstances can develop.

In establishing the area organization, the plant is divided into logical, properly sized geographical areas. See Figure 13-4.

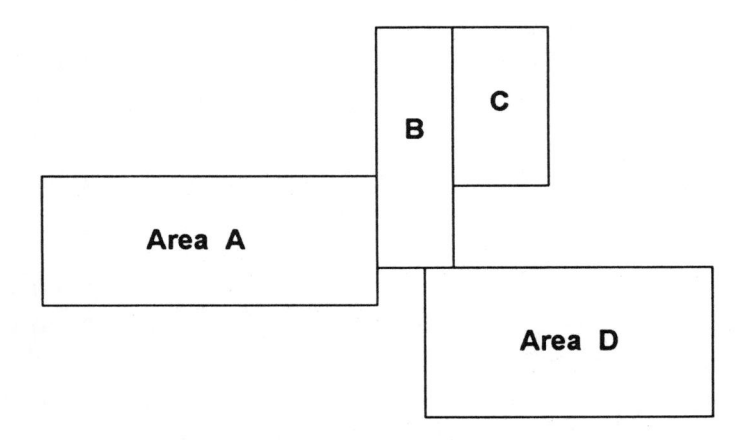

Figure 13-4. Each supervisor is responsible for all maintenance in a specific area.

The corresponding area organization then provides an area supervisor for each designated area. See Figure 13-5.

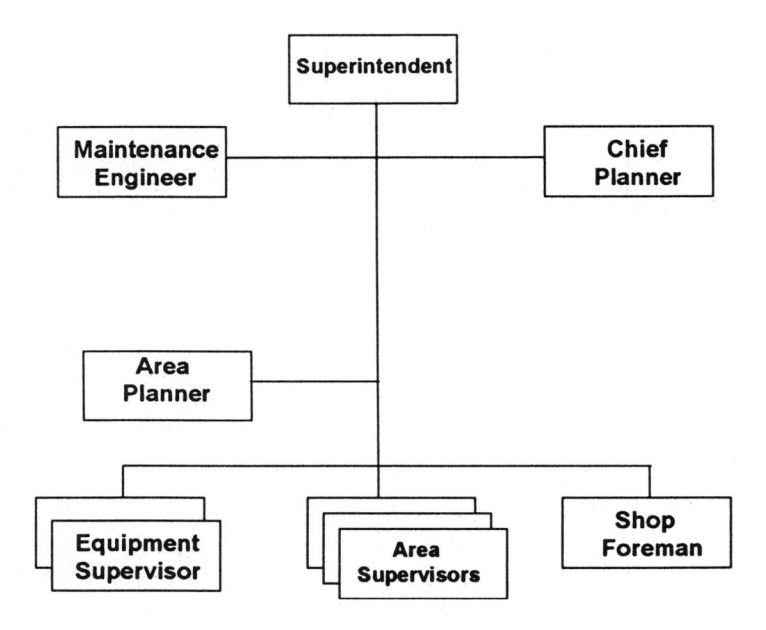

Figure 13-5. Area supervisors are assisted by planners plus shop and equipment support.

Craft Pool Operation - Many area organizations divide the work required in the area between an area crew that does the day to day maintenance and a craft pool. The craft pool supplies the additional personnel needed for a shutdown or major repair.

Examining Organizational Options 65

Once the craft pool personnel complete the work in any area, they are withdrawn for use elsewhere. In actual practice, craft pool personnel are left in an area for a whole week so that, in addition to completing the scheduled work, they can help the area reduce its outstanding backlog of work. This arrangement reduces travel time in moving pool personnel from one area to another. See Figure 13-6.

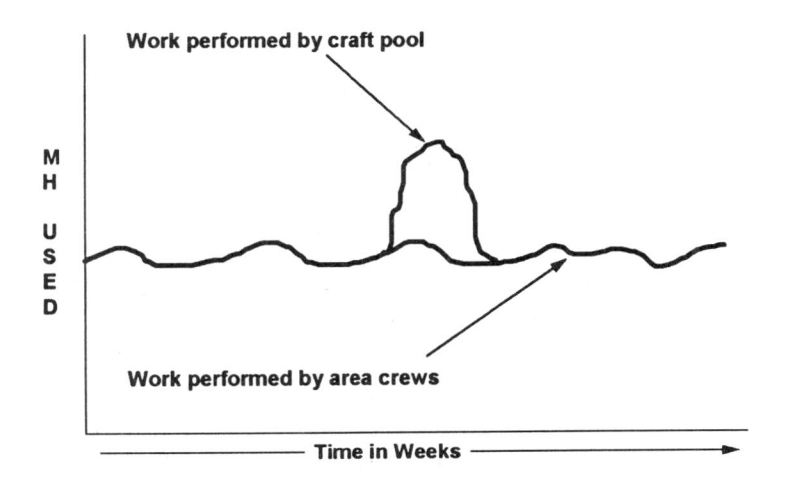

Figure 13-6. Craft pool personnel supplement area crews during peak work load periods.

Since regular area crew members are more familiar with the equipment in their area, the area supervisor may elect to use craft pool personnel on work other than the primary equipment during their time in his area. Alternately, he may mix them with his own personnel in accomplishing the work. Essentially, they are a temporary reinforcement to his regular area crew. Since the area supervisor is totally responsible for his area, he is forced to manage it. Thus, an information system that he may have ignored as a craft supervisor is suddenly important to him. It is not surprising to observe an area supervisor encouraging the pool personnel to finish up more quickly and get off 'his payroll'. Other phenomena take place as well, among them:

- Pool personnel with lesser craft skills are quickly discovered and trained.
- Craftsmen used to 'going along for the ride' are uncovered and motivated.
- Area supervisors with access to all resources are excelling as managers.
- Area crews are demanding and getting full cooperation from operators.

The combination of the area organization and the craft pool also improves planning and scheduling effectiveness. Areas bid on the craft pool at weekly scheduling meetings. The timing of the shutdown and the size of an area's backlog are elements judged in awarding manpower from the pool for the week following.

Measuring Schedule Performance - Also, at each scheduling meeting, performance for the previous week is reviewed. This ensures that a continuous measure of performance is carried out and progress is checked. Among the elements reviewed are:

- Compliance with the work schedule
- Compliance with the PM schedule (85% completion minimum)
- Reduction in backlog by craft
- Use of manpower (more planned, less emergency and unscheduled work)
- Planned jobs ready to go
- Status of non maintenance projects

Results of Craft Pool Utilization - After about six months of operation with the area organization and the craft pool, considerable information about the work force comes to light. First, it is discovered that after awarding pool personnel, there is a pattern in the number and types of personnel in the pool that are not needed. As this pattern continues, it signals that these are the personnel probably not required. Since, previous to the formation of the craft pool all personnel were assigned to craft supervisors, this confirms that their interests may not have included the best use of personnel. In fact, some craft supervisors may have been over staffed. Obviously, with a labor contract, these excess personnel cannot simply be discharged. However, for the first time, many mines know what the size and craft composition of their maintenance work force should be. Thus, they are in a position to make intelligent rather than arbitrary decisions regarding their disposition.

Second, all of the personnel whose craft skills are marginal are now well identified. The identification process is simple. Area crews simply don't want them in their areas as they cause more problems then they are worth. With solid evidence of their shortcomings, union officials are only too happy to support the re training of these personnel to keep them on the work force.

Third, the confirmation of personnel with a poor work attitude is seldom a matter of discipline as the other area crew and pool personnel soon exert enough pressure to straighten them out. If there is an incentive program, the peer pressure works even faster in solving this problem.

In most instances, about nine months after the area organization and craft pool has been in operation, the plant can confirm the size and craft composition of its maintenance work force.

Examining Organizational Options 67

At that point they are able to start applying equitable measures to downsize. See Chapter 16 - Downsizing the Maintenance Organization.

The area organization and craft pool are proving to be an important interim organizational step in the transition from a craft organization to a team. This is demonstrated in the next section which describes what plants are doing.

What Plant Organizations are Doing

New Organization Trends - In many industries, a pattern for new plant organizations is starting to take shape. Generally, they focus on the entire plant rather than on maintenance alone. Characteristically, those plants in the most difficult economic shape are moving the fastest, while those in a relatively comfortable situation are assuming a slower pace.

The organization theme is one of making a single person responsible for a total activity and assigning the necessary personnel to him or giving him access to them. Often operations and maintenance personnel are being combined under the same managers to flatten to the organizational hierarchy. This is similar to the idea of the area organization and craft pool arrangement described in the previous section. See Figure 13-7.

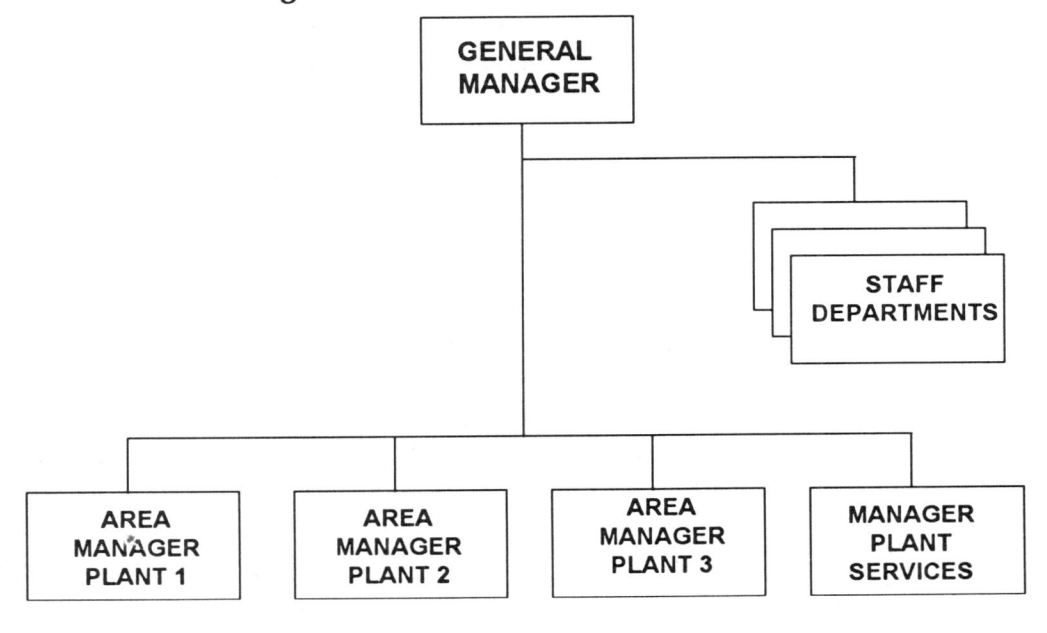

Figure 13-7. In the total plant organization each area manager controls both operations and maintenance. The plant services manager controls staff functions like engineering or planning. Regular staff departments like purchasing or accounting are often unchanged.

Within each area, managers are responsible for the total activity. Area managers are supported by section managers in which an area is equivalent, for example, to a smelter, a concentrator or the pit (mine itself). See Figure 13-8.

```
           AREA
          MANAGER
             |
    ---------+---------
    |        |        |
 SECTION  SECTION  SECTION
 MANAGER  MANAGER  MANAGER
```

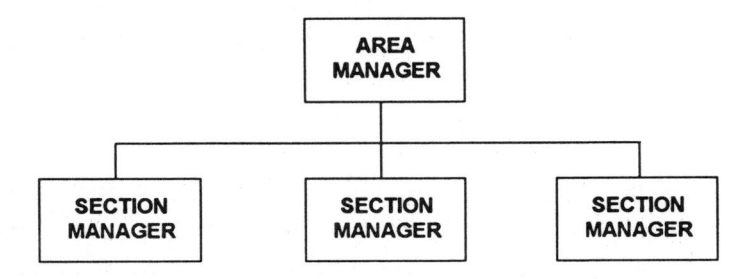

Figure 13-8. Section managers control sub activities such as: grinding, flotation or haulage (pit).

In turn, section managers control several coordinators. See Figure 13-9.

```
          SECTION
          MANAGER
             |
    ---------+---------
    |        |        |
COORDINATOR COORDINATOR COORDINATOR
 Lines 1-5   Lines 6-10  Lines 11-15
```

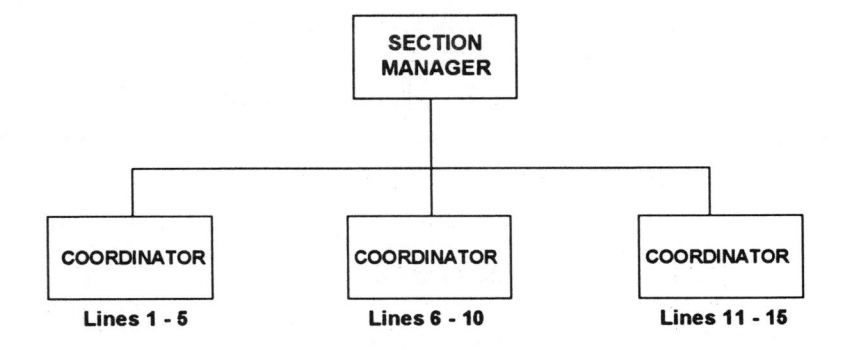

Figure 13-9. Typically, coordinators would manage several processing lines.

Each coordinator then:

- Controls all resources necessary to meet operational objectives
- Controls operators directly
- Controls craftsmen for day to day maintenance (as in area organization)
- Has access to additional craftsmen for shutdowns or overhauls (as in a pool)
- Has access to services like engineering, planning or lube services

It is important to note that certain aspects of these new organizations resemble the steps taken in the operation of the area organization and craft pool. This not only confirms that they work but, it is strong endorsement for the use of the area organization and craft pool as a transitional step, particularly in changing the behavioral patterns of former craft supervisors and their craft crews.

Examining Organizational Options

Note the organization of a typical concentrator grinding section below. See Figure 13-10.

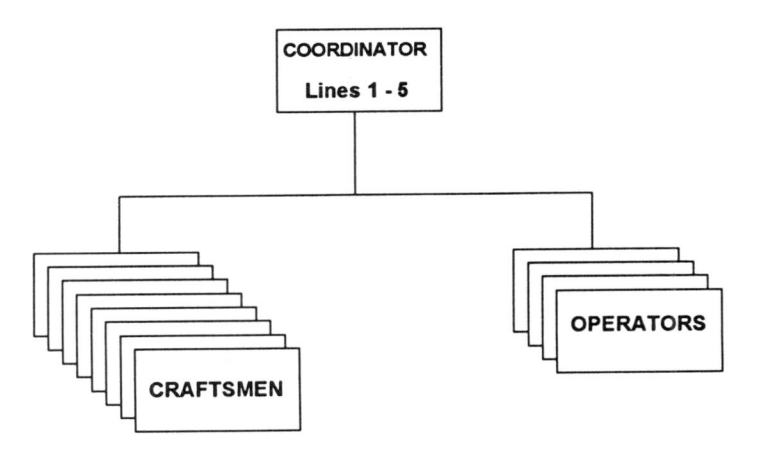

Figure 13-10. The organization for day to day operations provides for control of both operators and maintenance personnel by the coordinator.

To better support areas and sections, the plant services group controls all of the staff activities like planning and engineering as shown below. See Figure 13-11.

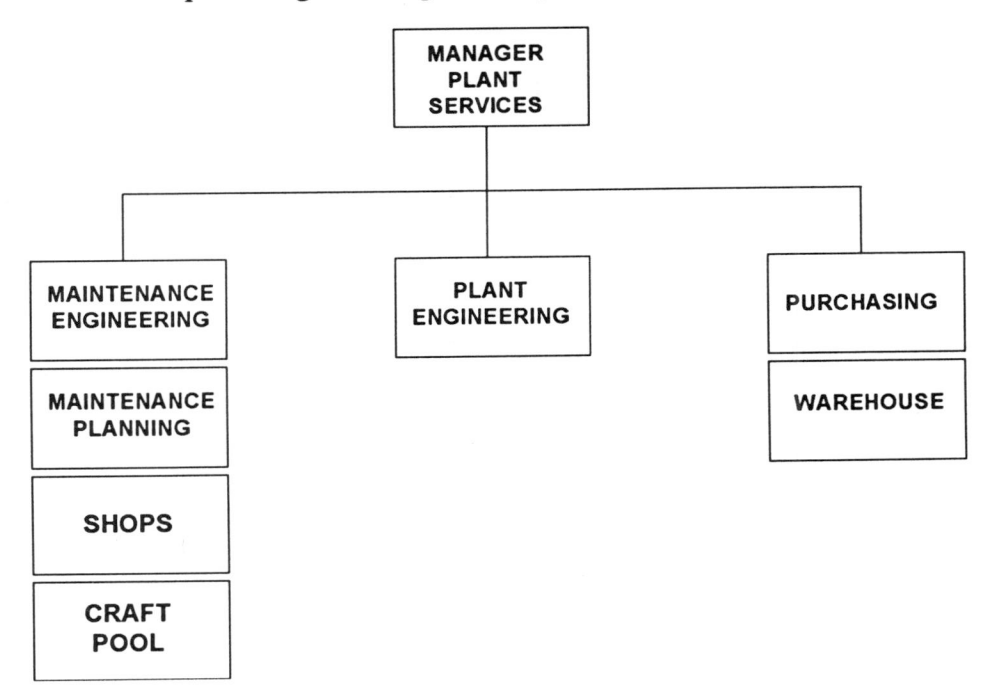

Figure 13-11. Note that purchasing and warehousing are placed under the plant services manager as an alternative to having them report to a general manager (as in Figure 13-7 On page. 67).

Typically, the staff agencies being placed in plant services include:

- Maintenance Engineering
- Predictive maintenance services like vibration monitoring
- Lubrication services
- Maintenance planning
- Shops
- All crafts not performing day to day maintenance (pool)
- Plant engineering
- Quality control
- Safety
- Training
- Environmental control and Labs

Already benefits are accruing in better planning and control of activities like lubrication and predictive maintenance. Larger plants, especially where they have a maintenance engineer, are profiting from the organization of a tight maintenance support group headed up by the maintenance engineer. Typically:

- Planning is now carried out against a strict criteria.
- Predictive maintenance results are being analyzed.
- Lube personnel are no longer 'at large'.
- The craft pool is well informed and organized.
- Specialized training is taking place.

These steps are improving productivity and performance. See Figure 13-12.

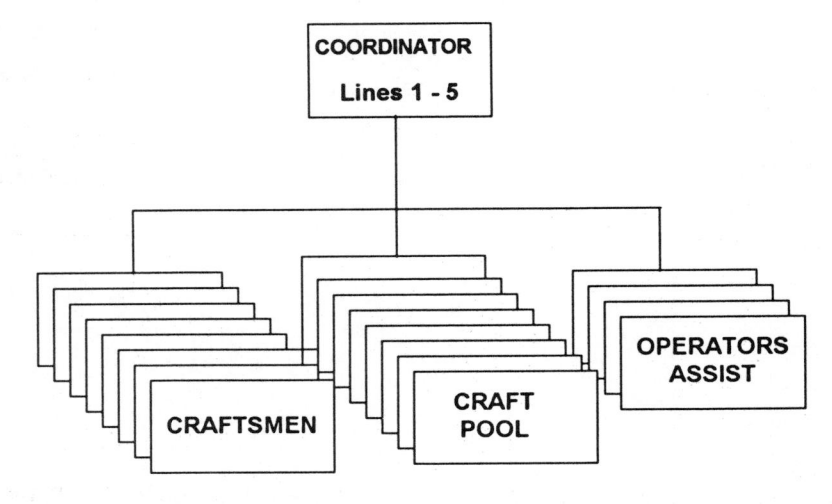

Figure 13-12. A coordinator using craft pool personnel during a shutdown to improve productvity.

Examining Organizational Options

Below, the road building organization is shown first in their normal organizational configuration and then with craft pool personnel added to assist in an overhaul. See Figure 13-13 and Figure 13-14.

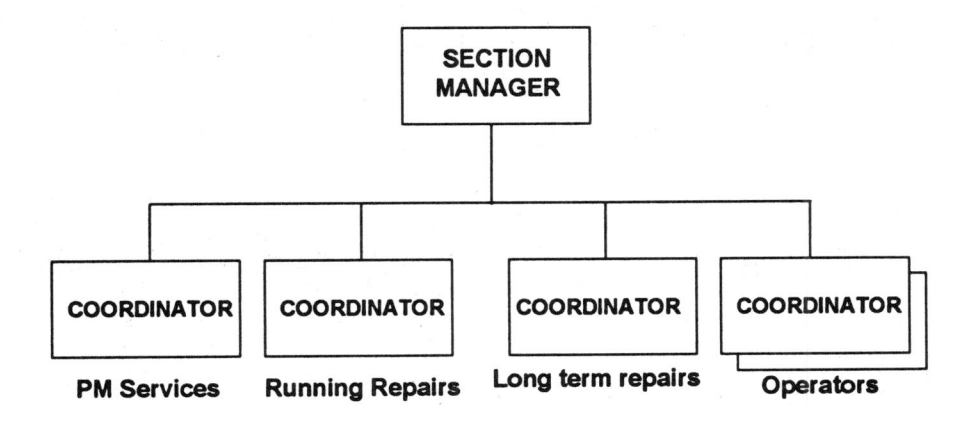

Figure 13-13. Note the division of the garage into PM, running repair and long term repair bays.

As the craft pool personnel are added they are assigned to work on an overhaul being carried out in a long term repair bay. See Figure 13-14.

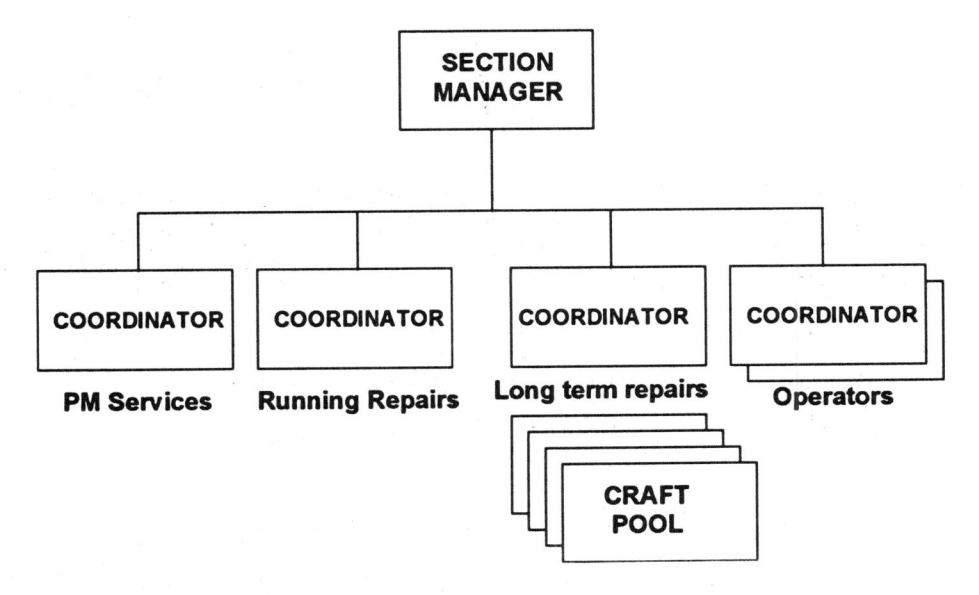

Figure 13-14. For an overhaul, craft pool personnel remain in the area until the overhaul is completed.

Changes in Maintenance Planning - One of the most dramatic improvements being brought about by these organizational changes is in maintenance planning. Previously, it was not unusual to see planners doing everything but planning. Generally, they were used as line supervisors saw fit. To often, they became clerks or parts 'gofers' or were used in administrative tasks like figuring out vacation times or preparing reports that a faulty information system couldn't produce. In the new posture they are often tightly controlled by the maintenance engineer with much better results. Typically, this central control has produced significant results. Planners now:

- Plan only major jobs
- Are assisted by material coordinators (from material control department)
- Develop:
  - Task lists
  - Material Lists and . . .
  - Tool Lists for major, repetitive jobs
- Have access to craft pool personnel

There have also been significant changes in maintenance, among them:

- Mechanical and electrical skills are under common supervision.
- Ex maintenance supervisors now control both operations and maintenance.
- Professionalism of craftsmen is getting major emphasis.
- Skill assessment and training are being carried out.
- Computer skills for all are being pushed.
- Maintenance engineering is becoming more prominent.
- Planners plan only major jobs.
- Material coordinators are assigned to help planners.

With operations, the same type of benefits are accruing. Typical:

- Ex operations supervisors now control both operations and maintenance.
- Operators are being trained on the maintenance program.
- Computer skills are being pushed.
- Operators are doing simple PM and many minor repairs.
- Craftsmen are training operators on PM and repair skills.
- Section managers run weekly maintenance scheduling meetings.
- Coordinators run daily coordination meetings.
- Coordinators are accountable for cost and performance.
- All are insisting on better information quality and timeliness.

Examining Organizational Options

Monitoring Progress in Organization Changes - A critical aspect of the organizational change is to recognize that personnel will be in a state of flux as the changes are being made. Therefore, a method of monitoring progress and performance levels attained by the changes is an absolute requirement. See Figure 13-15.

Also see Chapter 12 - Evaluating to Improve Performance and Chapter 42 - Conducting Effective Evaluations. Appendix G - Maintenance Performance Evaluation, is helpful as well. Chapter 15 - Making Effective Organizational Changes provides more detail in the conduct of the change.

However, the following steps for carrying out organization changes are pertinent at this stage. It is important in the successful implementation to:

- Step 1 - Conduct an 'as is' evaluation of the total plant organization.
- Step 2 - Correct as many problems as possible before moving forward.
- Step 3 - Develop a reorganization plan and get management concurrence.
- Step 4 - Consult with union officials, brief them on what, why and how.
- Step 5 - Inform small groups until all are informed (use chain of command).
- Step 6 - Form advisory teams to make recommendations.
- Step 7 - Install changes in test areas before going plant wide.
- Step 8 - Conduct periodic evaluations to check progress and find problems.
- Step 9 - Announce gains or problems, and seek advise from advisory groups.
- Step 10 - Carry on to conclusion and continue progress measurements.

Figure 13-15. At this stage the most important aspect of the change is to seek and utilize the participation of all personnel who will be involved.

The help of the personnel who are involved is needed to make the change successful. Thus, their education in understanding the what, why and how must begin as soon as the organization changes necessary are identified. [1]

Summary - Plant maintenance organizations have had difficulty emerging from the 'mind set' created by long association with craft organizations and difficult labor situations. Many are making the transition to more effective organizations by adopting area organizations to create a more team like environment.

Other plants are moving forward to a team organization covering the entire plant. In the new organization configurations, many long standing problems of control, productivity and performance are being solved.

1. Images copyright, New vision Technologies, Inc.

14

Implementing Successful Team Organizations

The Magic of the Word - Team

The reality of smaller, more productive maintenance work forces with 'self directed teams' achieving prodigious amounts of quality work at vastly reduced costs has eluded many industrial organizations.

Urged on by industrial managers who visualize astounding Japanese like achievements, maintenance organizations are struggling mightily to emulate them, often unsuccessfully and with considerable frustration.

Advisers, consultants and corporate specialists abound and eagerly offer their services on how to do it. They will:

- Train discussion facilitators
- Show how to conduct meetings
- Demonstrate how to get captive groups to list 'interesting suggestions' like painting stairs yellow and expanding lunch rooms

Unfortunately, too few such advisors understand industry or even the maintenance environment within which the advice is offered. As a result, very few substantive suggestions emerge on the one topic it's all about, improving productivity and performance.

Somehow, the promise of smaller maintenance work forces operating without supervision, working efficiently to yield high quality work at lower cost has not materialized in many instances.

Here are 10 points to consider.

One - A team probably won't succeed unless employees think they need it as badly as management wants it.

The Japanese have succeeded with the team concept, in part, because they are the homogeneous society we aren't. For example, try visualizing a crew of burly millwrights doing their morning calisthenics with their supervisor or explaining a union picket line to visiting Japanese businessmen. [1] See Figure 14-1.

Figure 14-1. Team means different things to different people.

Better quality work is not necessarily an exciting enough objective in America. But, a need to improve that threatens plant closure suddenly makes teams possible. This explains why an unprofitable plant sold to its employees is often made profitable in record time.

Two - Level With Employees.

Employee productivity, improved performance and cost reduction are among the expectations of management for the maintenance team organization.

However, management also knows that the cost of maintenance does not go down unless the number of people doing it and the frequency at which it is done is reduced.

Employees also know these facts. Therefore, a credible explanation must be offered on why a team is thought necessary. If the maintenance work force must get smaller to reduce costs, then preserve some jobs by displacing outside contractors with current maintenance employees. If the work force simply must be reduced, explain the options of early retirement.

Implementing Successful Team Organizations 77

Three - Straighten out the Maintenance Program.

The principal cause of the failure to successfully implement a team organization is the absence of a well defined maintenance program within which the team is expected to operate.

A self directed team operates well if they are presented with well identified, ready to go jobs, at the start of each shift. Their strength is in sorting out who does what and getting on with the work. ₂ See Figure 14-2.

Figure 14-2. A well defined system must exist before team members can do their thing.

Conversely, if the team is bombarded with requests for emergency repairs it responds to the most life threatening and then retreats to the sanctuary of their lunch room to let the dust settle. While many team members appreciate that continuous equipment condition monitoring can prevent emergencies, it is simply not their responsibility to create the monitoring program. The maintenance engineer must create a viable program into which the efforts of the team fit.

Organizations aspiring to implement a team successfully, must bring the team to life within the framework of a well defined maintenance program. Conversion of the traditional organization to a team is more difficult than starting new.

Four - Don't forget the traditional supervisor.

The biggest surprise is resistance from the maintenance supervisor, presumably part of management. He sees himself becoming obsolete. Because his future is threatened and uncertain, he fights to preserve the status quo or delay the questionable future that faces him. [3] See Figure 14-3.

Figure 14-3. The supervisor must be told what his new job will be in the future team organization.

When converting a traditional organization, assure the supervisor of his future task in the team organization.

Five - Recruit the new team carefully.

Generally, organizations that recruit directly into a new non union organization stand the best chance of success. Their personnel directors are able to spot and reject the 'storm-trooper' type in favor of the potential team player. As a result, the team starts out with fewer handicaps. Despite this, many 'frustrated contractors' are hired as craftsmen. Soon, preventive maintenance services that help guarantee reliable equipment are displaced with questionable projects encouraged or even solicited by these folks. Recruiters should look for personnel with good craft skills who exhibit the potential for team membership. However, to these qualifications must be added:

- An appreciation of how the 'total' maintenance program is carried out
- How it fits within the managers' production strategy (his plan for using maintenance services to achieve profitability)

Implementing Successful Team Organizations

If prospective team members appreciate the framework of maintenance management, their contribution to a successful team is much greater. By starting with the right type of people, the plant can avoid taking on difficult behavioral changes that are required to make the team functional.

    Six - Decide on the salaried status carefully.

The maintenance team may be made up of former hourly workers whose status may now be changed to that of salaried employees. These new salaried employees may be looking for the 'perks' (like flexible lunch periods) that other salaried employees seem to enjoy. They must realize that their work status and shift assignments simply do not permit them the same latitude afforded other salaried personnel. Equipment maintenance needs do not recognize individual preferences. When the need arises, there must be rapid, dependable response. [4] See Figure 14 - 4.

Figure 14-4. Awarding a salaried status to teams must consider potential problems it can trigger.

To help them make the transition, provide ground rules to clarify their expectations. Ensure personnel know that:

- Equipment maintenance requirements set their work schedules.
- Vacation periods or lunch hours may lack the flexibility of those provided to office workers, for example.
- As workers, as well as decision makers, they may have constraints on what decisions they can make. They can specify the order in which they do work providing it is coordinated with operations but, they cannot decide whether or not to do certain work requested by operations.

    Seven - Establish team decision making parameters early.

A is qualified to make decisions on how to perform work. However, if for example, they are asked to decide how to manage maintenance or which shifts they will work, it may ask more than they can deliver. Neither the decision nor the result will be satisfactory.

Determine in advance the boundaries within which the teams can make decisions. The team can:

- Schedule established PM services after coordinating with production but, they cannot specify what services are to be performed. They could however, recommend changes to existing programs.
- Modify equipment only as prescribed by maintenance engineering. But, they can recommend how the modifications might be made.

Eight - Determine how craft skills will be evaluated and remedial training provided to improve lagging skills.

A team can become a hiding place for a worker who needs skill training as well as a 'whistle blower' for someone who doesn't carry his load. Consequently, there must be a method of determining training needs and carrying them out. See Figure 14-5. 5

Figure 14-5. Provide continuous skill training but, also establish procedures for evaluation of skills and provision of remedial training when it is found to be necessary.

Nine - Give the team firm work control procedures, as most team members have not had the training or the experience to develop their own.

Within a framework of reasonable work control procedures, such as daily work assignments and determining which type of work requires planning, most teams can be successful. To illustrate:

At a plywood plant, maintenance personnel on back shifts operated without maintenance supervision. This arrangement thrust craft team members into a situation requiring that they call 'their own shots'. Fortunately, they had received considerable training on the overall maintenance program and, as a result, saw themselves as the direct maintenance contacts on the back shifts. With this education, they saw the total picture of work control and were able to competently 'manage' work requests, unscheduled repairs and handle 'surprises'. They did this with praise and compliments from their production counterparts.

Implementing Successful Team Organizations

Ten - Organize maintenance and establish a working relationship with operations that fosters team utilization and development. For example:

In a meat packing plant, each 6 man production crew was assigned one mechanic. The mechanics realized their success would depend on working as a team with the operators on the crew. They not only made it clear to operators what services they would perform but, when they must be done. To solidify the team working relationship, mechanics tried to be helpful to operators.

If an operator needed a break, for example, a mechanic relieved him. When the operator returned to the unit, the mechanic provided pointers on checking equipment during operation. In turn, if there were heavy repair jobs to be done, the mechanic could count on the operators for help. No one told them to do it, they just did it. This type of maintenance - operations cooperation made the team successful. [6] See Figure 14-6.

Figure 14-6. Mechanics and operators who work together create a healthy team environment!

Summary - Self directed teams can be beneficial to both the organization as well as to team members. The team will become successful more quickly if it is developed within the framework of a well identified maintenance program. Sensible guidelines on team decision making and skill training should be applied as the team is being developed rather than after it is discovered these are problem areas.

Conversion of traditional organizations will be more successful when the, about to be displaced, supervisor is given an honest picture of his future role. Most important to the success of a team is the creation of an environment in which the team can succeed, including solid reasons for employees to want the team as much as management does.

1 - 6. Images copyright, New Vision Technologies Inc.

15

Making Effective Organizational Changes

Implementing Successful Organizational Change

"It must be considered that there is nothing more difficult to carry out, nor more doubtful of success, nor more dangerous to handle than to initiate a new order of things. For the reformer has enemies in all of those who profit from the old order, and only luke warm defenders in those who profit from the new." [1]

Little differentiates the problems of organization change today from the challenge implied in these words when they were written in 1532.

Any manager who has made a significant organization change can testify that the attitudes toward the changes can range from enthusiastic support to stubborn resistance. Thus, winning the support of individuals so that the necessary changes can be made often becomes the managers primary implementation concern. He knows he must successfully reduce resistance to change so that he can obtain support for desired changes. Therefore, a fundamental step in successful organizational change is the understanding of resistance to change.

Most personnel react in one of three ways to changes affecting them:

- Socially - Crew members resist as a group as all agree to slow down.
- Psychologically - Individuals alter attitudes toward work by complaining of working conditions.
- Behaviorally - Individuals or groups change work patterns by requiring helpers if they suspect cut backs.

Misunderstanding, insecurity, inflexibility and fear of loss of power or authority all breed resistance:

- Craftsmen are fighting TPM implementation because operators will do maintenance work even though TPM would free them to do more skilled work (misunderstanding).
- A craft supervisor opposes teams because he has not been informed of his future job (insecurity).
- The new information system is forcing supervisors to discontinue the familiar paper work order system (inflexibility).
- The chief planner will report to the maintenance engineer to give better focus to planning (loss of power).

Each level of acceptance or rejection is accompanied by identifiable reactions. See Table 15-1.

Table 15-1
Characteristic Behavior to Organization Change

Acceptance
- Enthusiastic cooperation and support
- Cooperation
- Cooperation under pressure from management
- Acceptance
- Passive resignation

Indifference
- Apathy; loss of interest in the job
- Doing only what is ordered
- Regressive behavior

Passive resistance
- Non learning
- Protests
- Working to rule

Active resistance
- Doing as little as possible
- Slowing down
- Personal withdrawal
- Committing errors
- Deliberate sabotage

Overcoming Resistance to Change

Management's success in affecting change will depend on the skill exercised in utilizing techniques to reduce resistance to change. Including:

Persuasion - If rewards or bargaining are used as a means of persuading personnel to support change, managers must know the actual causes of resistance in order to focus their persuasive efforts on fears, such as loss of pay or status.

Security - An effective measure for reducing resistance is to create a feeling of security. If the work force must be reduced, management should make places for the men who will be laid off. This can be done by reducing overtime, cutting back on the use of contractors or retraining personnel for other jobs in the organization. It may even be necessary to assure personnel at the outset that none will be laid off but, that redundancy will be covered with attrition.

Understanding - Normal communications are often impaired during periods of organization changes. Therefore, management must ensure full understanding of the changes and their consequences. Face to face discussions are best, since employees questions can be answered directly.

Time - If well planned, organization changes rarely need to be made suddenly. Employees need time to condition themselves to the upcoming adjustments. If time is short, increased resistance can be expected and should be weighed against the value of early initiation.

Flexibility - Trial runs have several advantages, particularly in the flexibility they afford. Persons involved in the changes will have opportunities to test the new systems before making judgments. Supervisory personnel can collect more data. Persons who are expected to resist can, during a trial run, more objectively judge the value of the changes. Management, also, can test the new systems and the methods of implementation before committing itself, and make modifications as indicated. Trial runs reduce threats to those affected and, thus lessen resistance.

Participation - Of the methods available to management to reduce resistance to change, personnel involvement is the best. If changes are to be forced on maintenance employees, expect resistance. But, given the opportunity to have a say in their own destiny, they will generate effective recommendations instead of resistance.

In one industrial organization, for instance, the personnel department needed a new classification plan for maintenance craftsmen and it was difficult to come up with an acceptable one. A committee made up of superintendents and supervisors was appointed and five meetings were scheduled, two weeks apart.

At the first meeting, a personnel representative explained the problem and the basic principles of personnel classification. At the second meeting, the committee roughed out the types of jobs and the levels of difficulty involved, and created four tentative personnel classifications. At the third meeting. they were able to define the specific entrance qualifications for each classification, and reduced the total number of classifications to three. The fourth meeting was held with union representatives who had already been briefed and were in agreement. Meanwhile, the 'grapevine' was used by non participating employees to make suggestions to the committee. The final classification plan was presented at the fifth meeting and approved on the spot by the personnel department.

When the personnel department scheduled face to face discussions with the craftsmen, they found that there was no need for explanatory meetings. The new plan was well understood and had been fully accepted.

Employees need a feeling of responsibility. They can get it through participation. Advice should be asked of the persons involved and action taken on recommendations with merit. Personnel who make suggestions, and then see that their recommendations are included in the new procedures, have a part ownership in the change. Thus, they will work hard to convince others of the value of the whole change. Therefore, participation should be a commanding element in the method of implementation. Often, the degree of participation dictates the success or failure of the reorganization.

Minimizing Organizational Change Problems

Examination of the current organization can reveal needs essential for planning for the new organization as well as implementing it. For example:

Objective - Organizations have objectives, unique to their own operation. These must be analyzed before initiating organization changes. If changes will create new and wider objectives, the proposed organization must be able to support them. Thus, new organizational concepts, like teams, may require different objectives to accommodate the alteration of the traditional relationships with operations for example.

Philosophy - Management philosophy includes the beliefs, attitudes, values and supportive actions that affect the running of an organization. Management philosophy can generally be evidenced in the way the organization operates.

- Does it exhibit a friendly, easy going atmosphere with a minimum of formal rules ?
- Or, does it seem cold, formal, and constantly under pressure ?

Changes must be evaluated in terms of the philosophy of the organization. If changes are not compatible with the philosophy, conflict can result. Thus, the objectives of the reorganization must be in agreement with management philosophy.

Key Activities - Each organization has a concept based on established facts and circumstances and, often, key activities are related to them. Some are easy to identify:

- The principle activity of a maintenance crew is to keep equipment in a safe, effective operating condition.
- There must be timely detection of equipment problems to ensure adequate time to plan work.

Sometimes key activities are not so obvious. For example, although the best source of information on equipment should be the supervisor who maintains it, data obtained from him are often carelessly collected and measured, frequently distorted, and, hence, of limited use. When supervisors collect these data reliably, many maintenance problems are reduced. Thus, the supervisors effort to collect quality data is a key activity.

Through analysis, established activities can be shifted into new organizational status and prominence. Thus, organizational planners must identify key activities to optimize their effects on the future organization. Typical:

Information Needs - Management is the control of activities, either directly or through subordinates. The tools used to shape decisions for control are objectives, strategies, policies, organization, formal planning and clearly defined programs. Because information powers these tools, systems must be designed to provide every level of management with the data needed for decision making and control.

Analysis of the Current Organization - The present organization should be examined carefully to identify the sound characteristics that should be retained. This study should ask:

- Is the present structure clearly defined, understood and adhered to?
- Are managerial processes such as planning, decision making and performance appraisal arranged for?
- Have future manpower needs such as training and recruiting been covered?
- Has leadership considered the willingness of key personnel to alter behavior to accommodate changes?

Findings of the analysis will indicate the manner in which the organizational change should be carried out. For example, if reorganization can be phased in over several years, it may be better to develop supervisory talent than recruit new personnel.

Adherence to Management Principles - After the preliminary planning, it is necessary to review applicable principles of organization. Typically:

- The organization must reflect its objectives.
- The organization must coordinate people, activities and efforts.
- Duties, responsibility, authority and relationships must be defined.
- Superiors are responsible for the acts of their subordinates.
- Final authority rests with mangers.
- Lines of authority must be established to subordinates
- Responsibility must have corresponding authority.
- Reorganization is a continuous process and should be planned for.

Leadership During Change

Generally, the originators of the change (usually the staff specialists) will have the least objectivity and the most optimism about the change. They are the change agents and not the implementors. They will be impatient about timing and implementation and will also resist efforts to modify the original concepts. They may even consider suggestions as interference.

Supervisors and superintendents will be the most objective, since they must deal with the actual implementation and ensure its benefits are secured.

Clearly, it is important that all participants be as unified as possible in their desire to accomplish changes.

Making Effective Organizational Changes

As an organization is restructured both individuals and groups are affected by changes in responsibility, levels of supervision, delegation of authority, size and characteristics of work groups, and individual job assignments.

At the start of any change process, the sensitivity of individuals and groups must be taken into account. Intended changes which are properly announced, no surprises, always objective and positive, never personal or critical, have an excellent chance for acceptance. It must be remembered that personnel at all levels are essential to the implementation procedure. Therefore, the following steps for carrying out organizational change should be considered:

1 - Start by conducting an 'as is' evaluation of the maintenance organization, its program and the environment in which maintenance operates. (In examining the environment look at the quality of support provided to maintenance by activities such as material control or accounting as well as operations. These are the activities on which maintenance depends yet, they are not controlled by them.) This step will help you to judge the capability of the organization to take on changes as well as assess their readiness to do so.

2. - Classify the results of the evaluation into problems requiring fundamental change or improvement and those problems that can wait. Also identify areas that were satisfactory. Correct as many problems as possible before moving forward. The most serious problems must be corrected before starting into the organizational change to preclude them from undermining the change effort. The evaluation must be done. An organizational change attempted without first evaluating to find out the current situation is like stepping into a 'mine field blind folded'.

3 - Develop the maintenance re organization plan and get management concurrence. You must determine whether the actions to be taken by maintenance fit within the plant manager's future organizational plans. The plant manager may have other plans for key personnel in maintenance. He may even want to transfer some into operations or vice versa. Maintenance and operations may even be merged.

4 - Consult with union officials, brief them on what, why and how. As the leaders of the maintenance hourly personnel their comprehension of the changes will help to dispel considerable potential resistance.

5 - Work through small crew groups, using the chain of command until all are informed. Answer individual questions and ensure that all get answers to the questions they pose. By using the chain of command, the line supervisors who must ensure that the implementation is successful get involved. They must, of course, be well informed on the what, why and how. Since they will be talking with their own personnel, they can translate questions and concerns into practical answers while warning of areas of potential resistance.

6 - Extend participation by meeting informally with groups that can recommend practical methods of implementation in their areas. For example, ask them to list the specific types of minor repairs that would be made by operators and to prescribe the training they will give to operators. Always link a responsibility of the group with any suggestion they make. This will keep recommendations practical and commit them to getting them done.

7 - Install changes in test areas before going plant wide. Empower the test area to modify the original plan with a better one. Thus, participation and action are combined.

8 - Conduct a follow up evaluation to uncover problems that people are having with the new organization. Since few organizational changes will go through without some modification, there must be a method for finding out what the problems are and obtaining suggestions to correct them. Such an evaluation also measures progress from the initial evaluation (step one above).

9 - As specific problems are encountered, seek advise from the advisory groups formed in step six (above) to gain further participation.

10 - Carry on to the conclusion of the implementation process punctuating it with periodic evaluations to verify progress while providing personnel with a known method for reporting problems and offering recommendations.

Summary - Organizational changes generate people resistance and the conversion of resistance energies into implementation support mark successful change. By understanding resistance and how to reduce it, managers can assure that the organizational changes will be properly implemented.

1. - Machiavelli, *The Prince*, 1532.

16

Downsizing the Maintenance Organization

Downsizing Maintenance

Downsizing maintenance to meet economic needs or even to assure the survival of the plant is now a fact of life. Unfortunately, successfully trimming of the work force creates the additional problem of how to reduce the maintenance work force in the least harmful way to the plant. Traditionally, management has felt that the easiest way to cut maintenance costs is to arbitrarily reduce the work force. See Figure 16-1.

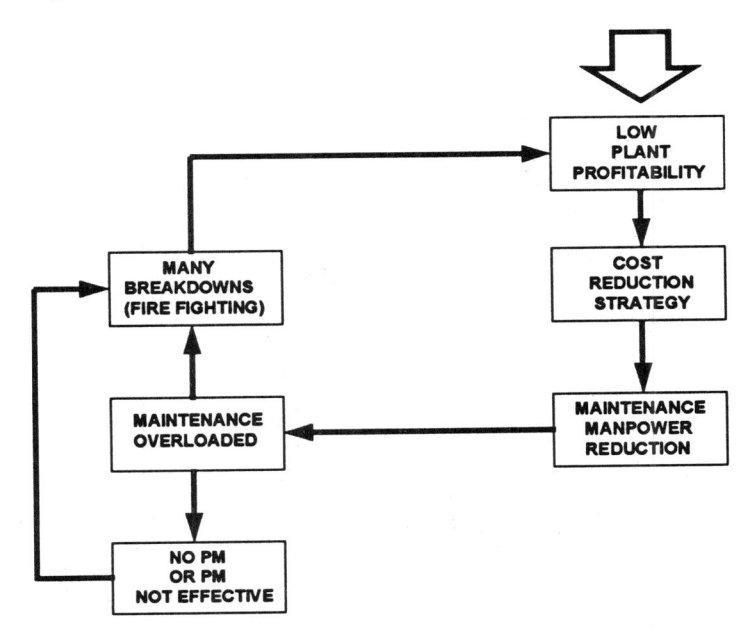

Figure 16-1. Arbitrary work force reductions start a chain reaction of further damage.

To avoid arbitrary manpower reductions and keep the maintenance work force effective, maintenance must command certain facts to be able to successfully defend the minimum levels to which it can be reduced. Maintenance management must establish:

- What constitutes the work load and how it was measured
- Where the work load can be reduced with the least harm to plant operations
- How to measure and then improve labor productivity

The need for efficient maintenance work force reductions has been created by today's economic pressures. In the past, mines were less concerned about work load and productivity. Measurements were avoided because of the possibility of adverse reaction. Now, maintenance organizations are being forced to measure the work load and verify productivity to ensure their maintenance crews are at effective levels.

In those instances in which work force reductions are necessary, there must be a plan that will allow maintenance to reduce the work force with a minimum loss of efficiency. Generally, plant management, maintenance, and labor unions can work as a team to resolve difficult situations. Most unions now accept the economic facts of survival and agree to work force reductions, providing they are fair to its membership. In most cases, management is keeping faith with long-term employees in terms of retirement benefits and other rights they have accrued through the years. By working together jobs, profits, and overall company welfare can be protected.

Typically, the maintenance workload includes:

- Basic Maintenance - Preventive maintenance services, unscheduled repairs, emergency work and planned, scheduled maintenance
- Fixed Assignments - Janitorial or custodial services, administrative jobs, utility work plus buildings and grounds service
- Capital Projects - Construction, installation of new equipment and equipment modification

When work force reductions are being considered, each of these maintenance actions must be examined and it determined whether:

- A particular job can be deferred
- The work has to be done so often
- The work has to be done at all

Downsizing the Maintenance Organization

Generally, lubrication, preventive maintenance services, and planned, scheduled maintenance should not be reduced. Rather, these actions should be emphasized.

While, emergencies and unscheduled repairs are a poor utilization of manpower, emphasis on PM services and regularly scheduled maintenance can minimize them.

Janitorial and custodial work can be cut back if the plant can tolerate a lower level of service. Administrative jobs can often be combined and clerical tasks can often be performed using a computer.

Utility work should not be cut back because it is needed to preserve assets. But buildings and grounds services can usually be curtailed without problems. For example, the grass need not be cut as often. Capital projects, especially new construction jobs, can often be deferred. But, equipment modifications to ensure maximum production or to ease maintenance tasks should not be delayed.

The Impact of Productivity

Maintenance labor control is a major challenge for maintenance and improving productivity is a must.

In one processing plant, a work sampling survey revealed that of the total time a mechanical crew was available for work, 32 percent of the time was spent working, 23 percent was devoted to travel, 26 percent was waiting or idle time, 16 percent was spent on clerical tasks and the remaining 3 percent was spent in 'other' activities.

By combining electrical and mechanical crafts into an area crew and moving the crew into the area where most of the repair work requests originated, waiting times were cut in half and productivity was substantially boosted. In addition, an examination of projects showed that a significant amount of construction and modification work could be deferred. The combined increase in productivity and project work load reductions enabled maintenance to consider reducing the work force.

The question then became: What strategy can be used to reduce the maintenance work force to assure the least disruption to the plant ?

A step by step approach to work force reduction involves measuring maintenance workloads and crew productivity, and studying labor turnover and absenteeism trends. This example indicates how a 300 man work force can be adjusted to a 245-man force with minimum disruption. See Figure 16-2.

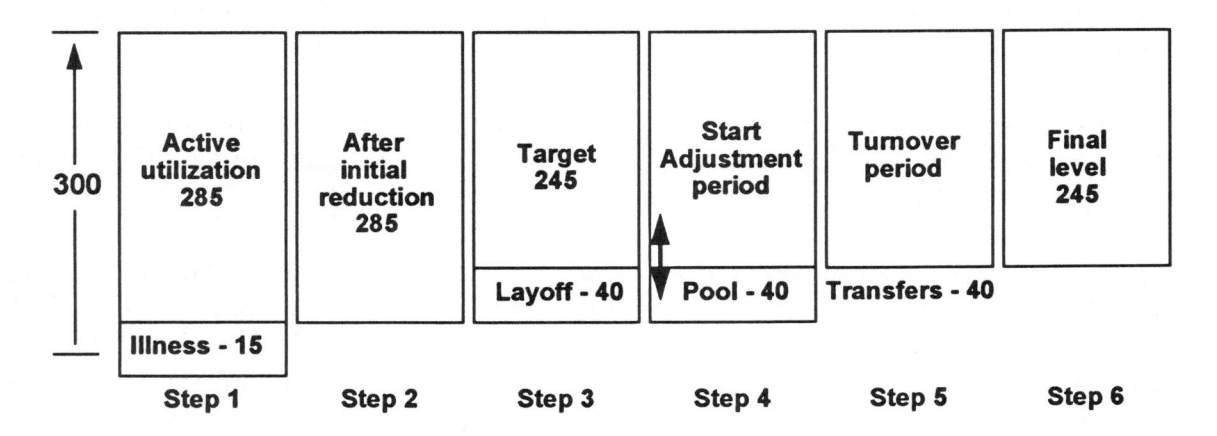

Figure 16-2. A step by step strategy was used to minimize disruption to maintenance.

Step 1 - Personnel who have been away from work for several months (usually because of illness) are identified. These workers are unlikely to return to the active work force but, they must be classified according to the specific action recommended in each case. For example, should the individual be urged to retire at the normal retirement age ? Should he be encouraged to retire early, with financial inducements acceptable to both the company and the individual ? Or, should retirement (with satisfactory financial arrangements) be forced because of medical problems ?

Step 2 - Personnel on permanent disability are reduced based on individual cases (-15).

Step 3 - The exact number of skilled and unskilled personnel required to meet workload levels are determined (245). Maintenance also identifies the skills and the number of men that can be eliminated from the work force without hindering the maintenance operation (-40).

Step 4 - A personnel pool is formed of employees who, although trained in skills valuable to the company, must be targeted for layoff. These men remain in the pool until normal turnover makes room for them on the regular work force. The pool gives flexibility at a time when the whole work force is in a state of flux.

Downsizing the Maintenance Organization

Step 5 - As attrition in the regular work force occurs, the men who leave are replaced from the personnel pool, not by new employees. As personnel transfer from the pool to the regular work force, a few new employees may be required to ensure a work load versus work force balance.

Step 6 - The adjusted work force level, the number of maintenance men who can satisfactorily perform the work at the desired maintenance labor productivity level is reached (245). Men with the right skills have been slowly and systematically worked into the regular work force as other personnel retire and leave.

Based on projected improvements in productivity, the work force level can be reduced. Work force level is affected by elimination of unnecessary jobs, productivity increases, and workload variations. Basic changes in the workload can influence the projected work force reductions. See Figure 16-3.

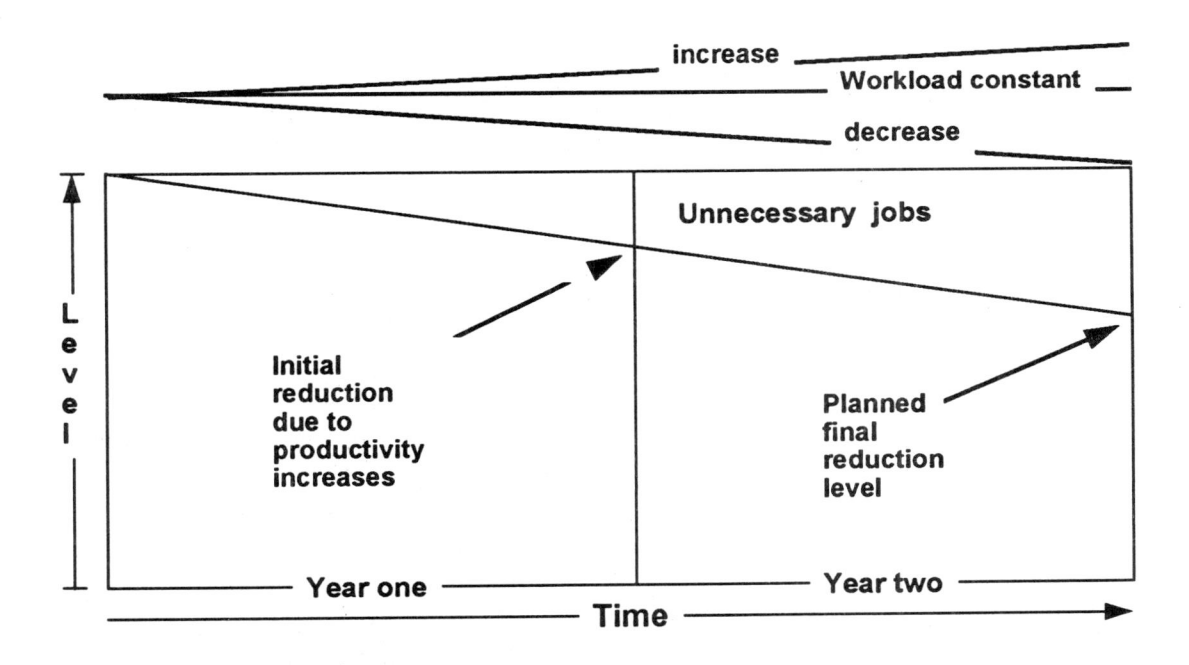

Figure 16-3. Over an extended period, reductions are based on projected work loads.

For example, planned reductions can be carried out if the work load remains constant. However, if the workload increases, the planned work force reductions will have to be restudied, because more personnel will be needed to meet the increased work load. And, if there is a decrease in the work load, there should also be a larger work force reduction.

Steps That Facilitate Work Force Reductions

Traditional Maintenance Shortcomings - One has to look no further than at the things that plant maintenance traditionally does not do well to identify the things that, if improved, will greatly facilitate the task of work force reduction while improving overall maintenance performance. Consider this approach:

Step 1 - Evaluate maintenance, including its productivity and establish a performance 'benchmark'.

Step 2 - Examine the organization. Eliminate wasteful craft organizations. Combine electrical and mechanical crews under common supervision. Organize a strong maintenance engineering effort to ensure that PM is carried out properly. Install a craft pool to boost productivity.

Step 3 - Require maintenance to spell out its program in detail. What will they do and how will they do it? Have them educate all plant personnel.

Step 4 - Demand weekly reports of PM compliance since little planning will happen unless PM services first find the problems in advance to allow planning to take place.

Step 5 - Require that maintenance planning develop standard task lists, material lists and tool lists for every major job repeated at a frequency of six months or less. This will streamline parts procurement or component rebuilding and assure uniform, quality work

Step 6 - Scrutinize all non maintenance work to determine whether it is necessary, feasible and properly funded. About 40 % will be deferred or disapproved further reducing the work load.

Step 7 - Make operations responsible for the cost of maintenance. They will stop abusing equipment, demand that PM be done on time, make equipment available for work promptly and insist that major repairs be competed on time.

Step 8 - Require that every person in maintenance be able to use the work order - information system competently. Discontinue all systems and procedures not incorporated with the primary information system.

Step 9 - Oversee the weekly operations - maintenance scheduling meeting to ensure that work was properly planned, resources allocated and the previous weeks performance checked. If improvement is needed, take corrective steps.

Step 10 - Evaluate and measure performance relentlessly.

Summary - Smaller, more efficient maintenance work forces are necessary for survival. Careful planning and control with continuous evaluation will yield a satisfactory end result when proper attention is given to individuals, whether they stay or go.

17

Essential Duties of Key Maintenance Personnel

Who are the Key Maintenance Personnel ?

With the advent of teams that will require craftsmen to receive, organize and perform work, and then account for what they did, they must be added to the list of key maintenance personnel, who include:

- The superintendent who is responsible for the overall success of the maintenance program and the effective utilization of the work force
- The maintenance engineer who develops and implements services to ensure equipment reliability and maintainability
- The maintenance planner who organizes major jobs so that, upon execution, work is completed more effectively and efficiently
- The craftsman who does the work and must execute his work assignments diligently and produce quality work

The most critical person in the traditional maintenance organization is the supervisor. He is responsible for getting the work done, on time, with quality results and, often, under trying conditions. He controls craftsmen directly and is responsible for everything they do or fail to do.

A review of the supervisors situation gives perspective to the definition of essential duties of key maintenance personnel, since:

- Tomorrow's team members will pattern their behavior on the supervisors who directed them before the teams were formed.
- The supervisors who help launch the teams will leave their imprint on future work control, productivity and decision making.

Supervisory Problems to Be Faced - Today's maintenance supervisor is caught in the conflict of cutting costs and work force reduction while being required to repair equipment that is often pushed to the failure point. A maintenance supervisor who is really in command of his crew and his work load produces an atmosphere of confidence. Tangible results are measured by declining costs, reduced downtime, fewer emergencies and less overtime. See Figure 17-1. [1]

Figure 17-1. The supervisor is responsible for everything his crew does or fails to do.

The maintenance supervisor who is in trouble exhibits the reverse of these circumstances. Some are easy to spot, typical:

- Overtime is excessive
- The actual work load isn't known
- Productivity is poor
- There is little PM and less planning

Most maintenance supervisors were once craftsmen. As plant expansion provided for advancement, the span of time between leadman, foreman, supervisor and superintendent did not allow mastery of each job level before moving on. Thus, supervisors and superintendents found themselves with responsibilities they could not handle. They became the victims of an environment thrust upon them. Typical:

- An expanding plant will undergo an increase in its maintenance work force. Don't assume that an ex craftsman can adequately manage 30 - 60 personnel without being trained to do so!
- New equipment and processes require new technology. Don't assume the 'old timers' will pick up new techniques as they go!

Demands placed on the maintenance supervisor create an environment that would challenge more experienced, better educated managers. Yet, how does the supervisor see it? Does he feel:

- 'Dead-ended' with no opportunity for advancement
- Without a line of progression to assure promotion
- Blocked from promotion because he alone can make the 'equipment run again'

Essential Duties of Key Maintenance Personnel

How do craftsmen see the supervisors job? As the primary source of supervisors, the better craftsmen will step forward if they find it attractive. If not, don't expect good candidates. If a supervisor wants to go back to his tools, you have a real problem. Many maintenance supervisors are considered indispensable. After years at the plant, there are few crises that the supervisor has not encountered several times. See Figure 17-2. [2]

Figure 17-2. Do craftsmen see and appreciate the demands of the supervisors job?

If the plant does not have a means of measuring maintenance performance, he is measured on his response to emergencies. Soon he gears his whole effort to emergencies. He is soon elevated to 'folk hero' status as he bails the plant out of each new crisis. The crisis need not have occurred if there had been some type of PM effort. Now, the supervisor covers his odds with extra people and stashes 'bootleg' parts so they will be handy for the next crisis. When managers do not concern themselves with maintenance supervision, they contribute to these problems. Yet, today, plants are forming teams which eliminate the maintenance supervisor if favor of coordinators, for example, who 'control' both operations and maintenance. Most likely, the problems that beset maintenance supervisors will simply be inherited, unsolved by these coordinators. They, in turn, will struggle, to solve them, unless they are solved before the teams are formed.

Plants anticipating the benefits of team organizations, must first solve the fundamental problems of the traditional maintenance supervisor. Managers must make maintenance part of the production strategy. They must open up the sources of maintenance supervision to reduce 'in breeding'. Operations supervisors, for example, who have leadership skills can be effective in maintenance because they can make their crews perform. Supervisor selection must include management potential that can carry the candidate beyond the supervisor's job. It should be made clear to all who aspire to supervision that the supervisor's job is a step in a clearly spelled out line of progression, beyond the confines of the maintenance department. The technically oriented supervisor must be trained to broaden his administrative and managerial skills.

The Principal Duties Outlined

Supervisor - The principal duty of the maintenance supervisor is supervision. He controls his crew to carry out the work assigned him. See Figure 17-3. He carries out PM services, unscheduled and emergency work on his own initiative. Major planned and scheduled jobs appear on an approved weekly schedule. He focuses the crews efforts on these jobs while building other work around them. He is vigilant in his response to any emergencies. [3]

Figure 17-3. The supervisors primary task is supervision with focus on work control.

The supervisor instructs and trains his crew members, arranges their vacation time and counsels and disciplines them as necessary. He prescribes work methods and procedures. He ensures the timely completion of work by adequately supervising it. He is responsible for the conduct of PM services, the control of labor, including overtime, and the procurement of materials for unscheduled and emergency work.

Superintendent - The maintenance superintendent is responsible for the entire maintenance function. He performs this through general supervisors or first line supervisors. He uses maintenance engineers in a staff capacity to ensure the total maintenance program has continuity and that all new construction or installations can be supported by maintenance. Maintenance planners are used to develop plans for resource use on major work that must be scheduled. He defines the maintenance program and verifies that it is consistent with the overall plant operating plan. [4]

Essential Duties of Key Maintenance Personnel 101

In addition, the superintendent, ensures that all maintenance personnel are educated on the maintenance program and, as required, delegates specific aspects for execution by other supervisors. He coordinates closely with operations to ensure that the program meets their needs. He jointly schedules major activities so they interfere least with operations while making the best use of maintenance resources. He trains maintenance supervisors and planners to ensure they can control work properly and carry out planning tasks effectively. See Figure 17-4.

Figure 17-4. The superintendent has total responsibility for all maintenance activities.

He coordinates with staff departments such as purchasing or accounting to ensure they understand the maintenance program and that maintenance personnel utilize staff services properly. He is also concerned with budgeting and long range planning such as future overhauls. He anticipates future requirements and takes steps to ensure the work force is well prepared to handle them when the time comes.

The general supervisor controls two or more first line maintenance supervisors. Alternately, he may be responsible for several areas or shops. Normally, he controls the planning activity which supports his supervisors or areas of responsibility. His time is equally divided between direct field supervision and administrative duties such as planning, cost control, union business, vacation planning, etc. His time focus is the current week into the week following.

Maintenance Engineer - The maintenance engineer is a staff person responsible to the superintendent for ensuring that equipment and facilities are properly installed, modified correctly, well maintained and performing effectively. Specific responsibilities include: assessment of newly installed equipment to ensure its maintainability, availability of parts, maintenance, instructions, prints, adequacy of design and correct installation. See Figure 17-5.

Figure 17-5. The maintenance engineer ensures equipment reliability and maintainability.

In addition, he monitors on going work to ensure that sound, craftsmanship like procedures are being followed. He also observes the adequacy of stockage of correct parts in proper quantity. Further, he monitors the quality of parts used to ensure good performance and recommends action to correct inadequacies. He reviews repair history and costs to determine repair, rebuild, overhaul or corrective maintenance needs. 5

The maintenance engineer also reviews the conduct of PM services to ensure they are conducted on time, according to standard. He reviews equipment inspection deficiencies uncovered as a source of corrective maintenance. He also helps to develop standards for individual major jobs, procedures for performing work, cost levels and the quantity of resources used. Periodically, he makes cost and benefit reviews to determine make or buy actions. He also develops recommendations for training. As required, he prescribes methods for non destructive testing and the use of predictive maintenance techniques.

He constantly reviews the information system to ensure its adequacy and uses it to review the work load and backlog with the view of making recommendations to change work force levels.

Planner - The maintenance planner provides direct support in planning major actions such as overhauls, rebuilds or non maintenance work, like construction, prior to execution by maintenance supervisors and crews. These planning actions include job organization, labor estimating, identifying and obtaining materials and coordination of work with production or engineering. See Figure 17-6.

Figure 17-6. The planner organizes major jobs to help boost productivity. Well planned jobs can reduce the amount of labor required by up to 15 %, thereby reducing elapsed job time.

Once work is scheduled, the planner allocates labor according to priority and monitors the execution of on going major jobs.

Essential Duties of Key Maintenance Personnel

In addition, the planner monitors the conduct of preventive maintenance inspections. He also analyses repair history to convert results into planned maintenance work.

The planner confines his activities to planned, scheduled maintenance and project work. Unscheduled and emergency work are controlled by maintenance supervisors. [6]

Generally, once the current weeks major jobs are scheduled, he turns his attention to the planning of the following weeks work. On going programs like the PM program or the component replacement program guide his efforts in anticipating up coming tasks. By preparing task lists, standard bills of materials and tool lists, the planner can simplify the planning of most repetitive tasks and better ensure clear instructions.

Leadman - The leadman is regular maintenance craftsman who, in consideration of an hourly wage increment and any special talents, can be used temporarily to help a supervisor coordinate elements of major crew jobs.

The leadman should not be considered as a supervisor since he cannot be expected to discipline his fellow crew members.

Team Coordinator - The team coordinator is a craftsman who provides direction to a team in the receipt, assignment, control and accountability for the completion of routine repair work. He assures that it is accomplished expeditiously. The team operates within the framework of a previously well defined maintenance program and fits its activities in with the major planned tasks provided on the approved weekly schedule. Although team members may come from a variety of sources such as ex supervisors or craftsmen, they operate as co equal team members, often exchanging leadership roles to solidify team spirit. [7]

Summary - Maintenance personnel must know what is expected of them to ensure that they can carry out the maintenance program efficiently. Duties should be placed in the framework of the actual program performed to convey the overall picture of their combined activities. See Chapter 20 - How to Define the Maintenance Program.

Many plant organizations are forming teams. Therefore, the duties of the personnel who will make up future teams should be carefully defined so that the team is launched with a solid conceptualization of what needs to be done and how. Once the team becomes a reality, its members can more readily decide who performs or shares in which function.

1- 7. - Images copyright, New Vision Technologies, Inc.

18

Carrying Out Craft Training

A New Dimension

"If this plant were to close today", said a young plant repairman, "it won't matter what skills you have." He went on, "if you can't use the computer and don't understand the program, you aren't even in the job market".

Qualifications of Craftsmen - This is a sign of the times. Being a skillful welder or a certified electrician simply isn't enough. This change in the basic qualifications of individual craftsmen is the result of a need to operate with a smaller, more effective maintenance work force. Everyone must be able to contribute. The days of a supervisor 'calling all the shots' are disappearing. Tomorrow's maintenance departments will demand much more of the individual craftsman. Now he must add to his craft skills the ability to control work and operate within a team environment. Industry is faced with a whole new set of circumstances on how it must train craftsmen to prepare them for this challenge. While the traditional means of carrying out craft training remain, expectations of better performance add a new dimension to craft training.

Traditional Craft Training

Craft Training Program - A craft training program should have strong management backing. It must ensure that craft skills keep pace with current technology and that basic skills are kept up-to-date. The training materials and methods must be effective and, those giving the training must be well qualified. Training provided should challenge and motivate those being trained.

Personnel who are to be trained should meet well defined, published minimum standards of experience, education and demonstrated performance to qualify for the training.

Testing for Training Effectiveness - Realistic testing should be carried out to ensure that training is effective and those trained did learn. In addition, attendance and progress records should be kept. The program should require that progression from one level to the next be conditional upon demonstrating necessary skills. Higher skill levels should never be awarded on the basis of seniority and, when greater skill levels are achieved, personnel who qualify should be acknowledged with tangible rewards such as a skill level upgrade or a raise in wages.

The traditional craft training program usually includes:

- A general assessment of training needs
- Selection of personnel for training
- Emphasis on basic diagnosis and repair skills
- Limited technical training
- Safety and hazard training
- Training on the production process
- Specialized training on specific problems
- Training by manufacturers
- Limited training on the maintenance program

Training administration tries to assure that:

- Employees' suggestions for training are acted on.
- Techniques, such as video, match training needs.
- New employees are properly indoctrinated.
- Records are kept on training progress.
- Higher skill levels are awarded on successful completion of training.

Training results, if measured, focus on achieving:

- Less rework
- Fewer overtime hours
- Improved quality
- Timely job completion

The Challenge for Today's Craft Training

Training Yields Equipment Reliability - The profitability of a plant depends significantly on how well its production equipment is maintained. Equipment must be reliable so that the production levels established to ensure profitability can be met consistently. This equipment reliability must be guaranteed by the craftsmen who perform the maintenance. However, today's craftsmen must be able to do more than 'fix' things. They must meet a new challenge that requires them to:

- Control work
- Operate without supervision
- Understand the overall maintenance program
- Operate in a team environment.

In addition, the quest for plant profitability has created new circumstances.

Newer Equipment - With the advent of newer, more technically demanding equipment, the skills and abilities of the work force may not match the maintenance needs of the equipment. They must be upgraded.

Work Force Reductions - Downsizing, a necessary action to keep some plants in the black, compounds the challenge of carrying out effective maintenance. Younger low-seniority workers may be forced to leave, taking the latest skills on the newest equipment with them. Older craftsmen with lots of experience are being offered early retirement. Both constitute a dramatic shortfall in the skills and savvy required to perform maintenance adequately. To counteract these forces, a plant must have a quality skills training program. The program must determine the skills needed to maintain increasingly complex new equipment. Yet, the program must sustain current skills. Then, as downsizing takes its toll, skills must be transferred from those who go to those who stay. As the revised craft training program is developed, it must continuously be contrasted with the newly identified maintenance workload. That workload will identify the number of personnel authorized, by craft, after the downsizing. It will also dictate when they will work, shift by shift. The training program must meet new work load demands.

New Organizations - To this is added the task of team operation. Craftsmen will not only have to apply their repair skills, they must also control the work of their teams. This, by far, may be the greatest challenge for craftsmen because it is the least familiar to them.

The strong emphasis on basic skills provided in many existing craft training programs has not prepared most craftsmen to work in a team environment. Maintenance programs that only their supervisors understood are now theirs to master. Supervisors are no longer making their assignments, craftsmen must control the work themselves. Teams are here and craft personnel must be trained to work effectively in the new environment. [1] See Figure 18-1.

Figure 18-1. Craftsmen will no longer be able to rely on supervisors making decisions for them once they are within a team organization.

The Changing Work Force - Fewer personnel are choosing plant maintenance as a career. Also, those entering the plant work force are changing as well. Much of the now English speaking male work force will give way to more women, minorities, possibly immigrants, and fewer English speaking males. A shortage of plant maintenance personnel is forecasted. Therefore, the trend toward asking operating personnel to share in the maintenance task is becoming a necessity. To attract personnel to maintenance, a plant must shed the image of maintenance as 'fixers of things that break' to a perception of well qualified and trained craftsmen who improve equipment effectiveness and reliability. Apprenticeship programs must be established to provide opportunity for employees outside of maintenance to learn the skills required while examining maintenance as a possible career.

Train to Meet Technology Advances - Some plants have introduced new technology as a means of boosting competitiveness. But, those same plants have neglected the improvement of skills necessary to maintain the equipment containing the technological advances. For example, one plant invested in on board computers for its haul truck fleet to monitor over 80 different functions and warn of early signs of trouble.

Carrying Out Craft Training

However, they neglected to train personnel in how to interpret the data. As a result, the new technology has had little impact in improving maintenance performance and equipment availability.

By contrast, the successful plants have carefully determined in advance what new skills were necessary to utilize the new technology effectively. Then, they selected and trained personnel to achieve the maximum benefit from new technologies. Thus, new technologies must be accompanied by effective training to gain advantage from their use. [2] See Figure 18-2.

Figure 18-2. "I know exactly how to handle this, thanks to some good training."

Profile of Future Craftsmen - How maintenance craftsmen will be used in the future is getting attention as well. Prominent are multiskill and multicraft maintenance.

- Multiskill maintenance blends necessary craft skills allowing personnel to do a job from start to finish. It also refers to flexible trades, cross trades, and multicraft maintenance.
- Multicraft requires personnel to possess more than one trade. Unfortunately, multicraft is unpopular and, for that reason less successful or sustainable.

Multiskill - Multiskill maintenance is found in lean plant operations and among team or area maintenance organizations. Also, as plants implement total productive maintenance, multiskill maintenance may include operators performing routine maintenance on their equipment.

When multiskill is considered, a plant should focus on the desired gains rather than the process of implementing multiskilling. By identifying the gains it will reinforce the motivation for successfully implementing multiskill.

Among the gains are:

- Reduced equipment downtime
- Improved equipment performance
- More preventive maintenance completed
- Reduced waiting or idle time
- More effort on equipment improvement, modifications, predictive maintenance, training, planned maintenance and team development
- Reduced overtime work
- Reduced breakdown and crisis maintenance calls.

Work force reduction is often described as a potential benefit from multiskilling. However, it seldom happens. Craftsmen instead are shifted from breakdown maintenance to improving equipment performance. Thus, while the existing work force becomes more productive, it may not be reduced in size.

Multiskill maintenance implementation has barriers:

- Multiple unions try to protect their 'turf'
- Multiskill talents don't match plant maintenance needs
- Maintenance planning and supervision need realignment
- Managers may not understand benefits or strategies of multiskill use
- Training and qualification processes are inadequate
- Maintenance, union, and management are at odds
- Incentive programs not readjusted

Multiskill Implementation - Helping the maintenance work force understand multiskill maintenance should be one of the highest priorities of any multiskill transformation process. If the maintenance work force has been performing their jobs the same way for 20 years, they will ask, "Why change ?" Some maintenance work cultures have come to believe that high seniority jobs should be easier and less demanding.

However, multiskill maintenance works against this non competitive attitude. Therefore, multiskill implementation must include actions to change cultures, attitudes and behavior. Helping the maintenance work force and management understand multiskill maintenance is an educational as well as a commitment process.

The desired performance results must be linked to the objective of becoming more efficient, effective and competitive.

Carrying Out Craft Training

Concerns of Employees - Basic to any organizational change is a requirement that employees know they will still have a job when all the dust settles. Therefore, the plant must make assurances to employees:

- There will be no layoffs or pay cuts
- New technology requires multiskill

Assure them also that multiskill will benefit them as well as the plant:

- Primary skills have a place in a multiskill
- Multiskill not a jack of all trades
- Broadening primary skills emphasized
- Most maintenance people are multiskilled at home

The Plants Craft Training Strategy

Competitive pressures require plants to attain equipment effectiveness and reliability. Maintenance must deliver. Plant managers must realize that new technology alone, like on board computers, will not assure either better maintenance or productivity improvements.

Impact of Computers - New computer integrated technologies will have the biggest impact on maintenance job roles. The advent of digital electronics has caused changes in traditional maintenance job roles. First, application of craft skills will be affected. Typical, how is the data provided fixed monitors used to help diagnose a problem and the overall performance of equipment ? But, computers will have a tremendous impact on work control as well. Craftsmen who used to get a job assignment handed to them are now the assignors. They find themselves operating the work order systems that they resisted only last year with an 'over my dead body' attitude.

A New Maintenance Mission - Too many plant managers, supervisors and even workers base their ideas of maintenance on the 'fixers' mindset. In today's competitive world, there is little room for 'fixers'. Maintenance must be redefined as equipment reliability, because maintenance, especially multiskill maintenance, must concentrate on improving equipment reliability. Maintenance programs must be expanded to equipment management programs.

Managers must ensure that successful equipment performance is understood to be a collective effort of operations, maintenance and supporting staff departments like warehousing, plant engineering or accounting.

The productivity of maintenance workers must no longer be measured by 'wrench time'. Instead, a multiskilled reliability technician must be responsible for solving problems and troubleshooting, planning and expediting jobs. He must make equipment improvements while responding quickly to operations needs.

Education - Focusing on the results and benefits while educating management and the work force, can make the transformation to multiskill maintenance efficient and effective. Emphasizing work control and new skills training produces both short-term results and sustains long term performance. ₃. See Figure 18-3.

Figure 18-3. Training Methods for craft personnel must be related to both skills and work control.

Involvement - Plant management must involve the union or work force representatives in education and training. This way, their suspicions and tendency to resist will be replaced with knowledge and understanding. They will better appreciate the value gained and offer their support more willingly. Give employees an opportunity to be involved. Let them help define the program or identify areas in which they feel training must be provided.

Commit Money - A successful program has management support and budgeting. This commitment also measures management's understanding, without which not much can happen toward meeting the objectives of profitability.

Controlling the Training - Link skill training to specific job performance requirements and measurable results like improved equipment availability and reliability. Match the training needs of the work force and the requirements of new technologies. Combine education and training approaches to accommodate individual workers needs. Mix formal on the job training with peer coaching, guided self study courses, hands on simulation classes and shop work, or computer or video based programs.

Carrying Out Craft Training 113

Use other craftsmen as trainers whenever possible because they communicate in practical terms. Use skills certification and qualification to assure that skills and knowledge can be demonstrated upon completion of training programs. Establish and publish performance standards so that all understand what they must do to complete the training and be fully qualified. Treat knowledge gained as a skill and require the craftsman to demonstrate that he can do the work correctly. Always reward what is important:

- Improved performance
- Better productivity
- Greater equipment reliability
- Less downtime
- Lower employee turnover
- Reduced absenteeism
- Better teamwork

Establish a framework to sustain and improve team operation such as an advisory panel of craftsmen who must make it happen but, see it from a different vantage point than managers. 4. See Figure 18-4.

Figure 18-4. Get some practical advice from those involved.

The Right Kinds of Improvement - Don't be trapped into a one shot improvement effort. Avoid the consultant who promises that teams will solve all problems but, has no industrial experience. Instead, use professional trainers and consultants or workplace educators who can work with people and apply worker centered learning approaches. They will successfully develop the skills necessary to meet both the new requirements of maintenance and the ability to control work that is now being required of craftsmen.

Summary - Competitive industrial organizations are paying attention to the future of maintenance craftsmen. They appreciate that traditional repair skills must be upgraded to meet today's demands for profitability. Their craft training programs are adding the new dimension of work control and people skills required by the team environment. For those plants that have done it successfully, the rewards can be measured in better performance, greater productivity, lower costs, improved equipment reliability and, greater satisfaction of the craftsmen who have made it possible.

1. - 4. Images copyright, New Vision Technologies, Inc.

19

Conducting Supervisory and Staff Training

The Phantom of the Training Program

Urgent Need to Train Supervisors - Great concern is frequently voiced over the urgent need to train craft personnel. Yet, little is being done to train the superintendents, supervisors and planners who control them. Over 90 % of today's maintenance supervisory and staff personnel came up through the ranks from the hourly worker level. Along the way, the only exposure many received to prepare them for their current jobs was 'trial and error' in the 'school of hard knocks'. Too often, as they moved from the craft ranks, they found little to guide them as to what they were to do as supervisors and planners or how. Once in salaried positions, their worst suspicions were confirmed. They were on their own. No training was provided. Typically:

- There were vague job descriptions written primarily to justify a salary level
- Few written procedures existed on how maintenance was to be conducted, much less how they were to perform their jobs

The training needed to convert them into effective supervisory and staff personnel wasn't conducted. The inevitable result can be seen in many plant maintenance organizations today:

- Supervisors still work with their crews rather than lead them
- Planners are 'parts chasers' and wonder why little gets planned
- Superintendents add 10 % to last year's budget hoping no one will notice that budgeting isn't their strong point

Training on the Maintenance Program - Correction of these problems is only possible through the development and conduct of an effective training program for supervisory and staff personnel. However, before that can come about, a plant must acknowledge that a problem exists. For those plants that have assessed it, their most startling finding is the confirmation that the person most responsible for getting the work done, the supervisor, has had very little training. He is seldom trained on the maintenance program and, in most cases, has had no training on how to be a supervisor. That same supervisor is then charged with carrying out a program he may not understand. He is also expected to explain it to crew members. Often, he may not be able to do so. Against this unhappy revelation, we find that the supervisor is considered a key component in implementing new organizational schemes like total productive maintenance or teams. Who is kidding whom ? How can anyone expect that the long neglected education of the supervisor will suddenly and magically produce a leader who will make today's challenge of reorganization and downsizing successful ?

If tomorrow's organization is to be a team, its members will pattern their behavior on what their supervisors have demonstrated or taught them. If the guidance was poor, many aspiring team organizations will fail or struggle mightily before they achieve desired team performance.

What to Do

Find Out What They Need to Know - The first step is to find out what supervisors actually know about supervision, their organization and their maintenance program. This is difficult because, as adults, they dislike being questioned. A more viable alternative is exposure to the right training accompanied by an assessment of what they have learned. Older supervisors have a wealth of experience and judgment but, may wonder why some things work well and others don't. In their case, training may be a confirmation of the judgments they can now make after years of 'trial and error' operation. On the other hand, younger supervisors don't have a lot of experience and their training should focus on what works best and why. Then, as they apply their new knowledge, the training guides their actions so they avoid the 'trial and error' approach. Moreover, the tremendous technological advances made with equipment will no longer permit supervisors to go about maintenance using 'trial and error' as a method of controlling work. Thus, there is justification for filling this training void. After training has been completed, there must be a performance assessment to determine how effectively supervisors and planners are applying what they have learned.

Conducting Supervisory and Staff Training

The Right Training

A balanced picture of how maintenance supports the total mining operation is necessary as a foundation for effective control of day to day maintenance. Therefore, maintenance supervisors and staff personnel must know how their efforts fit within the plant's production strategy. They should know, for example:

- The timing of the annual plant shutdown
- Whether the production plan will permit the overhaul of critical equipment
- If operators will perform limited maintenance
- The amount of non maintenance work permitted
- The status of future organizational changes
- How maintenance costs fit within the total profit picture

Within this framework, maintenance supervisors and planners can better focus on the details of their own responsibilities. These responsibilities cover several areas including:

- The organization
- Duties of key personnel
- The maintenance program
- Information
- The role of operations
- Support from staff departments
- Performance goals
- Evaluation and improvement

In turn, each area is sub divided into practical aspects that are needed to carry out their tasks efficiently. Specific practical and useful training is necessary because supervisors are responsible for getting the actual work done. They don't want a lot of 'what ifs'. They want to know how they can put what they learn to work immediately to make their jobs easier and more effective. Few will simply accept training without asking why something works better than something else. Therefore, the people doing the training must know exactly what they are talking about. Supervisors and planners will be a critical audience. Training for supervisors and staff personnel should include: organization, duties of key personnel, the maintenance program, information, the role of operations, support from staff departments, performance goals, evaluation and improvement techniques. Each of these topics is discussed below.

Organization - The supervisor must be in agreement that the current organization is the best posture for carrying out the tasks he and his crew must perform. Therefore, he must know something of the choices available as well as their strengths and weaknesses. If, for example, the supervisor knows that the craft organization is merely the organization chosen because the union won't tolerate anything more efficient, he is unlikely to go out of his way to create organizational efficiencies. On the other hand, if he is aware of the merits of each type or organization and allowed to organize more efficiently, he will be more likely to do so. Therefore, training on organizations should include:

- The types of maintenance organizations
- The pro's and con's of each type
- The degree to which different organizational types can be blended
- Organizational change

Duties of key personnel - Listening to the questions posed by supervisors and planners in public seminars reveals concern over who should be doing what and how they do it. Typical:

- What, exactly, is the planner supposed to do ?
- Why hasn't our superintendent told us what our program is ?
- Who's responsible for getting our PM program on track ?
- Why is our maintenance engineer mostly on projects that have nothing to do with maintenance ?

Certainly, the duties of key personnel must be defined. But, it is not satisfactory to have the personnel manager create vague job descriptions only to justify a salary. Duties must be spelled out exactly in practical, realistic terms. However, before the duties of any key maintenance personnel can be defined, the program into which their efforts fit must be defined. This step yields the necessary total picture. When this is done, the duties of key maintenance personnel can be defined effectively, including the:

- Superintendent
- General supervisor (if any)
- Maintenance engineer
- Maintenance technicians or specialists (as opposed to craftsmen)
- Maintenance planner
- Material coordinator

Conducting Supervisory and Staff Training

In addition, the duties of selected hourly personnel should be defined to guide them in supporting the maintenance effort. Included might be:

- Lead men
- Lead hands
- Fill in supervisors

Teams - If there is a team organization, the duties of team members in a self directed team must be spelled out. You must keep in mind that as ex craftsmen, most have only craft skills and little or no experience in any form of work control. While, it is logical that they will learn to work together after 'settling in', the anguish they will experience at the outset can be avoided with the guidance that definitions provide. Team organizations also operate with co-ordinators. Therefore, the duties of the coordinator should be defined to avoid simply replacing the supervisor with a coordinator. To assure that a 'fix it as usual' approach does not prevail in the supposed 'productive team' environment, it is best to rotate coordinators. Thus, the definition of the coordinator's duties takes on greater importance. Yet, since various incumbent coordinators may bring different leadership and administrative skills to the job, leave some flexibility in the description. But, make certain that what is developed is well documented so there is reasonable uniformity and understanding across the plant. See Appendix E - Duties of Key Maintenance Personnel.

The Maintenance Program - The biggest single deficiency of most plant maintenance organizations is a failure to adequately define their maintenance programs. As a result, personnel in maintenance are uncertain of what they are to do and how. In addition, operations is in a quandary as to how they are to support the program. Similarly, staff departments like warehousing raise questions on how they can best provide essential services to maintenance. Thus, there is little question that the maintenance program must be well defined. However, what of the supervisors and planners ? Expecting them to operate effectively without first defining the maintenance program is similar to telling them "good luck getting to the other side of the 'mine field' "! Obviously, it is grossly unfair to them and it can't be permitted to happen. Essential to the supervisors success is a solid education on the maintenance program. In turn, he must be able to explain it effectively to his crew members who will help him carry it out.

Information - Supervisors without information shun responsibility while those with information seek responsibility.

Universally, we want supervisors who seek responsibility. Thus, the proposition that supervisors don't need information is contrary to the maintenance performance improvements that plants seek. It follows that if a supervisor lacks sufficient information to seek responsibility, two possibilities exist:

- The needed information does not exist
- It exists but, the supervisor doesn't know how to access it

The more likely situation is the later. This suggests that the type of information required by supervisors and planners must be identified and training provided on how they are to use it effectively. That information, at a minimum consists of:

Labor Utilization - A weekly picture of how effectively labor was utilized. The supervisor can then determine whether PM has successfully reduced emergency repairs while allowing maximum manpower to be used on scheduled maintenance.

Cost - The total cost of maintenance performed by unit and component, with month and year to date summaries. This will permit the supervisor to spot troublesome equipment and give emphasis to its repair.

Repair History - The chronological listing of significant repairs made on key units of equipment allowing chronic, repetitive problems to be spotted. Failure trends reveal underlying causes allowing corrective action to minimize or eliminate them. Supervisors use repair history to pinpoint the problems initially identified by high costs. They also use it to guide craftsmen in the way to do current jobs based on previous equipment repair patterns. Planners use the life span of components recorded in repair history to help them determine the timing of future replacements.

Status of Major Jobs - The status of selected, planned major jobs from inception to completion and a summary of job cost and performance upon completion. Used by planners to manage major jobs and by supervisors to assess their crew's performance when the job was their responsibility.

Backlog - The degree to which maintenance is keeping up with the generation of new work and the ability to adjust the size and craft composition of the work force as the workload changes.

Conducting Supervisory and Staff Training

Used by the planner to assess the degree to which the backlog is controlled and, by the supervisor to judge progress in backlog reduction and changes needed in crew size and composition.

Selected Performance Indices - Trends in performance are observed through the use of meaningful indices:

- Does the supervisor have at least 60 % of his time available for supervision ?
- Has the productivity of the crew improved with its new organization ?
- Has the labor cost to install each dollar of material improved ?
- Has an 85 % PM compliance resulted in more planned work ?

Collectively, these elements of information are basic to adequate supervisory control and good planning. After verifying that the information is available, accurate, timely and complete, supervisors and planners should be trained on how to obtain and utilize it effectively.

The role of operations - What should the maintenance supervisor expect from operations ?

- When the weekly schedule is approved by operations managers, should the maintenance supervisor have to beg to get a haulage truck into the garage for service or should he see it parked waiting for service at the appointed time ?
- If a conveyor drive motor is to be replaced, should maintenance craftsmen first have to clean up the area before they can begin work ?
- If total productive maintenance has been implemented should the supervisor have to remind operators to do routine daily maintenance ?

The answer to all three questions is no. However, why leave it to conjecture ? Specify circumstances like these and let supervisors know where they stand.

Support from Staff Departments - What help is forthcoming from staff departments ?

- Will the warehouse deliver materials to the work site ?
- If special cost data are needed, will accounting provide them ?
- If the vacation policy is poor, will employee relations correct it ?

As with the clarification of the operations role, the specific support that the supervisor can expect from staff departments should be identified and the supervisor advised as his training is carried out.

Performance Goals - What is expected of supervisors in terms of cost control, crew performance, use of overtime, schedule compliance etc. If these goals are made clear, few supervisors will have to ask, "just how do you expect me to do that ?" The plant will have to provide the training if it expects good results.

Evaluation and Improvement - An evaluation is the first step of improvement. Therefore, supervisors should be trained in what is important and advised that they will be evaluated to determine how well they are carrying it out. If training has been done competently, supervisors will know what they are to do and how. They will see an evaluation as a checklist against which they can assess their progress. By contrast, if training is poorly done and supervisors are operating by the 'seat of their pants', an evaluation constitutes a threat.

Performance Appraisal - Every supervisory training program must be accompanied by an opportunity for on the job training during which the learning can be applied. But, the success of the training will be evaluated in actual job performance. Performance areas should be related to material on which training was received. Thus, if a planner was trained in project control, then more projects should be completed in less time. Similarly, if a supervisor was trained in work control his crew should achieve better productivity and schedule compliance.

Summary - The proper training of maintenance supervisory and staff personnel is being overlooked. This oversight has resulted in poor work control and ineffective execution of maintenance programs. Yet, other reasons compel better training. Supervisors are expected to bridge the gap between today's organizations and tomorrow's presumably more efficient downsized maintenance teams. Unless dramatic improvements in their training are forthcoming, they can't deliver. Therefore, a plant must assess its maintenance supervisory and staff training needs. Thereafter, a program to meet these needs must be carried out with opportunity for on the job training. Performance evaluations must determine the degree to which the training has improved current supervisory performance and prepared supervisors to assist in the implementation of new, more effective organizations.

III

Tuning Up the Maintenance Program

20

How to Define the Maintenance Program

A Well Defined Program is the Secret to Performance

A Major Deficiency - The most frequent, major deficiency of plant maintenance organizations is an ill defined or undefined maintenance program. As a result, the maintenance personnel who are to carry it out are confused. Operations wonders how they can best support maintenance and vital staff services like warehousing second guess what maintenance wants.

By defining its program, a maintenance organization can ensure better performance of its work force and gain the essential support of operations, management and staff organizations.

Advantages

By defining the maintenance program properly and educating plant personnel on it, maintenance can gain significant advantages. They can, for example:

- Confirm an existing program.
- Develop a new program.
- Educate personnel.
- Encourage participation in change.
- Ensure a well informed work force.
- Improve performance.
- Assure quality work.
- Gain productivity.

The process of program definition requires that each program element be considered. Therefore, program definition should not be the effort of a single person. Rather, it should be a composite action taken by representatives of crews, supervisors and staff personnel, all guided by the superintendent. The superintendent must provide solid guidance and monitor the process to its conclusion.

Once the program definition has been documented, the educational process must be carried out. Education must include everyone in the maintenance work force, all operations personnel, plant management and staff. The education should be given by maintenance personnel so that pertinent questions can be answered promptly and correctly. These instructors should be aware that their preliminary definition may have omitted certain aspects requiring greater clarification and, as a result, recommendations should be welcomed and encouraged. When explanation of the program definition is being given in maintenance, adhere to the 'chain of command' so that those responsible for work control (supervisors) will be addressing the same personnel that will carry out the program. The maintenance superintendent must monitor the educational process and provide a means of measuring how effectively the total process has aided performance.

How to Define the Maintenance Program

Program definition begins by establishing the sources of new work. They include:

- The results of PM services, particularly inspections
- Requests from operations
- The review of repair history
- Recommendations from maintenance engineering
- Forecasts prescribing major tasks
- Safety needs
- Environmental requirements

Next, the personnel who will initiate new work are identified. These would include: operations, maintenance personnel, plant engineering, various management personnel, staff departments and safety personnel. As this group is identified, make certain that consideration is being given to how they will initiate the work. For example:

- Does the work order system adequately handle non maintenance work ?
- Is provision made for verbal orders ?

How to Define the Maintenance Program

Program Definition Techniques

Since program definition must consider every aspect of the maintenance program from work request through planning, scheduling, coordination, assigning, reporting and completion, it cannot be done by a single person. Rather, the development team should include crew members, supervision, planners and maintenance engineers or technicians. The team should be given complete guidance by the superintendent since he will review the recommendations made. Thereafter, it is his program to execute properly. As the program is being developed, the superintendent must monitor progress against some reasonable time table.

Once the preliminary program is drafted, it should be reviewed carefully by maintenance leadership. After necessary modifications, it is presented to all maintenance personnel, using the regular 'chain of command'. As each supervisor briefs his crew, it is useful to have senior supervisors listen to the discussion. They should observe whether he understands it well and is committed to making it work. A statement, for example, in which the supervisor starts by saying, "I'm required to tell you about this", suggests a lack of commitment. Rather, he should indicate that he wants their input to ensure it is a complete, effective program. He should also explain how they are going to help him carry it out.

Following the presentation of the program within maintenance, the superintendent should affect the education of operations, staff and management personnel. He should explain it himself to senior managers and delegate the task to his key people as they explain it to their peers in operations or staff departments. As explanations are made, changes recommended by various listeners should be welcomed. Those with merit should be included giving the contributor a sense of participation. On the other hand, if a recommendation is improper, it should be explained why it should not be included.

Once the educational effort has been completed, the maintenance superintendent should monitor the changes in the performance of maintenance as the result of a better defined program and its impact. See Appendix G - Maintenance Performance Evaluation as an effective way to carry out a self evaluation.

The most effective technique for presenting the maintenance program is the use of schematic diagrams which depict individuals and their tasks. This allows personnel to observe the overall interaction as they carry out the program. Flow charts are less effective because they omit this interaction.

Defining the Program

Initiating New Work - Starting from the simplest method of reporting problems to the more complex conduct of replacing major components, each has its place in the program definition. Therefore, as program definition is initiated, each program element, action, system or sub system should be covered. See Figure 20-1.

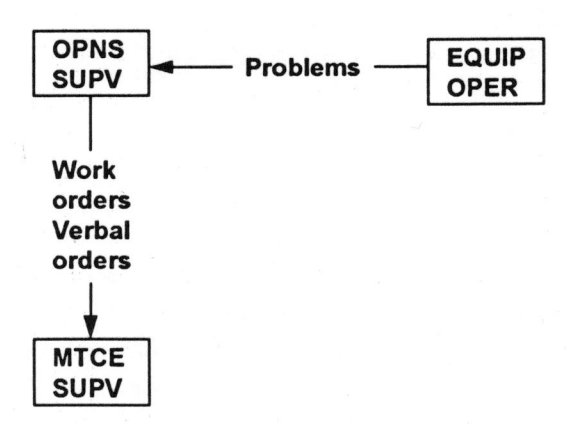

Figure 20-1. Operators report problems to their supervisors who, in turn, request help from the maintenance supervisor using standard work order procedures.

Next, maintenance crew members providing informal help to operators should have a simple way of reporting what they do. Often, much of this assistance is worthy of repair or cost history. But, if treated too informally may be omitted. Generally, exception reporting works best. Provide a simple criteria such as what is repair history and what is not. See Figure 20-2.

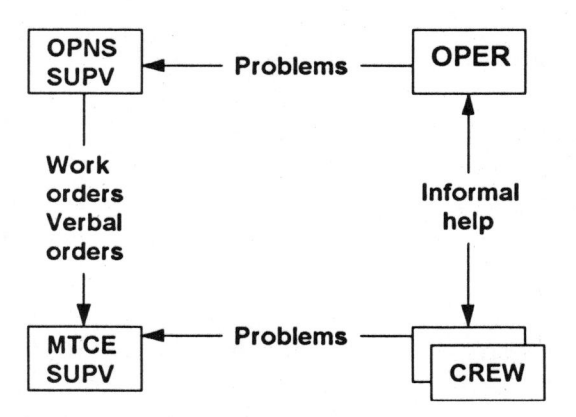

Figure 20-2. Equipment Operators get informal help from maintenance crew members who, in turn, advise the maintenance supervisor.

How to Define the Maintenance Program

Preventive Maintenance as a Source of Work - The PM program identifies services due. Through the scheduling procedures it determines when they are to be done. While the PM program itself serves to identify services, the more important work from the PM program is the new work generated from inspections and condition monitoring. See Figure 20-3.

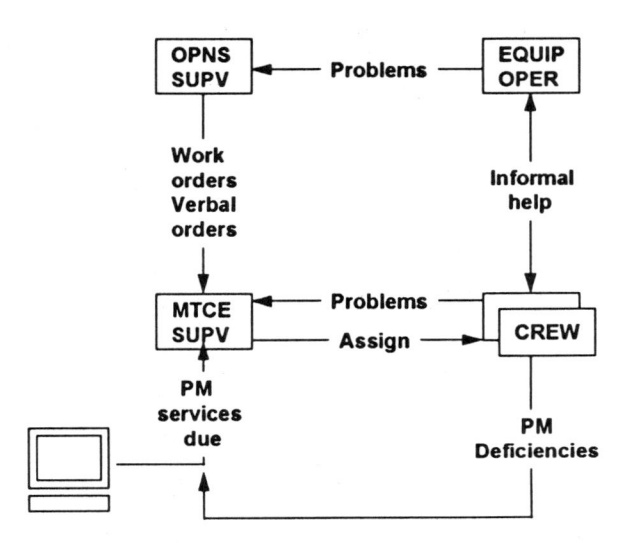

Figure 20-3. PM services are assigned to crew members, who report deficiencies found.

Component Replacements - The component replacement program is another source of new work. Each replacement is planned and scheduled and its actual condition is verified by a PM inspection. See Figure 20-4.

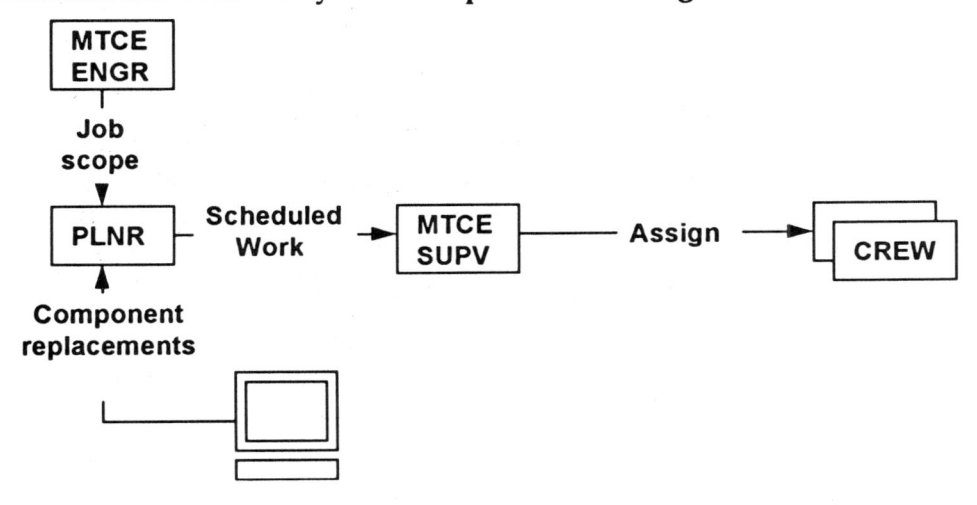

Figure 20-4. Unlike PM services that are scheduled routinely, component replacements are planned and scheduled.

Defining the Preventive Maintenance Program

Services Due - The PM program identifies static (while down) and dynamic (while operating) services each week. See Figure 20-5.

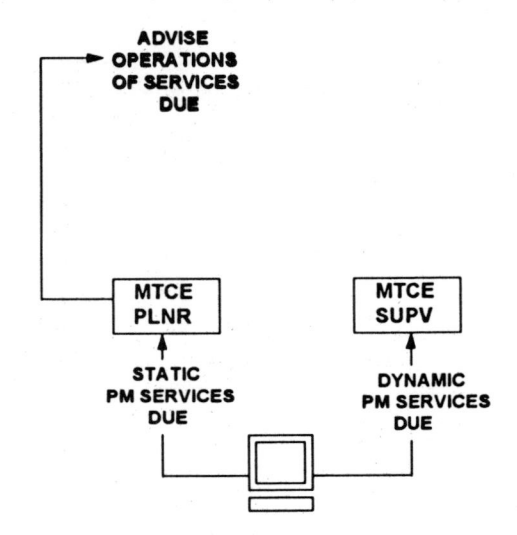

Figure 20-5. As the PM program identifies services due, the planner handles static services while the supervisor fits dynamic services into his daily schedule.

Next, the PM program must provide a means of assigning services. See Figure 20-6.

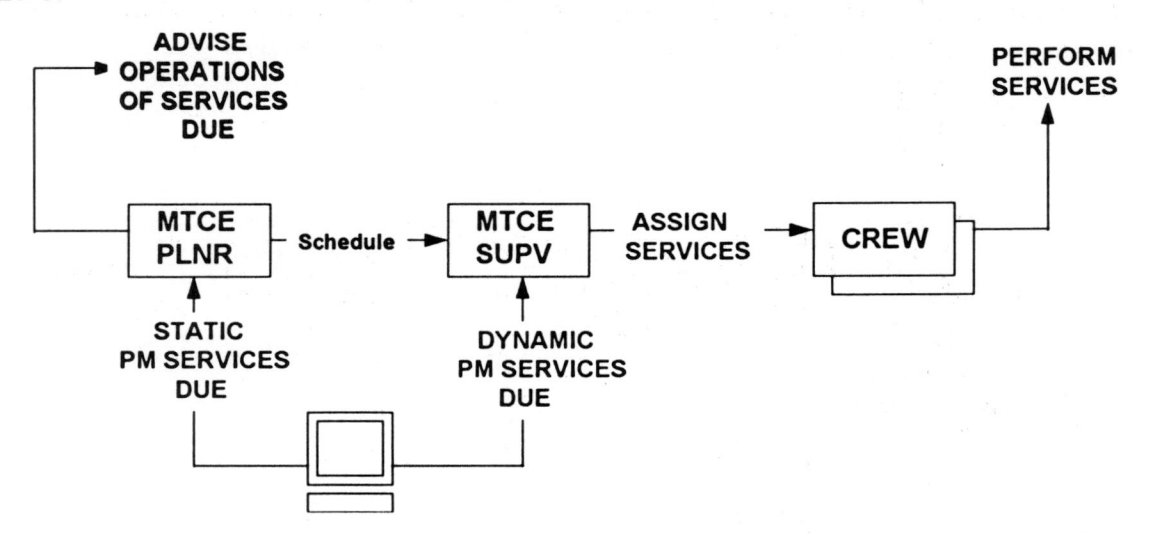

Figure 20-6. Static services are scheduled via the weekly schedule. Dynamic services are assigned on the supervisors daily schedule.

How to Define the Maintenance Program 129

With the team organizations, it is important to spell out the operator's role. See Figure 20-7.

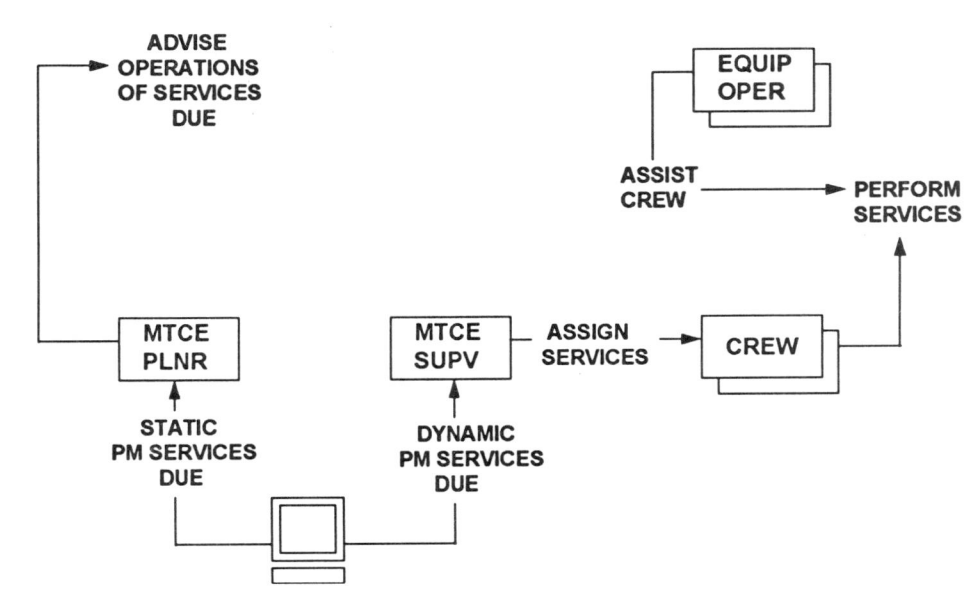

Figure 20-7. Operators assist crew members in PM services and perform some services independently.

The operations supervisor should be advised of deficiencies. See Figure 20-12.

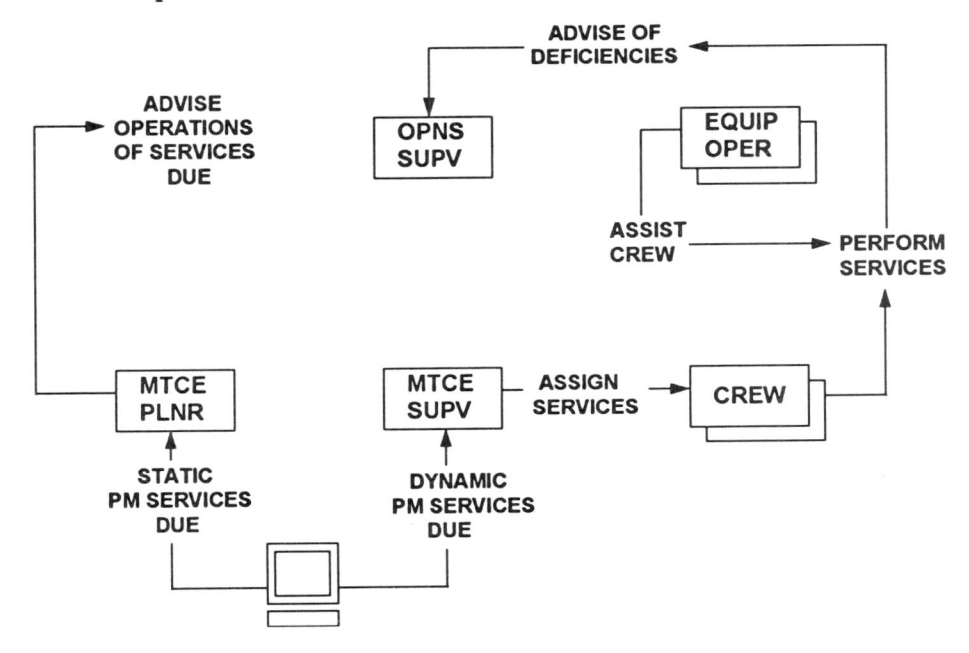

Figure 20-12. Advising the operations supervisor of deficiencies found, makes him aware of the value of PM and ensures his crew participates actively.

Decisions must be made on how to handle the deficiencies uncovered by the PM inspections. See Figure 20-9.

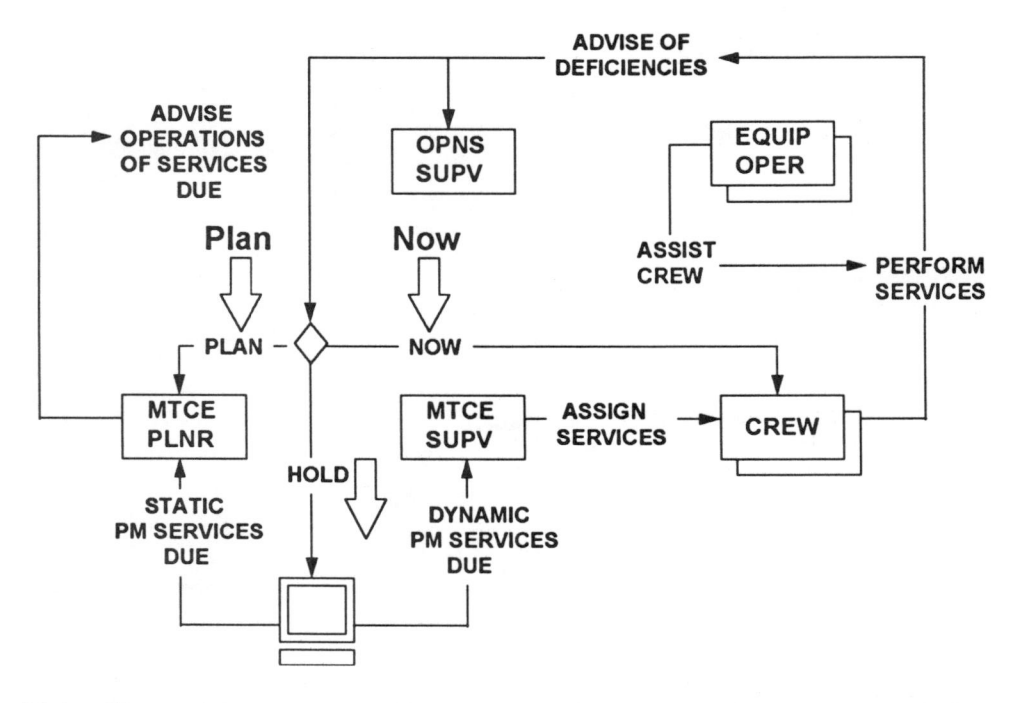

Figure 20-9. The maintenance supervisor reviews deficiencies to determine their future work status: planned, emergency or unscheduled.

Verifying the overall PM program - The maintenance engineer would review the overall PM program to determine whether it is complete, the service frequencies are correct and it is producing the desired results. See Figure 20-10.

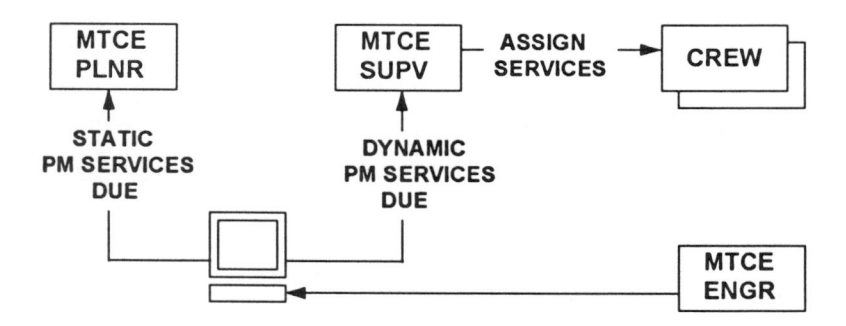

Figure 20-10. The PM program is reviewed by the maintenance engineer (or superintendent) in conjunction with the responsible planner and supervisor. Changes are made as required.

How to Define the Maintenance Program

Component Replacement Program

During the life span of much of the equipment used in mining, a major activity will be the replacement of its major components as they near the end of a wear period. In many instances, component replacements represent over 60 % of the work that can be planned and scheduled. Because it is a repetitive activity, major component replacements lend themselves to the use of standard task lists (how to do the job), standard bills of materials (that can be ordered in advance) and tools lists (so special tools can be reserved). Thus, the overall planning task can be greatly simplified. Consider the following steps in defining this portion of the overall maintenance program. See Figure 20-11.

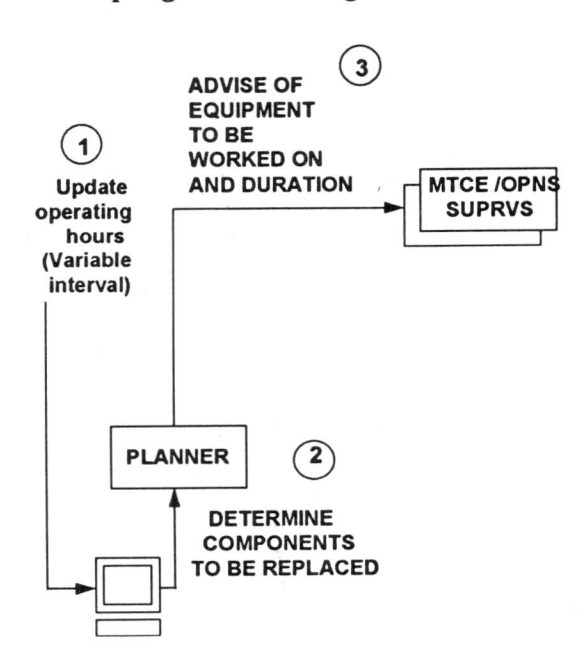

Figure 20-11. Step 1, operating hours are updated. Step 2, the computer identifies the units and components due for possible replacement. After verifying components condition via latest PM inspection, operations and maintenance supervisors are advised in step 3.

It is at this point that the use of standards pays dividends. Because a standard bill of materials exists, the planner is able to order the materials and have them on hand. The standard task list now allows the maintenance supervisor to confidently make assignments to crew members knowing that their job instructions are complete and quality, uniform work will result. Similarly, when crew members draw the tools they will find them waiting for them. Collectively, these steps make for efficient work. The latest PM inspection is used to determine that the component is actually due for replacement.

The balance of the component replacement activity is carried out. See Figure 20-12.

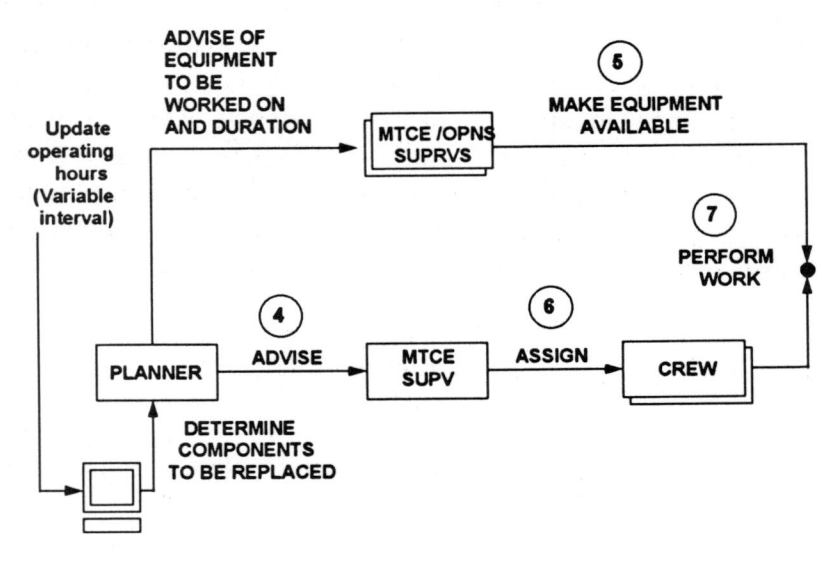

Figure 20-12. In step 4, the maintenance supervisor begins preparation for doing the work while, in step 5, the operations supervisor makes the equipment available. In step 6, the maintenance supervisor assigns the job to crew members who perform the work in step 7.

Once the work has been completed, labor is reported by the crew members and any additional stocked materials required are debited to the job. If special problems are encountered the crew members might alter the task list, material list or tool list.

Planning and Scheduling Major Jobs

Major jobs are derived from sources such as:

- Results of PM inspections
- Modification requirements
- Forecasts of future component replacements
- Decisions to overhaul a unit
- Non maintenance projects like the installation of a new conveyor
- Safety requirements
- Environmental needs

With the exception of component replacements, most planned jobs will be unique.

How to Define the Maintenance Program

Program definition should specify how unique major jobs are planned and scheduled. See Figure 20-13 and 20-14.

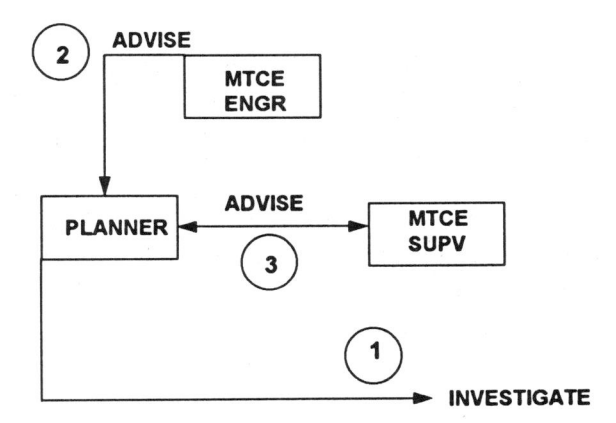

Figure 20-13. In step 1, the planner investigates the job in the field. With a good idea of what is needed, he is advised in step 2 on the scope of the job. Then in step 3, he confers with the supervisor whose crew will do the work to obtain his advice.

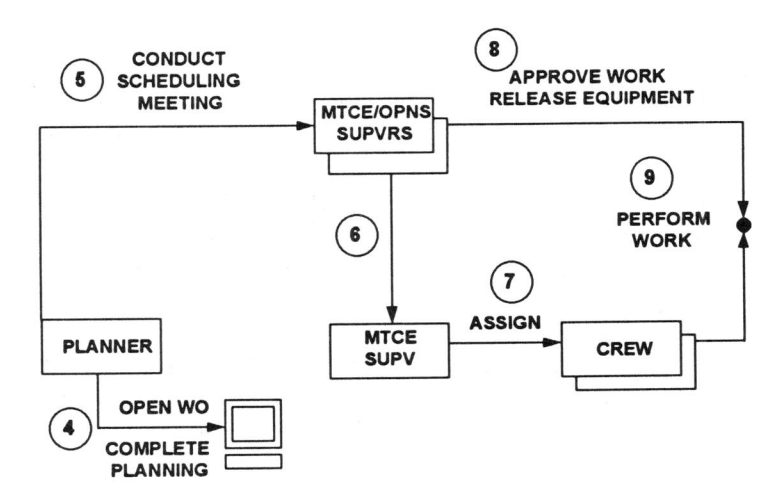

Figure 20-14. In step 4, the planner opens the work order and completes the planning steps. Then, in step 5, he participates in the scheduling meeting with operations and maintenance supervisors where they negotiate and approve the schedule. In step 6, the maintenance supervisor is notified. At the proper time he assigns the job to his crew, step 7. The equipment is released in step 8 and work performed in step 9.

On the completion of work, whether planned or not, there is a requirement to report the status verbally to the maintenance supervisor while reporting the use of labor and material formally on time cards and stock issue documents respectively.

Reporting work completion

The best source of information on work performed is the craftsman who performed the work. He should report labor and material field data on the time card and stock issue card respectively. In turn, these data are verified by the supervisor before the data enters the information system. Alternately, the craftsman might enter the data directly into the computer. See Figure 20-15.

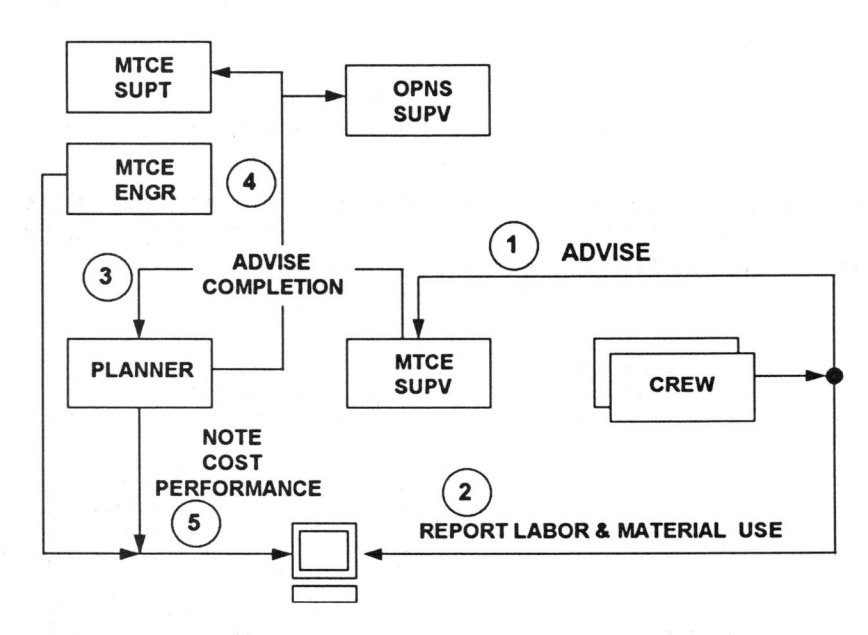

Figure 20-15. Step 1, crew advises supervisor verbally of job status. In step 2 they report formally. If work has been planned, planner advised of job completion by supervisor in step 3. Planner advises the superintendents in step 4. Step 5 planner and the maintenance engineer observe job cost and performance, .

All of the program elements are then assembled into a single schematic diagram and the legend is added explaining each step. This provides a single cohesive picture of the overall maintenance program. It is best to leave the diagram in a preliminary form during the following training sessions so that personnel who may wish to offer recommendations are less likely to feel that the diagram is final. The best program definition will result from the direct input of everyone's ideas. Moreover, as they see their input incorporated, they will gain a sense of ownership and actively sell it to others. See Figure 20-16.

How to Define the Maintenance Program

Typical Definition of the Maintenance Program

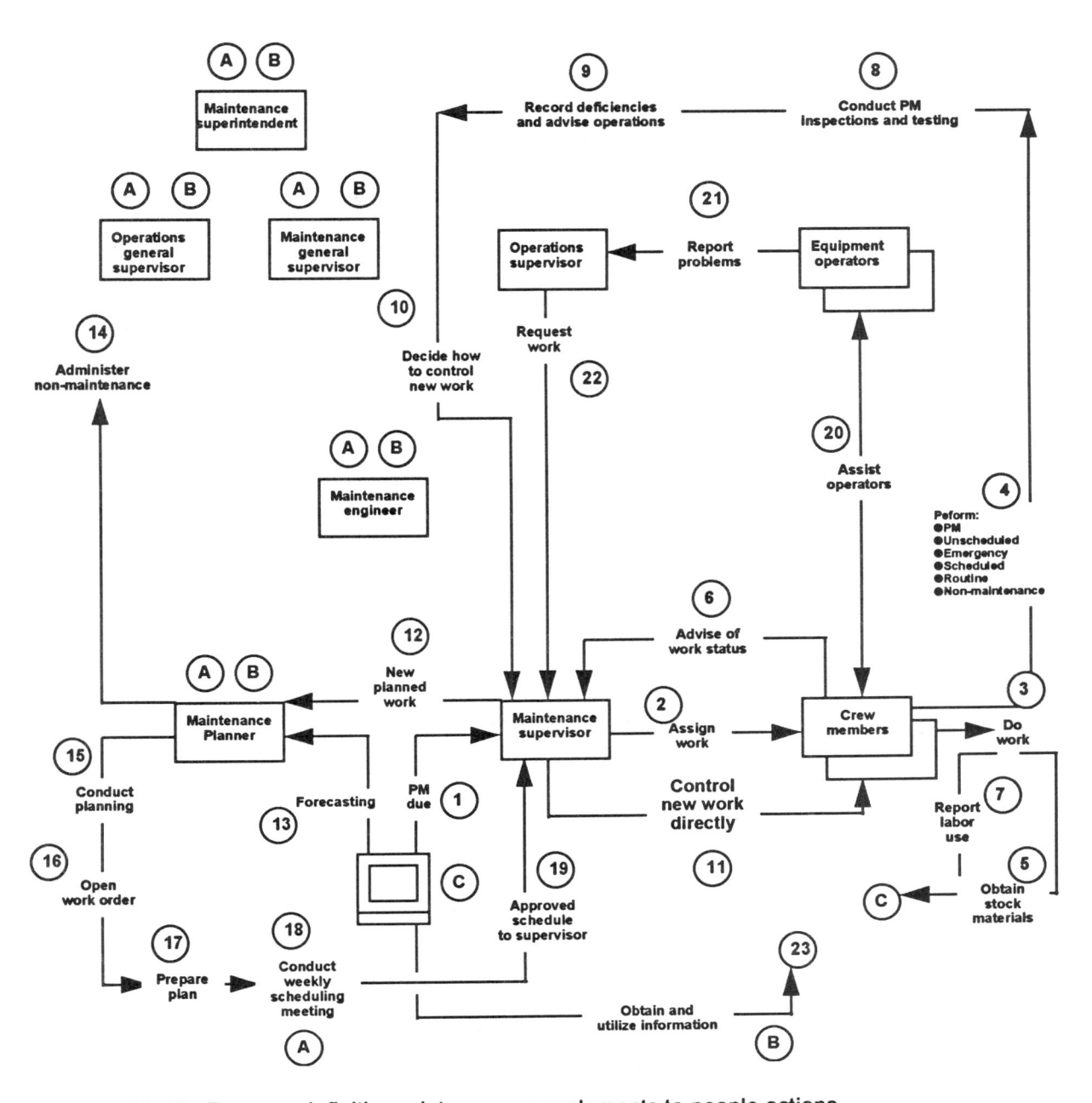

Figure 20-16. Program definition relates program elements to people actions.

The diagram on the previous page depicts how key maintenance personnel interact while using system elements to carry out the program. Concurrently, it shows the actions of key operating, staff and management personnel as they utilize maintenance services or support the maintenance program. The legend to accompany the diagram is shown below:

1 - The computer provides the maintenance supervisor with information on preventive maintenance services due. He, in turn, makes assignments to his crew members. Alternately, the planner might advise the supervisor of services due if equipment shutdown must be coordinated with operations.

2 - The maintenance supervisor makes work assignments to his crew members using the work order system. Since verbal orders will also be used, they should be a legitimate part of the work order system.

3 - Crew members receive work assignments and carry out work.

4 - Crew members (craftsmen) perform: preventive maintenance services, unscheduled repairs, emergency repairs, planned and scheduled maintenance, routine maintenance and non-maintenance work such as construction or equipment modification.

5 - In preparation for doing the work, or while work it is in progress, crew members obtain stock materials. Completed stock issue cards flow into the data base to adjust inventory levels and provide information to maintenance on stock material costs and usage (C).

6 - As they do work, or at its completion, crew members advise their supervisor of the status of work assigned to them.

7 - Upon work completion, crew members use the time card to report man-hours spent on each job. After verification by the supervisor, time card data flows into the information system where man-hours are converted into labor costs (C).

8 - PM inspections and predictive techniques (like vibration analysis) are used to monitor equipment condition and help uncover equipment deficiencies.

9 - When equipment deficiencies are found they are classified as work requiring planning, unscheduled or emergency repairs. The operations supervisor is advised so he:

How to Define the Maintenance Program

- Is aware of the inspection effort
- May anticipate the need for repairs
- Knows maintenance will follow through on the work needed

10 - As the maintenance supervisor is advised of work requirements he may:

- Make direct assignments to crew members especially if emergency repairs are required (using verbal orders, if appropriate)
- Forward jobs to the planner if they meet criteria for planning
- Hold jobs that can be done later in the computer

11 - Unscheduled and emergency repairs should be controlled directly by the maintenance supervisor since they require no formal planning. The general supervisor should be advised of the nature and corrective status of emergency repairs.

12 - Work requiring planning is identified and sent to the planner.

13 - Using the forecast, the maintenance planner determines the periodic maintenance actions that are due over the next several weeks.

- The planner verifies, through the latest PM inspection results, that the components identified do, in fact, warrant replacement
- Since most forecasted jobs will have been done many times in the past, they should have a standardized task list (what to do) as well as standard bills of materials and tool lists
- The planner also allows enough time to assemble materials or arrange for shop support such as fabrication of assemblies. He advises material control and shop personnel of needs in sufficient time for them to respond
- As materials or shop work become available, the planner is advised so he may carry on with planning tasks

14 - Some work to be performed by maintenance will be non maintenance projects such as construction, equipment installation etc. This type of work should be classified to distinguish it from maintenance work. It should also be evaluated to ensure it is necessary, feasible and properly funded. It should be prioritized because it will compete for resources that are also required for maintenance work. If significant non maintenance work is required, management should establish limits to avoid over commitment of maintenance resources.

15 - As each job is planned the planner would:

- Establish the job scope
- Conduct a field investigation or use an existing standard to develop a job plan
- Estimate manpower by craft and sequence into the job plan
- Identify materials, tools, rigging and support equipment needs
- Estimate job cost
- Establish the future week during which the work should be performed
- Establish job priority with the equipment user (operations)
- Obtain job approval

As required, the general supervisor or maintenance engineer would help the planner to develop the job scope of unique major jobs.

16 - The work order is then opened. This authorizes the procurement of materials, the ordering of shop support and the subsequent use of labor to perform the work.

17 - For jobs that are candidates for inclusion in the following weeks schedule, the planner prepares a preliminary plan. He verifies that, subject to operations approval, maintenance has the resources to perform the work.

18 - A scheduling meeting is conducted in which maintenance supervision presents the weekly plan to operations. Approval of the weekly plan by operations binds them to make the equipment available at the prescribed time and requires maintenance to make its resources available to do the work. At this point, the weekly plan becomes an approved schedule.

- During the scheduling meeting, compliance with the previous weeks schedule and performance on selected major jobs should be verified
- Preparations for selected important jobs that are scheduled to be done in several weeks should be assessed also

19 - The approved schedule is presented to the maintenance supervisors who will accomplish the work. The planner then coordinates with supervisors to arrange the timing of events such as on site material delivery. Daily coordination meetings are held between operations and maintenance to adjustment the schedule in the event of operational delays.

How to Define the Maintenance Program

20 - Crew members assist equipment operators on problems as required. In some instances, equipment operators may report problems directly as a result of inspections they make.

21 - Operators request help from maintenance crew members informally for small, simple problems. However, major problems are reported to the operations supervisor because greater resources and more downtime may be necessary to make repairs.

22 - Operations supervisors request necessary work using the work order system. Verbal orders are part of the work order system.

23 - The maintenance information system should provide information on:

- Utilization of labor (including overtime use and absenteeism)
- The status of cost and performance of selected planned jobs
- The backlog of pending jobs to help determine whether maintenance is keeping up with the generation of new work
- Repair history so that chronic, repetitive problems and failure trends may be uncovered and corrected, and the life span of critical major components verified
- The cost of maintenance against units and components for the current month and year to date

In addition:

- Operations should be advised of all jobs completed, unit by unit, each week
- Maintenance personnel should be able to research part numbers, drawings, wiring or assembly diagrams using the computer
- Management should be provided downtime data and performance indices like cost per ton or the labor cost to install each dollar of material (trend reveals improvement in productivity).

Summary - One of the most prevalent deficiencies of many plant maintenance departments is the absence of a well defined program. As a result, maintenance personnel are uncertain of how their actions must fit within the program. Similarly, operations may be uncertain how they are to support and cooperate with maintenance. The end result is a less efficient operation. Thus, a well defined program is necessary for an efficient maintenance effort.

The technique by which the program is defined is important as well. Great volumes written to try to clarify the program often fail simply because no one reads them. The best approach is the use of a schematic diagram with legend. It is easily reproduced, often on a single sheet of paper. Thus, as the program is discussed, constructive comments can be added, enriching the program. Once the program has been defined and generally agreed upon by key maintenance personnel an explanation should be given to all maintenance, operations, staff and management personnel. Once they understand the program, the help maintenance needs to carry out its program successfully will be forthcoming.

21

Examining Preventive and Predictive Maintenance Essentials

Avoiding Equipment Failures

The terminology associated with PM can be confusing. Yet, whether you are talking about preventive maintenance, predictive maintenance. non destructive testing or condition monitoring, they all have the same objectives of:

- Avoiding premature equipment failures and . . .
- Extending the life of equipment

Through inspection, the application of predictive techniques testing and condition monitoring, equipment failures can be prevented by finding problems before they reach the failure stage. With lubrication, cleaning, adjusting and minor component replacement (belts, filters etc.,) the life of equipment can be extended. Reliability engineering isolates the root cause of problems so they can be isolated, corrected through design changes and eliminated. [1]

Characteristics of PM services

PM services are routine, repetitive activities:

- Routine since the same check list or procedure is used with each repetition
- Repetitive because the service is repeated at regular intervals

They are also classified as:

- Static - Carried out only when equipment is shut down or . . .
- Dynamic - Able to be carried out while equipment is running

In addition, PM services are carried out at:

- Fixed intervals - Two weeks, six weeks etc.
- Variable intervals - 250 hours, 10,000 tons etc.

Overall, preventive maintenance should have a 'detection orientation' to:

- Find problems before they create a failure or stoppage.
- Utilize the time gained to plan and schedule the corrective work.

A good analogy is to compare a deteriorating piece of equipment with a stick of dynamite with a burning fuse. [2] See Figure 21-1.

Figure 21-1. The sooner the problem is found the longer the fuse.

Thus, the 'detection orientation' of PM assures that early discovery of problems will yield:

- Less serious problems
- Fewer failures
- More time to plan

There is an important relationship of PM with planning. It supports the idea of early detection of problems. Essentially, if there is not sufficient lead time between the discovery of a problem and the time when it must be repaired, there is not sufficient time for planning. Thus, successful planning depends on the proper application of the 'detection orientation' of PM.

Examining Preventive and Predictive Maintenance Essentials

Consider the relationship also of the 'detection orientation' of PM with the cost of repair and the amount of downtime to affect the repair. Essentially, problems found sooner are less serious and can be corrected at less cost and in less elapsed downtime. Conversely, problems found after the equipment has deteriorated substantially are more serious. Therefore, their correction requires more time and greater cost to restore them to proper operating condition. See Figure 21-2.

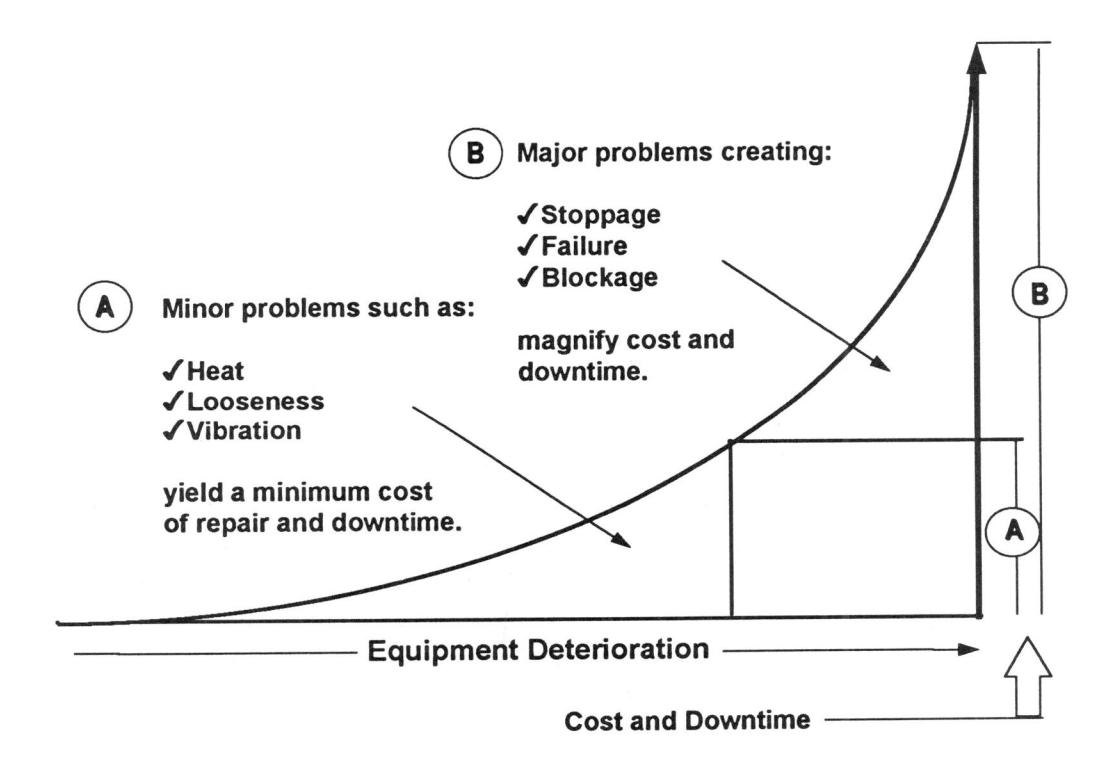

Figure 21-2. The level of deterioration dictates the cost and downtime of the repair.

A successful preventive maintenance program should help to:

- Maximize equipment availability.
- Minimize downtime.
- Prolong equipment life.
- Increase equipment reliability.
- Reduce emergencies.
- Increase the opportunity for planning work.

Of these, the most important is the opportunity for planning more work. More planned work yields better productivity, better quality work and requires less downtime to perform the work since it is better organized.

Types of Preventive Maintenance

There are two types of preventive maintenance services:

- Routine preventive maintenance
- Predictive maintenance

Routine preventive maintenance includes:

- Inspection
- Lubrication
- Cleaning
- Adjusting
- Replacement of minor components (belts, filters etc.)
- Testing and calibration

These routine preventive maintenance services are carried out through the direct physical activities of maintenance craftsmen as they perform visual inspections, cleaning, lubrication etc.

Predictive maintenance is a condition based system that measures an output from equipment signaling the level of deterioration of a component. Current signals are compared with the norm to enable analysts to determine corrective actions and their timing. Predictive techniques may be applied periodically or continuously. Especially critical components whose failure could be catastrophic are monitored continuously. Generally, sensors located at the critical components are linked with computers set to indicate a normal versus an abnormal condition. The indication of an abnormal condition signals the operator with an alarm so that action can be taken to restore the component to a normal operating condition. Continuous readings establish trends in the running condition of the equipment and their analysis leads to the interception of a worsening condition before more serious problems arise. See Appendix F - Predictive Maintenance Techniques for a detailed description of these techniques and their applications. Techniques include:

- Vibration analysis
- Shock pulse
- Resistance testing
- Infrared scanning
- Sonic testing
- Ultrasonic testing

Examining Preventive and Predictive Maintenance Essentials

The Impact of Preventive Maintenance on Program Execution

By observing the conditions before a PM program is initiated versus after it has been successfully established, the maintenance manager can develop a powerful argument for a quality program. The impact of PM on the execution of a maintenance program can be best illustrated by observing the before and after of:

- Deficiencies reported
- Time between reporting and failure
- Use of manpower

Deficiencies Reported - The importance of establishing an effective planning effort is demonstrated in the illustration below. Deficiencies found will continue to rise until about the 5th month after the PM program is started. If those deficiencies are converted to planned work, the equipment is less likely to be troublesome at the same frequency as before the PM program was started. See Figure 21-3.

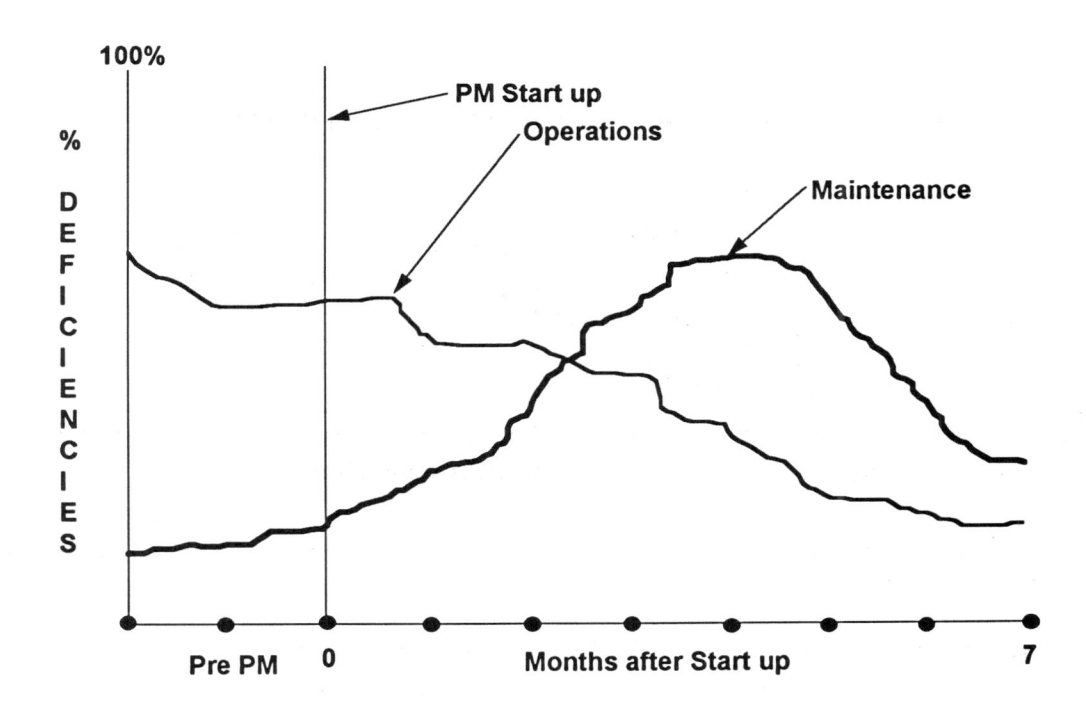

Figure 21-3. Deficiencies found by maintenance increase until the 5th month. Thereafter, if deficiencies are converted into planned work, fewer deficiencies will be found.

Planned work is done more deliberately. Thus, the frequency of repair will be reduced. As this happens, fewer deficiencies will be repeated. Conversely, if deficiencies are not converted into planned work, they will be 'rediscovered' and those personnel finding them will soon get discouraged. If this happens, the number of deficiencies will continue to climb after the 5th month and about the 7th month, the whole PM program will be abandoned. Thus, as any PM program is initiated, it will be important to set up effective planning to accompany it. Note that the number of deficiencies reported by operations also falls as maintenance takes the initiative in finding problems.

Time Between Reporting and Failure - The amount of time between the discovery of the deficiency and the time when the job must be done is also critical to the ability to plan work. See Figure 21-4.

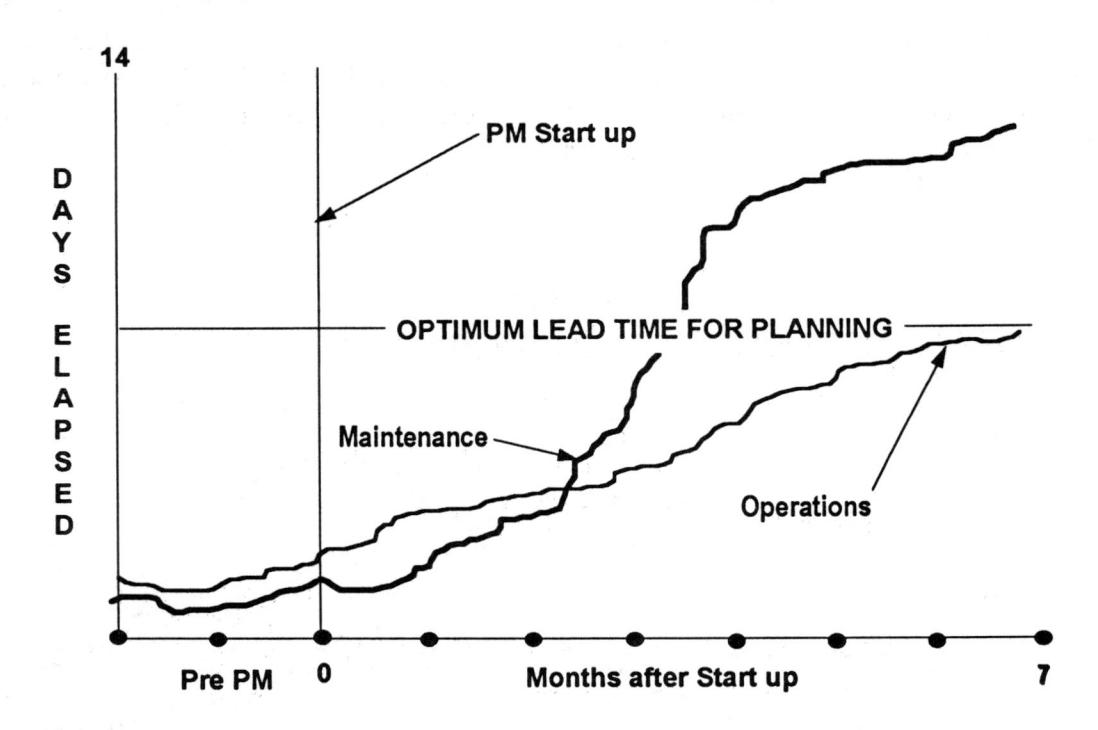

Figure 21-4. Within five months after starting the PM program, many deficiencies will be found soon enough to allow planning.

Generally, the optimum amount of lead time for planning a job is one week. Prior to the startup of a PM program, most deficiencies reported by operations were not known to maintenance until about 2 days before the equipment might fail. Thus, there was insufficient time to plan the job.

Examining Preventive and Predictive Maintenance Essentials

However, as maintenance starts the program, the significance of finding problems soon enough to be able to plan them becomes important. By the 5th month, providing the planning effort has been organized effectively, the benefit of well planned work is apparent in less downtime per job, better productivity and lower costs. With these improvements, it becomes much easier to convince operations of the importance of reporting problems as soon as they become aware of them. As a result, operations personnel become more aware of their responsibilities for monitoring equipment condition.

Use of Manpower - The PM program will also impact the way that manpower is used. Generally, as PM services are done more completely and on time, the amount of manpower used on emergency repairs will be reduced while more is spent on planned and scheduled work. See Figure 21-5.

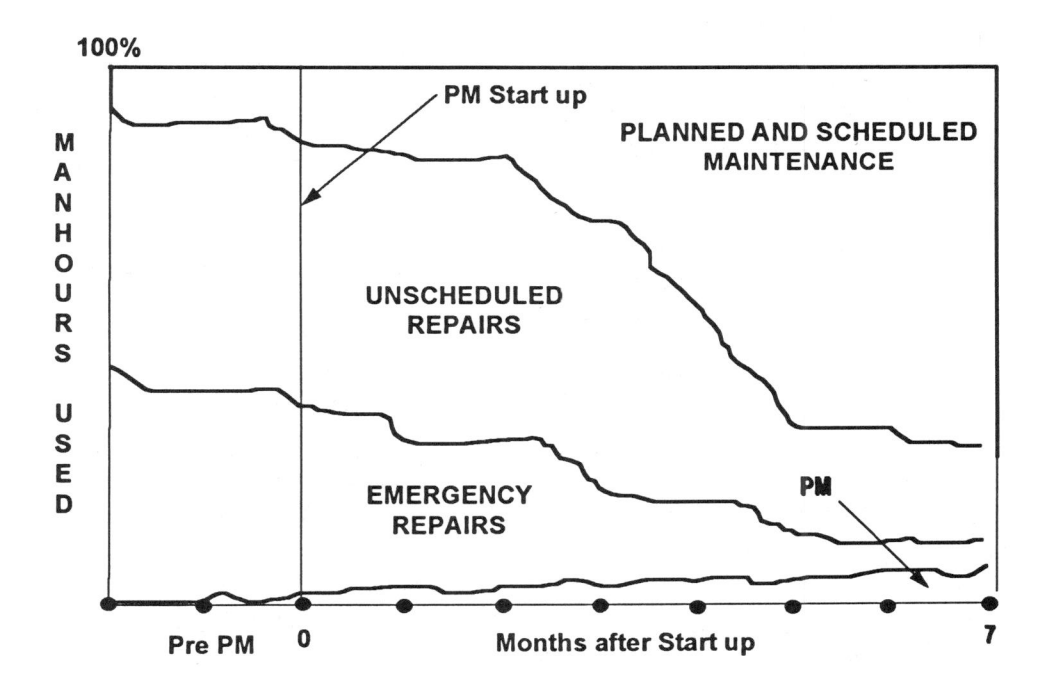

Figure 21-5. Optimum manpower used on PM yields a reduction in manpower spent on unscheduled and emergency repairs with an increase in manpower spent on planned and scheduled maintenance.

Generally, about 9 - 11 % of total maintenance manpower spent on PM services will assure an 85 % compliance with the PM program and the reduction of emergency and unscheduled repairs, As this happens, planned and scheduled manpower should reach a desirable level of over 50 % of all maintenance manpower.

Establishing the Preventive Maintenance Program

Establishing the preventive maintenance program consists of two steps:

- Organizing the program
- Operating the program

Once the program is in operation, experience will dictate modifications. The actions required in each step are shown in Figure 21-6.

Organizing

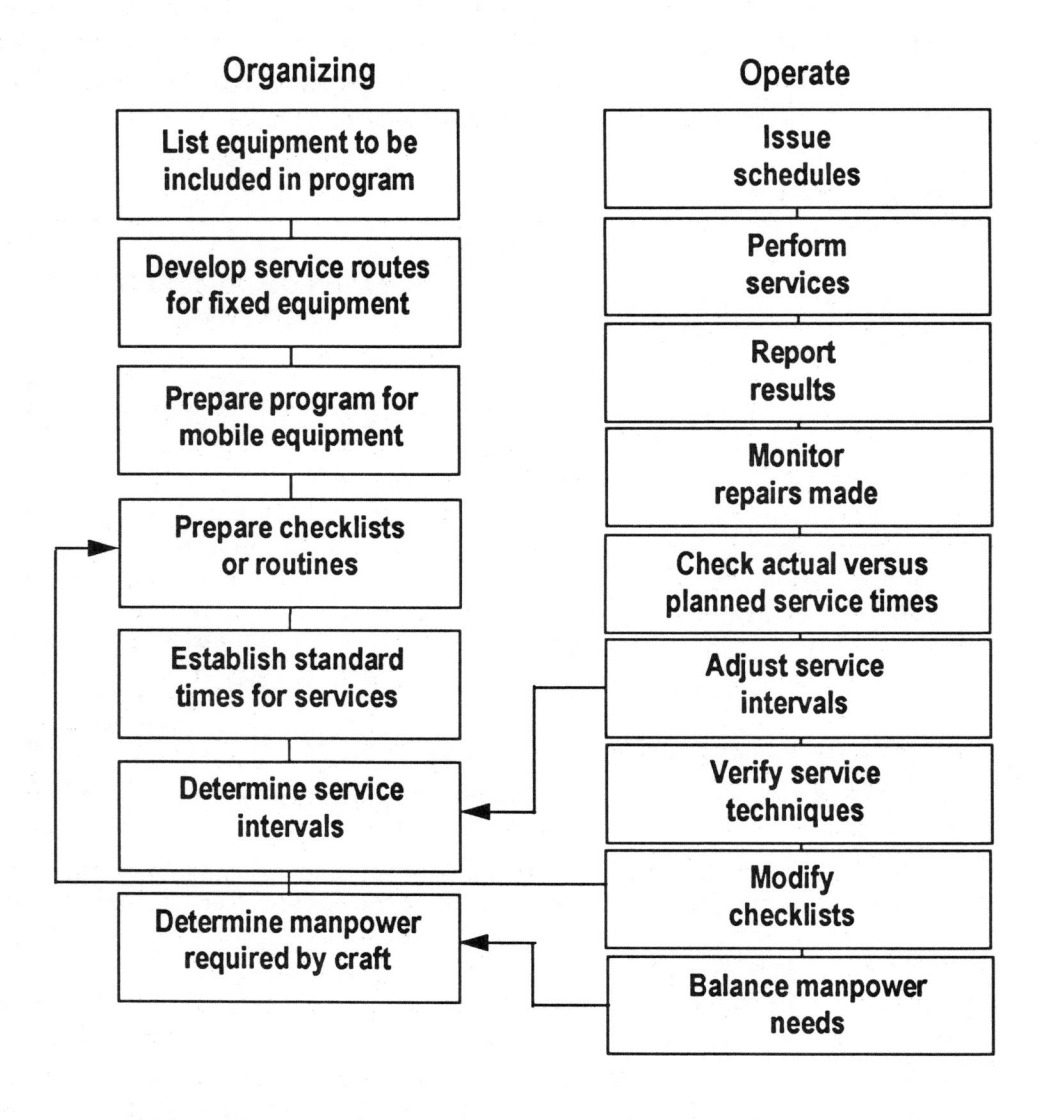

- List equipment to be included in program
- Develop service routes for fixed equipment
- Prepare program for mobile equipment
- Prepare checklists or routines
- Establish standard times for services
- Determine service intervals
- Determine manpower required by craft

Operate

- Issue schedules
- Perform services
- Report results
- Monitor repairs made
- Check actual versus planned service times
- Adjust service intervals
- Verify service techniques
- Modify checklists
- Balance manpower needs

Figure 21-6. As experience is gained in operation of the PM program, modifications will be made.

Examining Preventive and Predictive Maintenance Essentials 149

Determining Equipment to be Included in Program

Identify potential equipment problem areas that will deteriorate with continuous operation. Then establish PM tasks required to ensure equipment reliability. See Figure 21-7.

The impeller will deteriorate with continuous operation.

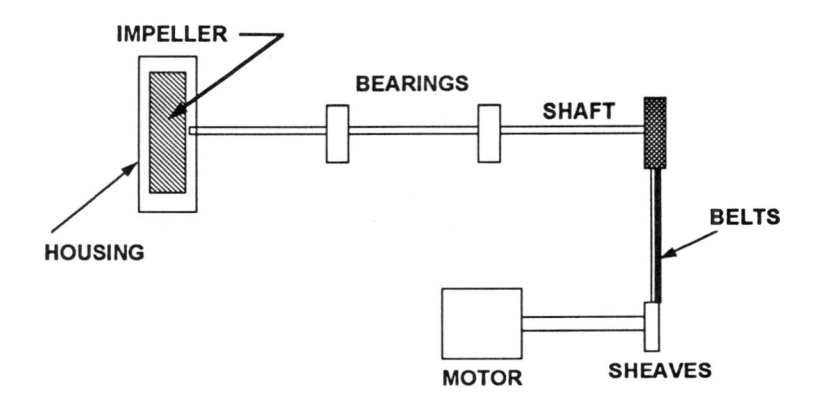

Inspect, test and lubricate to ensure equipment reliability.

Figure 21-7. Equipment criticality dictates PM action required.

Establishing the Preventive Maintenance Control Network

Operating the PM program requires a control network to schedule, assign and monitor services then measure program effectiveness. A step by step process is illustrated in Figure 21-8.

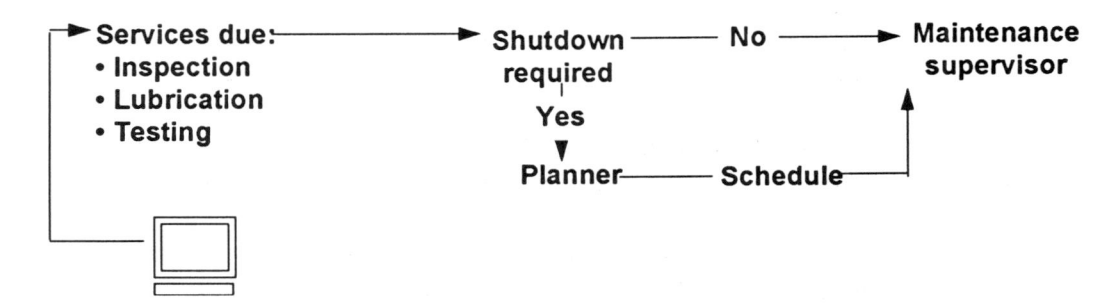

Figure 21-8. After services due are identified, determine whether equipment must be shut down. If so, the planner should schedule the services. Otherwise, the supervisor should assign the services routinely.

After receiving assignments, crew members conduct PM services, query operators, list deficiencies and advise the supervisor of them. Generally, the deficiency list is written out component by component and, recommendations of the crew members are provided suggesting whether the repair must be made immediately (emergency), can wait (unscheduled) or might require considerable materials (possibly needing planning). See Figure 21-9.

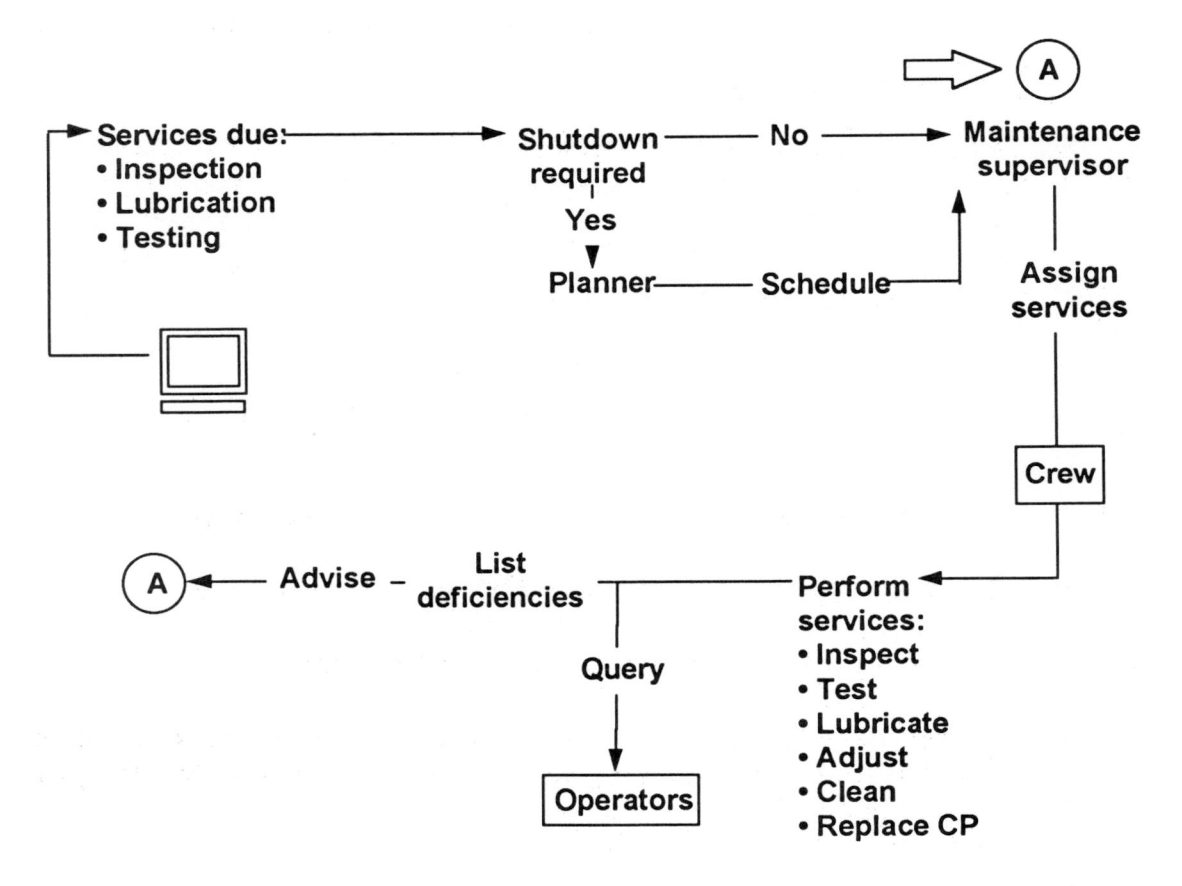

Figure 21-9. On the receipt of deficiencies from crew members, the supervisor must classify each deficiency and decide on an appropriate action.

Based on the advice of crew members, the supervisor will consider each deficiency and decide how it will be handled. He would, for example:

- Assign emergency jobs to crew members immediately.
- Enter unscheduled repairs in the computer pending the next opportunity to shut down the equipment.
- Take jobs that meet the criteria for planned work to the planner and discuss them with him.

Examining Preventive and Predictive Maintenance Essentials 151

The total picture of the preventive maintenance control network is then assembled. Notice that it has the same configuration as the schematic depicting the definition of the maintenance program except that it deals only with PM. Reference Figure 21-10. See also Chapter 20 - How to Define the Maintenance Program.

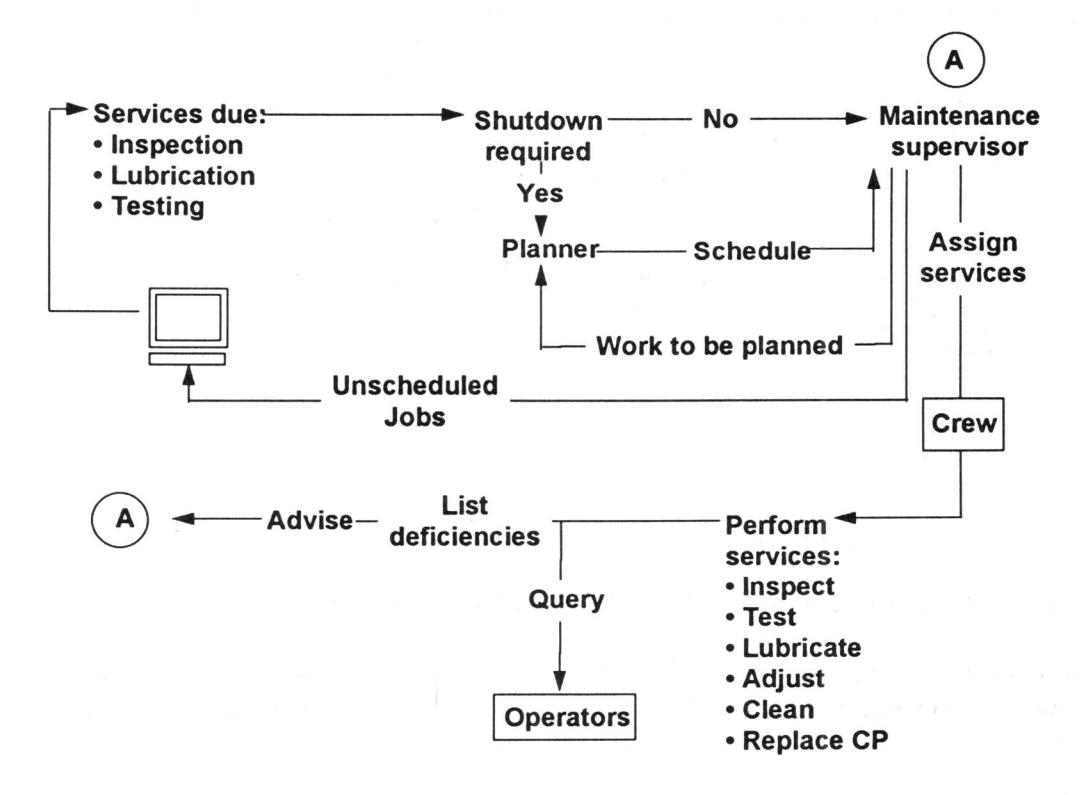

Figure 21-10. After classifying deficiencies, the supervisor takes action to assign, hold or have selected jobs planned.

Scheduling Procedure - Most preventive maintenance scheduling procedures are included in the computerized maintenance management system (CMMS). Systems determine services due based on fixed or variable intervals:

- Fixed interval every 2 weeks for example
- Variable intervals such as the accumulation of operating hours of a unit with services carried out at 250 hours, 500 hours etc.

Some CMMS systems contain the pertinent checklists, allowing them to be printed by the crew member at the time he sees his PM assignment on the daily crew assignment sheet.

It is desirable for the crew member to enter new deficiencies into the system as he completes the service. This will allow them to be easily classified by the supervisor or, if there is a team organization, other team members or the team coordinator.

The scheduling procedure for fixed equipment is a matrix. See Figure 21-11.

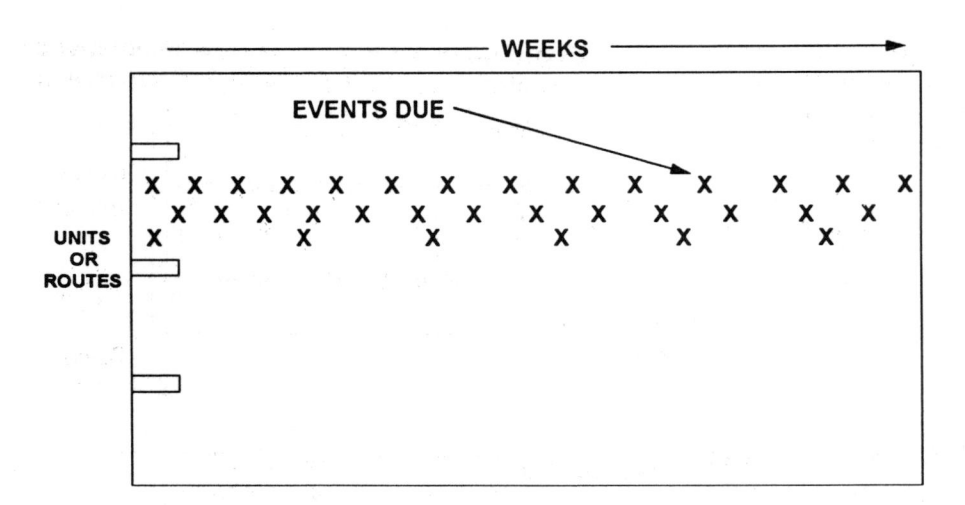

Figure 21-11. Week by week, the computer can identify events or services due.

The manpower required for all PM services annually, by craft, provides an estimate of manpower required for the total PM program. See Figure 21-12.

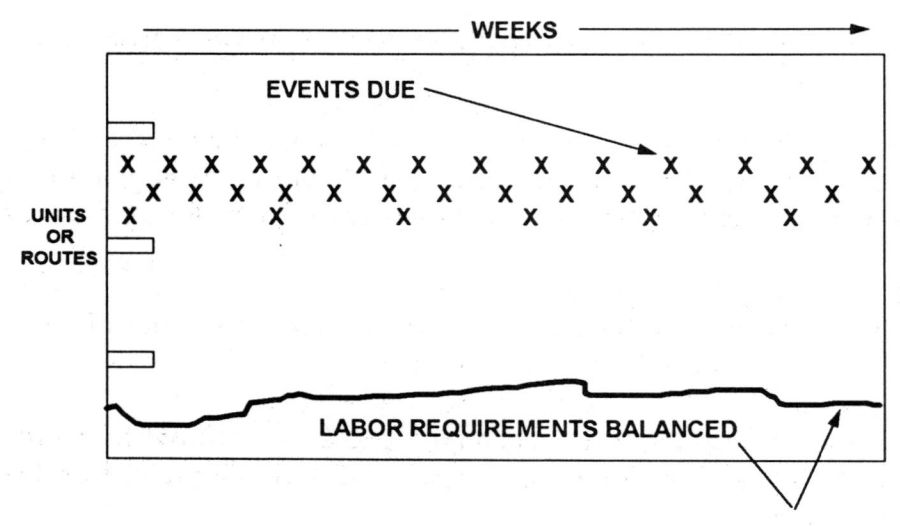

Figure 21-12. Manpower for the total program is balanced week by week.

The Successful Preventive Maintenance Program

The preventive maintenance program will utilize from 9 - 11 % of total maintenance manpower in performing required PM services. Deficiencies found as a result of performing inspections, testing or monitoring are classified as:

- Potential emergency repairs
- Unscheduled repairs (running repairs)
- Possible planned and scheduled maintenance

The manpower required to perform this work should be classified as emergency, unscheduled or planned maintenance, not PM.

Thus, one important index of the success of the PM program is the amount of manpower used and its impact on reducing emergencies while increasing manpower used on planned work. Next, the compliance with the program should be measured week by week. Generally, 85 % compliance is desirable and managers should specify this. PM compliance reporting should be included in the performance evaluation checked at each weekly scheduling meeting. If compliance is less than 85 %, the reason for non compliance should be determined. Many plants that have difficulty achieving good PM compliance should require operations to be accountable for making its equipment available to meet the schedule.

The success of the PM program can also be measured in tangible ways, including:

- More work is being planned.
- Components are lasting longer.
- Downtime is decreasing due to more planned work.
- The interval between repetitions of repairs is increasing.
- Emergency repairs are less frequent.
- Overtime is being reduced as emergencies decrease.
- There is less unscheduled work.
- More manpower is being used on scheduled work.
- Labor is being better controlled due to more planned work.
- Productivity of personnel is increasing.
- More total work is getting done.
- Overall costs are being reduced because jobs use less labor.
- Fewer personnel are needed or some may be shifted to other work.

The most significant impact of a successful preventive maintenance program is its potential for reducing cost through fewer emergencies and more planned work. ₃. See Figure 21-13.

Figure 21-13. Remember, successful PM always shows up on the bottom line !

Summary - The objectives of preventive maintenance to avoid premature equipment failure and to extend equipment life are achieved through the faithful application of routine PM services and predictive maintenance. However, successful PM depends also on an effective planning effort to convert PM deficiencies into planned work. Only then do many of the real benefits of PM accrue.

1. - Reliability - based maintenance, sometimes classified as preventive maintenance, is better classified as maintenance engineering. It appears in Chapter 25 - Conducting Essential Maintenance Engineering.
2. - 3. - Images copyright, New Vision Technologies, inc.

22

Conducting Effective Planning

Effective Planning Yields Better Productivity and Performance

Planned maintenance includes major jobs, which by virtue of their scope, cost and importance warrant advance planning to ensure that work is accomplished with efficient use of resources in the least amount of downtime. See Figure 22-1.

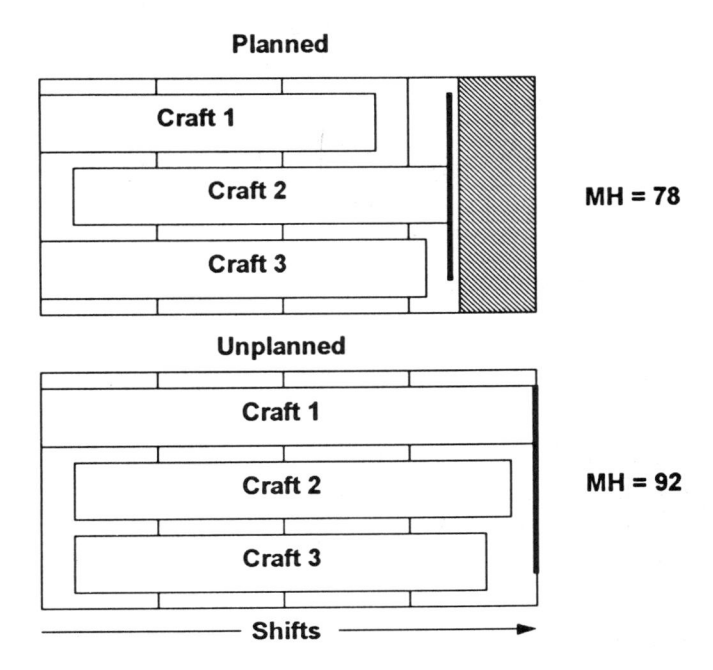

Figure 22-1. Two comparable major jobs, one planned and the other not will, on completion, reveal that the planned job was completed with 15 % less labor and, in at least 6 % less downtime.

Quality planning is made possible by ensuring that there is a solid, 'detection oriented' preventive maintenance program. That PM program will successfully uncover equipment problems far enough in advance to allow planning to take place. Planning creates its impact before work begins as well as during the execution of work and after it is completed. ₁. Se Figure 22-2.

Before work begins:

- The job scope is properly identified.
- A solid job plan is organized.
- Supervision knows what is expected of them.
- Crews can be advised of tasks in advance.
- Materials can be pre ordered.
- Tools can be reserved for jobs.

Figure 22-2. The planner should have sufficient time to get major jobs properly organized in advance.

As a result of good planning, the execution of a job will be improved by:

- Clearer job instructions
- Allowing the supervisor to be more effective
- Fewer job interruptions
- Better productivity of crew members

The completed job will be enhanced as well as a result of better job preparation and execution. Typically:

Conducting Effective Planning

- Less manpower will be used.
- Each planned job will cost less in labor.
- Overall costs will be reduced.
- Jobs will be completed in less elapsed downtime.
- Better quality work will be produced.
- There will be a longer interval before the work has to be repeated.

Criteria for Planning Work

To utilize the planner more effectively and establish a common understanding of what should be planned, a criteria should be developed to determine which jobs should be planned. This criteria should be clear and practical to administer. A criteria suggesting that every job requiring over 2 hours or costing more than $ 500 be planned is both simplistic and impractical.

Ultimately, the criteria should serve to clarify working relationships. For example, if certain PM deficiencies require planning, the supervisor would take them to the planner and the planner would be in agreement that the work should be planned. The criteria should have two levels. One, a listing of all conditions under which the job will be planned. Two, desirable conditions that suggest planning would be beneficial but, leave some discretion for both the planner and the supervisor. Typically, jobs will be planned when:

- Cost and performance must be measured.
- The job is done to satisfy a manufacturer's warranty.
- A standard of quality, cost or performance must be met.

Alternately, a job is planned if any 10 of the following 12 conditions are met:

- Work not required for at least one week.
- Duration of the job will exceed one shift.
- Two or more crafts needed.
- Requires crafts not part of the regular crew.
- Requires two sources of materials of: stock, purchased or fabricated.
- A coordinated shutdown is needed.
- Special equipment, like a mobile crane or power tools are required.
- Rigging and transportation will be needed.
- Drawings, prints and schematics are required.
- A job plan must be prepared.
- Work requires contractor support.
- The estimated cost exceeds $ 5000.

Planning Steps

The illustration below shows the sequence and timing of planning steps against a typical period covering eight weeks. Planning covers weeks 1 - 5. Planned work is presented to operations for approval of job timing during the scheduling meeting on week six. Operations agrees to make equipment available during weeks seven and eight. The planner would then monitor the work during execution and note job performance. As the weekly cycle repeats, the planner turns his attention to planning new major jobs. See Figure 22-3.

```
─────────────── Weeks ───────────────▶
┌─────┬─────┬─────┬─────┬─────┬─────┬─────┬─────┐
│  1  │  2  │  3  │  4  │  5  │  6  │  7  │  8  │
└─────┴─────┴─────┴─────┴─────┴─────┴─────┴─────┘
```

- **Identify work**
- **Investigate**
  - **Get advice from supervisor**
    - **Determine if standards apply**
    - **Confirm job scope**
    - **Make job plan and set up WO**
    - **Determine resources**
    - **Establish manpower by craft**

  - **Estimate cost, set priority and get approval**
  - **Establish preliminary time to do job**
    - **Open WO and order materials and shop work**
    - **Await receipt of materials**
         Confer with operations on job timing
         Arrange rigging, transport and tools
         Conduct official scheduling meeting   ◀─ **Perform work** ─▶
Monitor job execution and note cost and performance

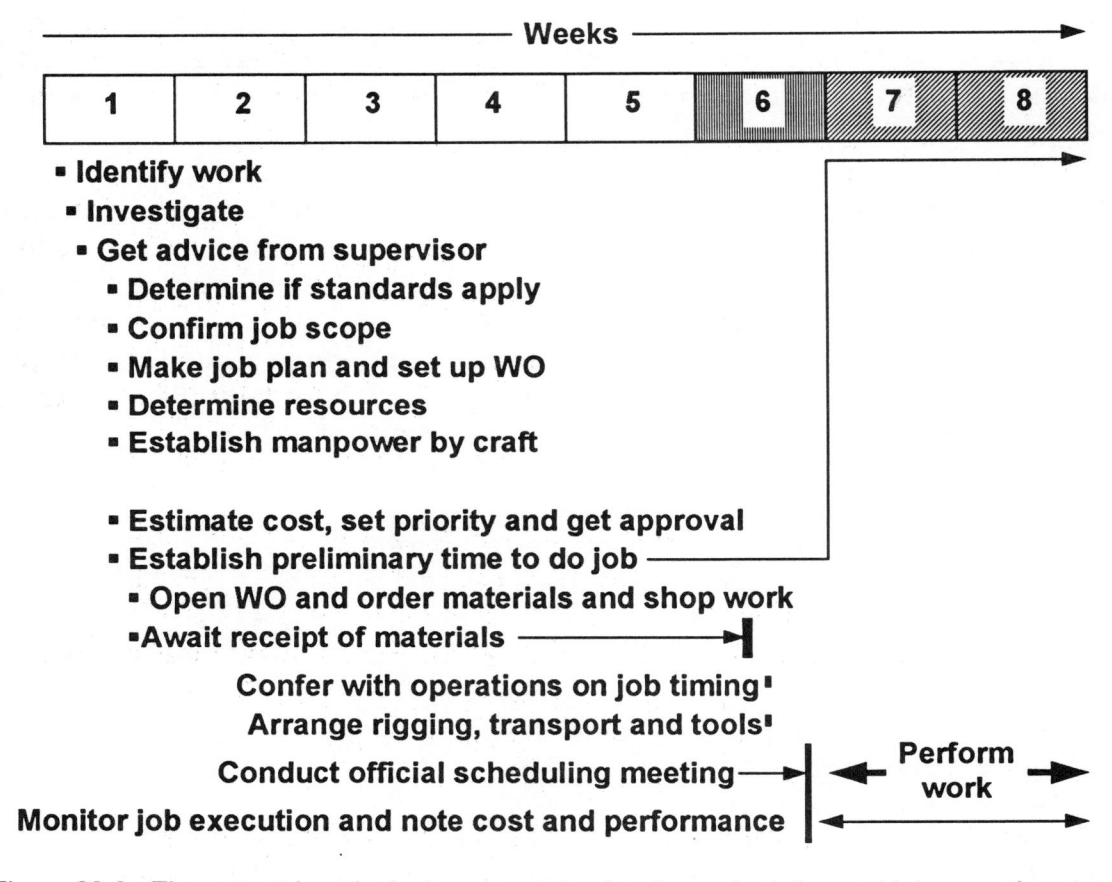

Figure 22-3. There must be a logical sequence to planning, scheduling and job execution steps.

The planning steps are explained in more detail below.

Identify Work - As new work is presented, the planner verifies that it meets the criteria for being planned. He then proceeds to determine the general nature of the job and, if required, establishes a time when he can investigate the job in the field.

Conducting Effective Planning

Investigate - The planner conducts a field investigation of the job if it has never been done before. If the job has been done before, he may consult previous work order details

Get Advice From the Supervisor - The supervisor who will perform the work is consulted along with crew members to obtain their views on how the job is best accomplished.

Determine If Standards Apply - If standard task lists, material lists and tool lists are available they should be incorporated into the planning. A less detailed field investigation and discussion with the supervisor and crew should be conducted in the event that changes have been made in the equipment or its operation.

Confirm Job Scope - Determine what is necessary to restore the equipment to safe, effective operating condition. This will constitute the scope and nature of the job and guide in planning it.

Make Job Plan and Set Up Work Order - The job plan is the step by step approach that will be used by the maintenance crew to perform the job. Usually, a bar chart is used to depict the order, sequence and duration of each step, the crafts involved and the completion time. Figure 22-4.

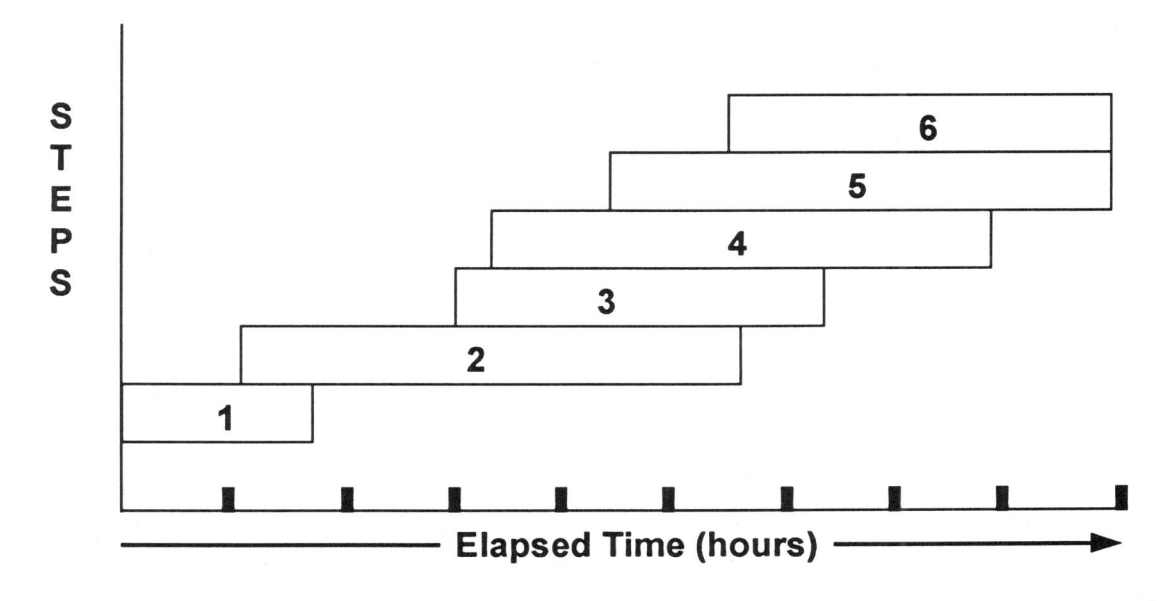

Figure 22-4. A similar bar chart would accompany the work order and be used on the job site to guide the supervisor and his crew through the job steps.

Determine Resources - Resources would be identified to include labor, materials, tools, equipment like mobile cranes, use of overhead crane, rigging, ladders, transportation etc.

Establish Manpower by Craft - For each step of the job plan, the craft manpower including the number of men and the sequence in which they are required would be identified. Manpower needs are then balanced according to the shift on which they will be available.

Identify Materials and Shop Work - Materials from stock, purchased materials and shop work such as fabrication, machining and assembly requirements are identified.

Estimate cost, Set Priority and Get Approval - The total estimated cost of all resources including labor, materials from all sources, shop work, the use of equipment, rigging and transportation are determined. Next, a job priority is established from the maintenance point of view. On job approval, the priority may be altered by operations. When the job is approved, it means that operations has accepted the need to perform the work, its estimated cost, the approximate timing and its priority

Estimate Preliminary Time to Do Job - Now the planner must tighten the timing of the job. The job cannot be started any sooner than the availability of all materials and shop work. The actual condition of the equipment will reveal how long the equipment can operate before failure or trouble may occur. The job must be accomplished sometime within this 'window of opportunity'. If this timing if significantly different from previous estimates, he should advise operations.

Open Work Order and Order Materials and Shop Work - Opening the work is the authorization to spend operating funds for the job. Only at this time can materials and shop work be ordered.

Await Receipt of Materials - A period of time will pass before materials and shop work will be available. During this period other jobs are being planned. The planner should monitor progress in assembling materials by observing the purchase order tracking system and work order status reports for shop work completion. Generally, stock materials would not be ordered until other materials are on hand. However, if there is concern that they may not be available, stock materials should be reserved.

Conducting Effective Planning 161

Confer With Operations on Job Timing - Just prior to the official scheduling meeting with operations, the planner will assemble a preliminary schedule of all jobs proposed for the week following. In order to make a preliminary allocation of manpower, he must have some idea of when required units of equipment could be made available. In this discussion, he might, for example, know that some materials will arrive at the last minute and may wish to have the corresponding unit available on Friday rather than Tuesday. $_2$. See Figure 22-5.

Figure 22-5. A preliminary meeting with operations before the 'official' weekly scheduling meeting permits a realistic plan of jobs for the next week to be prepared.

Conduct 'Official' Scheduling Meeting - The weekly scheduling meeting with operations is the opportunity for maintenance to present its plan for accomplishing major jobs and key PM services for the week following. Based on the production plan, operations will consider the needs for mobile equipment and the 'mix' they must have against the requirement to perform maintenance at the recommended times. Similarly, the shutdown of certain fixed equipment, lines or areas is considered. Necessary negotiations are carried out by the principal superintendents with the planner acting as an advisor. The meeting should conclude with an approved schedule. Thereafter, operations makes equipment available according to the schedule and maintenance makes its resources available. The objective is to accomplish the work with the least interruption to operations and the best use of maintenance resources. During the week when the scheduled jobs are being done, daily coordination meetings between operations and maintenance are held to make adjustments in the schedule as operations delays are encountered.

Monitor Job Execution and Note Performance - The planner should monitor the progress of each job during the week and on the completion of each job, note its cost and performance. 3. See Figure 22-6.

Figure 22-6. The work order status report permits the planner to observe the day to day accumulation of manhours and cost against each work order. On job completion, he can determine the job cost and performance.

In addition, the planner may wish to help coordinate on going jobs. This makes for a solid working relationship with supervisors and crews.

Summary - Well executed planning can have a significant effect on improving productivity, reducing the elapsed time to do jobs and assuring better work quality. As a result, planning makes a major contribution to cost reduction.

2, 5 and 6 - Images copyright, New Technologies, Inc.

23

Forecasting and the Component Replacement Program

Simplifying the Control of Repetitive Major Jobs

The replacement of major components represents over 60 % of maintenance work that can be planned and scheduled. As a result, the planning process, as well as the overall maintenance program, can be significantly streamlined with the inclusion of a properly organized and executed component replacement effort.

Key fixed and mobile equipment components such as engines, drive lines, pumps or conveyor drives lend themselves to this program.

Since each job has been performed previously, each such job can have:

- A standard task list (explaining what is to be done and how)
- A standard bill of materials (to help identify and ordering materials)
- A tool list (permitting them to be reserved for the job)

The task of controlling component replacements is most commonly called forecasting. Essentially, forecasting is the identification of the most likely future time when components may have to be replaced. It is important to recognize that forecasting must not be an automatic action in which, for example, a component is simply replaced after the accumulation of a certain number of operating hours or the passage of so many weeks. This type of 'guesswork' will either 'bankrupt' maintenance or cause maintenance craftsmen to question why 'perfectly good components' are being replaced prematurely.

Establishing the Forecast

The frequency at which major components are to be replaced is determined initially by establishing a 'mean time before failure' (MTBF). This MTBF is derived from repair history (the chronological listing of significant repairs and the record of intervals between previous component replacements). See Figure 23-1.

Conveyor Drive Motors

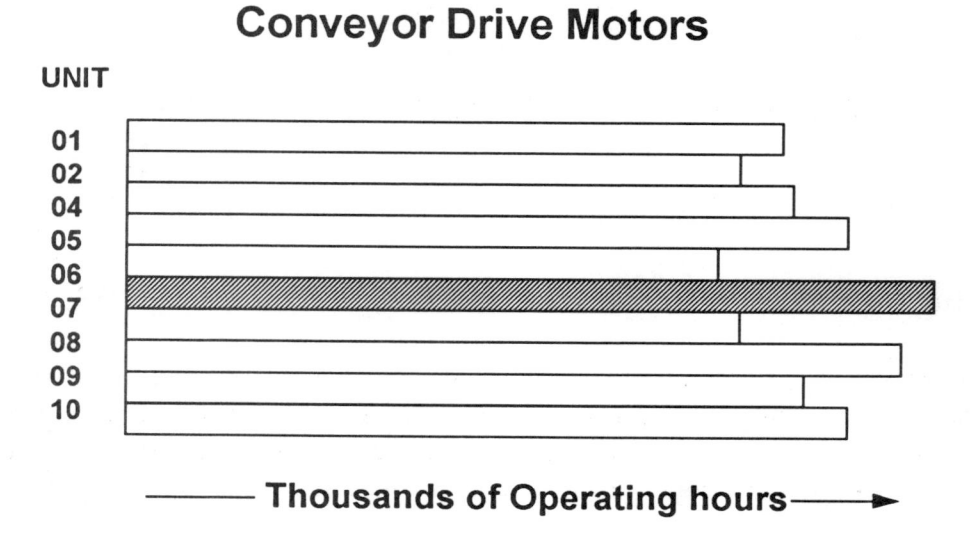

Figure 23-1. The 'mean time before failure' data of a critical component is developed from repair history data and used to establish the forecast.

Once the preliminary interval has been placed on the forecast, it can be used to help identify the possible future time when the component may have to be replaced. The initial uncertainty of the forecasted intervals must be compensated for by verifying the actual component condition before scheduling the replacement. Thus, the exact timing of the component replacement must be determined by the condition established during the latest preventive maintenance inspection.

While this procedure may seem tentative, it is merely an automatic step to better assure the future dependability of the forecasting procedure. In time, the forecasted intervals will be confirmed and the integrity of the program established.

It is important to acknowledge that the real benefit of forecasting derives from its simplification of planned work thus, encouraging maintenance to do more. The subsequent benefit is reduction in downtime for each job, lower cost and, ultimately, higher quality of work and less troublesome equipment.

Forecasting and the Component Replacement Program

The forecast provides a picture of all components to be replaced, unit by unit, and the approximate timing of their replacement. See Figure 23-2.

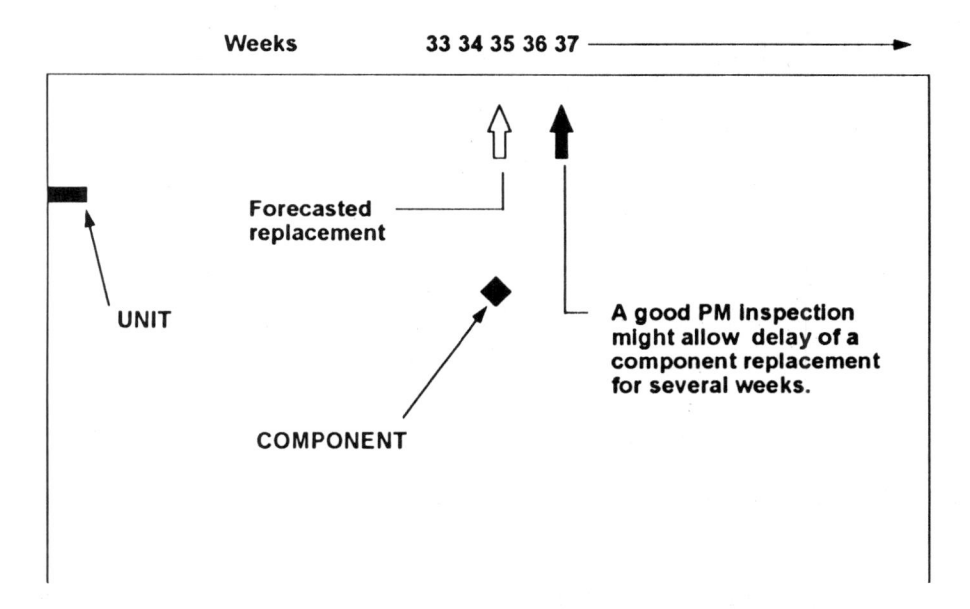

Figure 23-2. The latest PM inspection signals the actual condition of the component assuring it is replaced at the most appropriate time.

Each forecasted event includes standards on how to perform the job, as well as the materials and tools required. The essential purpose of these standards is to assure that the job is performed properly with each repetition. In addition, the existence of such standards provides a short cut to planning:

- Jobs do not have to investigated individually
- Materials are identified in advance

Not only does this simplify the planning but, it pushes the event from a planned and scheduled event to a scheduled event. Thus, it is not essential that the component replacement program be within the exclusive domain of the planner. With a competent forecast and good feedback from PM inspections, any supervisor or, in fact, any team could operate the component replacement program efficiently.

Consequently, the forecasting procedure could affect the way maintenance organizes and operates. In effect, the capability of the computer to gather repair history facilitates the identification of component replacement intervals. Next, the linkage of the work order with the inventory control program and purchasing provides for the rapid development of standard bills of materials.

In addition, the forecast helps to identify the manpower required to carry out component replacements. See Figure 23-3.

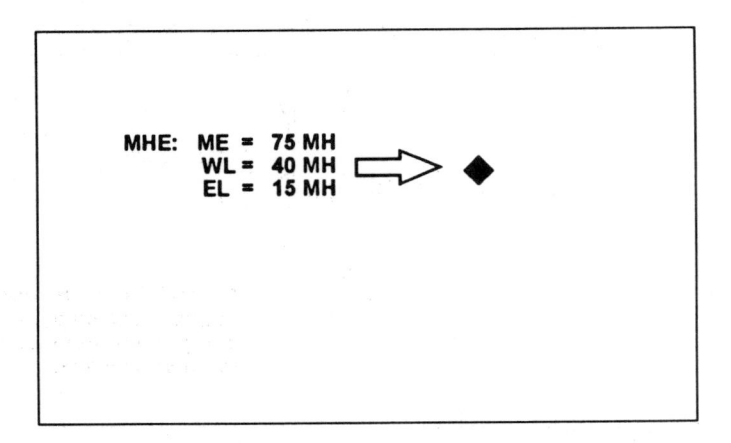

```
MHE:  ME = 75 MH
      WL = 40 MH  ⇒  ◆
      EL = 15 MH
```

Figure 23-3. The repetition of each component replacement confirms the number of man-hours by craft required to carry it out.

Collectively, these event provide a total picture of man-hours requirements for the whole component replacement program. See Figure 23-4.

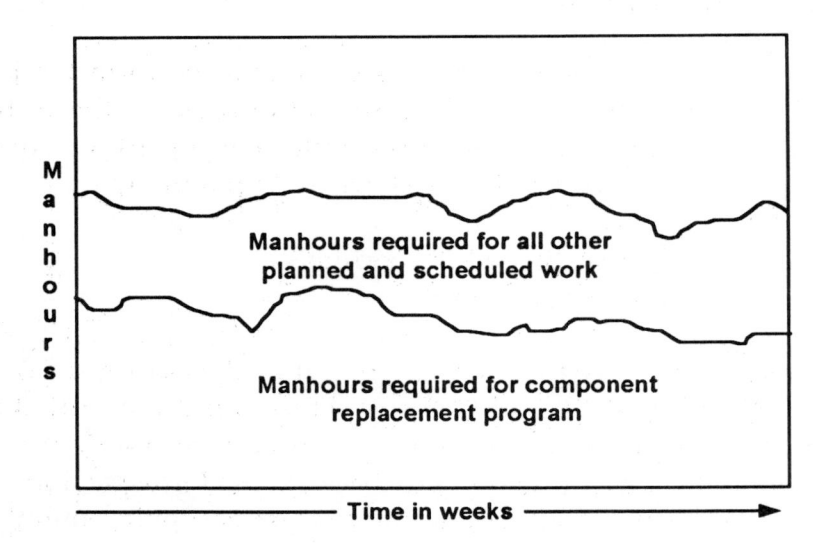

[Graph: Manhours vs Time in weeks, showing two curves — "Manhours required for all other planned and scheduled work" (upper) and "Manhours required for component replacement program" (lower)]

Figure 23-4. A complete component replacement program can represent over 60 % of all work that is able to be planned and scheduled.

Pulling together the elements of the forecast like material lists can be facilitated with the use of the computer. However, developing and assembling the task lists is not so easy, unless the help of individual craftsmen is obtained.

Forecasting and the Component Replacement Program 167

Typical Task list (Engine Module Removal Procedure)

1 - Degrease and pressure wash the module.
2 - Wash down truck with water hose.
3 - Move the truck into the shop. Check braking ability.
4 - Drain coolant into a water cart for future use.
5 - Drain engine oil into an oil cart for disposal.
6 - Disconnect main alternator, generator, blower and exciter.
7 - Remove all induction piping.
8 - Remove emergency shut-down housings.
- Plug inter cooler air inlet horns so as to exclude dirt.
- Rotate turbo compressor housings downward to exclude dirt.
9 - Remove exhaust piping at the turbo 'Y's'.
10 - Tag all hoses and wires upon removal to facilitate reassembly.
- Use aluminum tags and fasten securely.
- Ensure that all wiring harnesses and hoses are clear of module.
11 - Remove the blower ducting.
12 - Remove the hydraulic pump drive shaft.
13 - Remove the blower/exciter belts and belt tension idler.
14 - Remove the blower/exciter.
15 - Remove the left side rear exhaust cross-over pipe at muffler.
16 - Remove the blower inlet.
17 - Install engine lifting bracket.
- Use the 25 ton crane and the 6 ton come-along.
- Inspect lifting brackets. All bolts tight. No cracks in bracket.
18 - Remove module mounting bolts. Then remove module.
19 - Degrease and pressure wash truck frame.
20 - Inspect frame for cracks or damage.
21 - Inform welding of needed repairs.
22 - Inspect all hoses and wiring harnesses for wear or damage.
23 - Proceed to module installation sheets. Prepare new module.
- Do module build-up inspection.
- Prepare-installation parts (mounting pads, bolts, etc.)
- Check all air piping and connections (Items 32-39)

Task lists like the one illustrated should be prepared with the help of the craftsmen who normally perform the work. Many mines have them record the procedure and a transcript is made for them to edit later on. Then, other crafts personnel, selected by the first ones. try out the procedure adding items or modifying it. Once satisfied, the maintenance engineer would review the procedure and have it entered into the forecast data base for subsequent use.

Advantages of the Forecasting Procedure

Since each component replacement requires purchased as well as stocked materials. Two of the principal benefactors of the forecasting procedure are:

- Purchasing
- Warehousing

Typically, the purchasing agent can:

- Anticipate the number of components required annually
- Arrange to keep the right number flowing through the rebuild 'pipeline'

The warehouse can, on the other hand:

- Ensure the right number of associated spare parts are on hand
- Pull together 'packages' corresponding with forecasted tasks
- Arrange on-site delivery of parts
- Set aside tools needed from the tool crib

But, the planner, or a line supervisor, as well as a team member can:

- Apply standards to preclude planning jobs individually
- Accomplish much more planned work

Thus, the forecasting of major component replacements is a must for any maintenance organization striving to gain efficiency and better performance.

Summary - The forecasting of major component replacements is an effective way to streamline the planning task in plant maintenance as well as to assure less downtime, reduced cost and improved equipment performance.

24

Performing Effective Scheduling and Work Control

Scheduling to Obtain the Best Use of Resources

Objective - A principal objective of effective scheduling is to perform work when it interferes least with operations while making the best use of maintenance resources. Maintenance scheduling is a joint operations - maintenance activity. Maintenance recommends specific major tasks unit by unit together with suggested times. Operations responds with counter proposals based on the time they feel they can make equipment available and still meet their production goals. In a mining operation, for example, mobile equipment availability for maintenance is negotiated against the production plan which requires a certain 'mix' of mobile equipment. In the fixed equipment environment, usually a long term shut down schedule is provided to guide maintenance as they organize major tasks. At the conclusion of each meeting, there should be an approved schedule in which operations agrees to make the equipment available while maintenance agrees to make its resources available to accomplish the work. All planned major jobs should be included in the proposed weekly schedule along with preventive maintenance services that require an equipment shutdown in order to accomplish the work.

Meeting Chairman - It is best to have the superintendent of operations and the superintendent of maintenance present since binding decisions are required. The most effective meetings are those in which the operations superintendent conducts the meeting. He states the conditions under which operations commitments will permit him to make equipment available for maintenance. He then receives recommendations from maintenance as to which equipment they need to maintain or service, when and for how long.

Performance - It is always a good idea to check schedule compliance and performance for the previous week to ensure that it has been satisfactory. Schedule compliance is the percent of jobs completed versus those scheduled. Generally, 85 % is the target. Performance includes statistics such as the reduction in the backlog or the use of manpower on emergency versus PM or scheduled work. The purpose of checking is to note progress and ensure corrective actions are taken if needed.

Attendance - Attendance at the weekly scheduling meeting should include maintenance, operations, plant engineering, material control and safety personnel. The plant manager should attend from time to time to better judge the effectiveness and harmony with which the meeting is conducted. He should also listen as performance for the previous weeks schedule is discussed to observe cooperation and attitudes.

Coordination Meetings - Each weekly scheduling meeting should be supplemented by a daily coordination meeting to make adjustments in the schedule in the event of operational delays.

Maintenance Preparation

The maintenance planner is constantly preparing for the next scheduling meeting. He will start by assembling selected work orders for the week following and, after verifying the validity of the plan within maintenance, will assist in presenting it to operations at the meeting. See Figure 24-1.

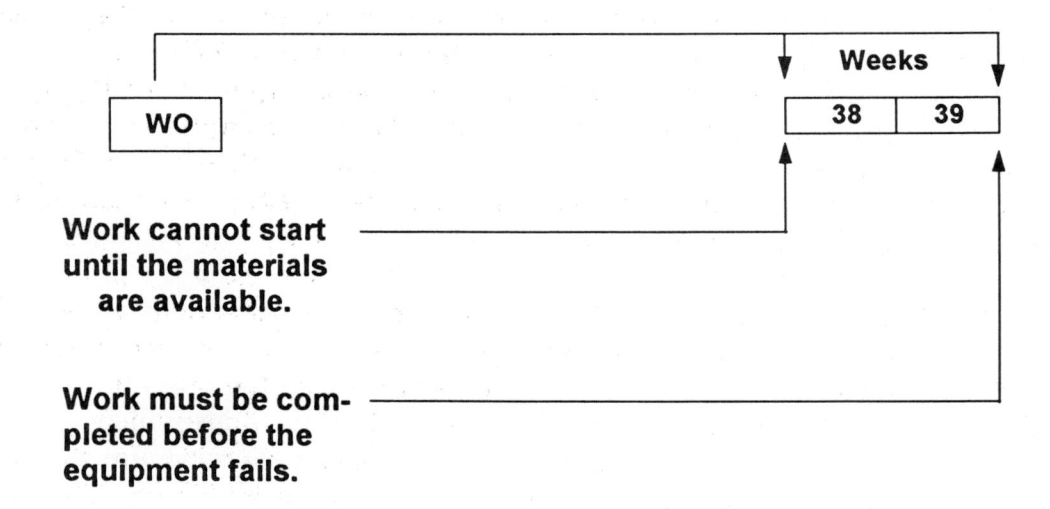

Figure 24-1. Each work order has a future time during which work should be done.

Performing Effective Scheduling and Work Control

By prescribing the approximate timing of each work order during the planning phase, the work order system can quickly pull the work orders for a particular week out of the general backlog. See Figure 24-2.

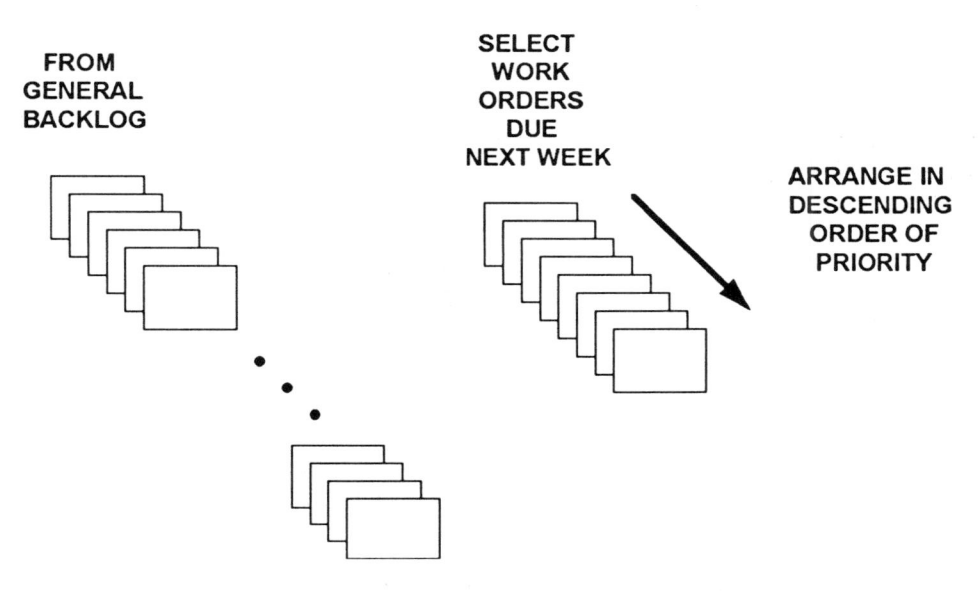

Figure 24-2. The work order system selects work orders due next week and arranges them by descending priority. The planner may then verify materials and allocate labor by priority.

The planner verifies resources availability for proposed work orders. See Figure 24-3.

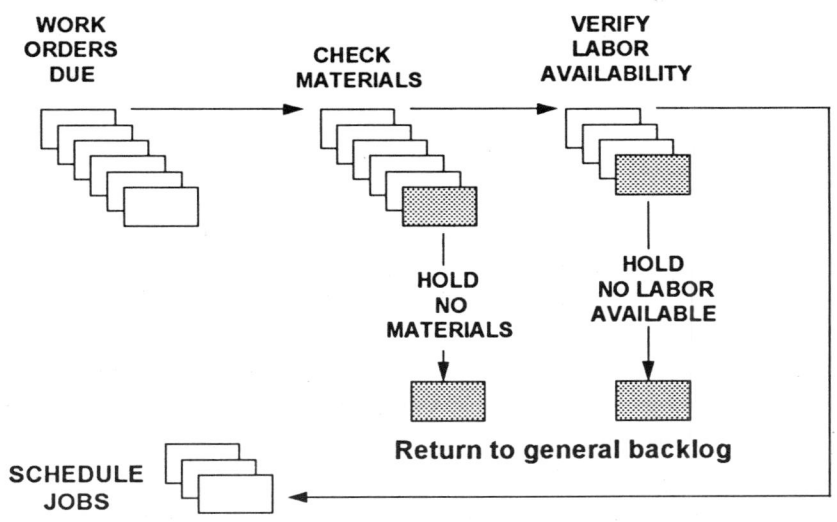

Figure 24-3. Materials and labor are verified before placing work orders on the preliminary schedule.

Conduct of Scheduling

Scheduling is divided into three phases:

- Getting ready for the scheduling meeting
- Conducting the meeting
- Follow up during execution of the schedule

The first two phases of getting ready for the scheduling meeting and the conduct of the meeting are shown in Figure 24-4.

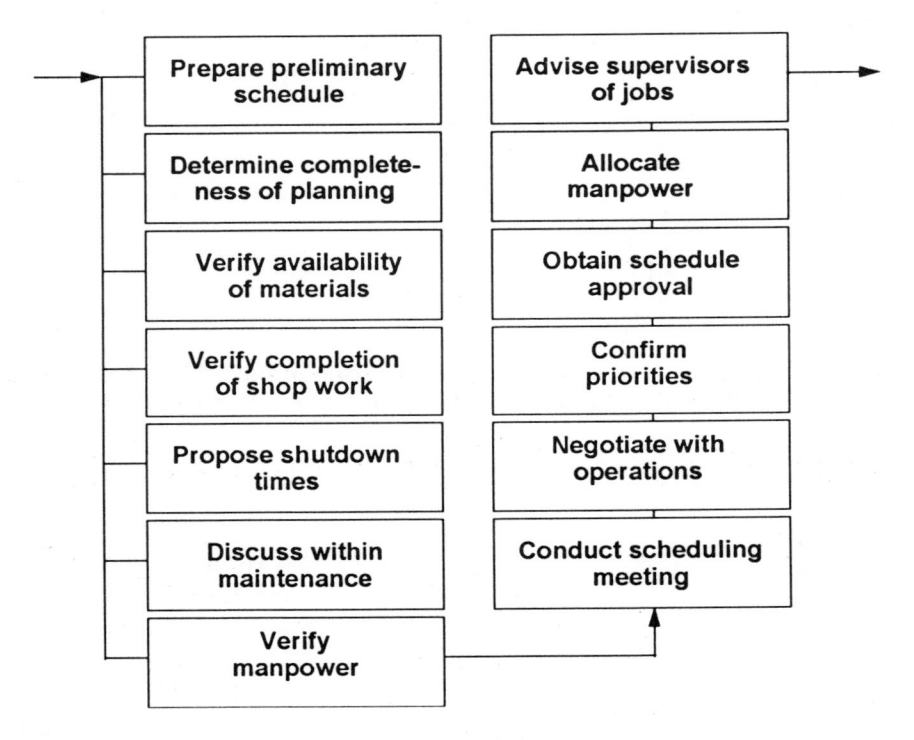

Figure 24-4. The left column describes steps taken to prepare for the meeting. The right column lists steps for conduct of the meeting.

Prepare Preliminary Schedule - The planner uses the work order system to help him select the candidate work orders for the preliminary schedule. This step is usually taken in the early part of each week because the scheduling meeting is often conducted on Thursday or Friday. Therefore, there will be several days to get ready. Since a planner may be working on from 75 - 125 work orders in various stages of planning and scheduling (some just starting and others on the schedule), it is important to have a work order system that facilitates the administration of a large volume of work orders.

Determine Completeness of Planning - Each work order would be reviewed to ensure that all of the necessary planning steps had been taken: job plan, labor, materials, tools, timing of shutdown etc. This review is necessary since the planner may not have looked at some work orders for several weeks.

Verify Availability of Materials - No job will be able to be completed unless the materials that the maintenance crews will install are available. Therefore, the planner should verify that the necessary materials are now on hand or will be before the work is scheduled. Primary attention should be given to direct charge materials, those being shipped to the plant. These are the materials over which the planner has the least control. Hopefully, during the planning phase, he allowed enough lead time to ensure that the supplier could deliver on time. Stock materials are less critical. Generally, a check should be made to verify that stock materials will be available. If there is some doubt concerning the availability of stock materials, the planner should reserve the materials in advance. This would require the warehouse to reserve certain quantities of specific parts just as if they had been issued. However, since the job might be deferred, it is best not to actually remove the items from the warehouse shelves until the related work order has been scheduled. Some mines provide a secured holding area in which the parts are held until needed. In other instances, the warehouse personnel may deliver the materials to the job site. In each instance, stock materials are charged to the work order number associated with the job.

Verify the Completion of Shop Work - Shop work such as machining, fabrication or assembly requires the same lead time considerations as do direct charge materials. Enough time must be allowed to ensure that the shop work can be completed in time. Again, the work order system is useful in permitting the planner to track the status of each work order that the shop is working on.

Propose Shutdown Times - To ensure that the most practical preliminary schedule is presented to operations, specific shutdown times should be discussed with operations in advance. This will permit the planner to start assessing labor requirements. Since craft personnel are not available equally on all shifts, special attention must be paid to making a careful allocation of labor. This cannot be done adequately unless operations gives some initial indication of when they can make equipment available.

Discuss Within Maintenance - The planner, as a staff assistant, cannot determine that the preliminary schedule is fully acceptable to maintenance, no matter how carefully he has thought it through and prepared it.

This step must be taken by the maintenance superintendent who is ultimately responsible for getting the work done. The superintendent should review the proposed schedule to ensure it is realistic based on his knowledge of overall plant operations for the coming week. He often has special knowledge of upcoming actions that the planned does not. Also in the process of reviewing the proposed schedule, he should ask pertinent questions to assure himself that the work can be done if operations approves the schedule. Certainly, basic questions about the availability of materials and labor must be asked during the review. As necessary, other personnel like general supervisors, other planners or the maintenance engineer might be included in the review.

Verify Manpower - A trial balance of manpower availability by craft, shift by shift, should be made to ensure that the necessary manpower will be available if the proposed schedule is approved by operations.

With the completion of the steps leading up to the scheduling meeting, the meeting is held to obtain approval of the schedule.

Conduct Scheduling Meeting - The objective of the scheduling meeting is to reach joint operation - maintenance agreement on the work to be done, the equipment on which it will be done and the timing of the work. Operations must commit to making its equipment available and maintenance must commit to making its resources available. Thus, decision makers must attend the meeting. Since, the proposed schedule must be negotiated within the production plan or a shutdown schedule, the most logical chairman of the meeting is the operations superintendent or his equivalent. Similarly, the maintenance superintendent, or is equivalent, should present the maintenance recommendations, in the form of jobs to be done and their proposed timing. While planners are present at the meeting, their role is to help adjust the proposed schedule in the event that certain equipment cannot be made available at the times that operations originally thought during the preparation of the preliminary schedule. Since the planner is intimately familiar with these details, he is the best person to make the adjustments. Moreover, when the maintenance superintendent presents the schedule, attendees at the meeting know they are listening to a decision maker who accepts the responsibility for seeing that the work is done. In addition, the superintendent, having had opportunity to review the schedule in advance is confident that the plan is a good one. Attendance should also include material control and safety personnel. The plant manager might attend periodically to help judge the effectiveness and harmony with which the meeting is conducted.

Negotiate with Operations - Negotiate is the best word. It is what actually happens. The proposed jobs are weighed against what operational circumstances will permit.

Confirm Priorities - As the negotiations progress, priorities may change. Units originally thought to be available at a certain time are now not available and the timing of the proposed work must be shifted. It is at times like this that the planners intimate knowledge of resources and their availability at a certain time pays dividends.

Obtain Schedule Approval - With negotiations complete, agreement must be reached. This agreement is schedule approval. On approval there is now a contract between operations and maintenance:

- Operations is to make equipment available at the scheduled time
- Maintenance is to make its resources available to accomplish the work

Allocate Manpower - With the approved 'contract' (schedule) at hand the planner can now allocate manpower to verify that it will be available.

Advise Supervisors of Jobs - The planner can now notify supervisors and crew personnel of the work they will be doing next week and its timing. This step is especially important in helping crews to better anticipate what is coming up and prepare for it.

Before examining the steps associated with schedule execution, it is useful to consider the scheduling meeting as a management forum for verifying performance. For example, the assembly of key operations, maintenance and material control personnel presents an outstanding opportunity to assess how well the previous weeks schedule was carried out. For example:

- Were all PM services completed ?
- Did all major jobs get done ?
- If operations failed to produce equipment, find out why
- If material control did not deliver, determine why not

Similarly, other related performance indices might be checked. Among them:

- Was the backlog reduced last week ?
- Are fewer emergency repairs being done as a result of better PM ?
- Is manpower being used more productively with more planned work ?

Often, a scheduling meeting begins with a look at the forecast for the next few weeks to alert attendees to major tasks that are coming up.

After the scheduling meeting, the planner should accomplish the following steps:

- Distribute the approved schedule so that all concerned parties are aware of the major jobs that are to be done and their timing
- Explain the tasks to individual maintenance supervisors so they can begin preparation
- Confirm the shutdown times with individual operations supervisors to ensure there are no misunderstandings
- Confirm the use of mobile equipment such as mobile cranes to ensure they are available at the prescribed times
- Confirm the on site delivery of materials or ensure that materials are made available from designated holding areas

As the new week gets underway the planner will:

- Monitor progress of on going work
- Coordinate the use of equipment between jobs
- Coordinate material delivery to ensure it gets to the right place on time
- Participate in daily coordination meetings where adjustments are made in the timing of events to accommodate operational delays

Summary - The weekly scheduling meeting is an opportunity for operations and maintenance to negotiate the shutdown of equipment for necessary PM services and major work. They are to schedule it to interfere least with operations while making the best use of maintenance resources. During this meeting, performance during the previous week can be noted and, if needed, corrective actions taken. Many organizations find that a daily coordination meeting is essential but, do not conduct weekly scheduling meetings. By not doing so, they fail to look ahead far enough to be able to get ready. Mines that do this should consider extending the Thursday daily coordination meeting by 15 minutes. Then, operations should ask the question: What are we doing next week ? The results will be worth the effort. Maintenance usually knows and they are delighted to tell operations. As a result, operations is better informed and thus, more helpful in cooperating with the maintenance effort.

25

Conducting Essential Maintenance Engineering

Assuring the Reliability and Maintainability of Equipment

Objective - The objective of maintenance engineering is to ensure the reliability and maintainability of equipment. These actions are built on a combination of preventive maintenance, predictive maintenance, assessment of equipment performance and the monitoring of new installations and rebuild procedures

These tasks have been prominent in the mining and mineral processing industries for many years. They have now been re labeled as 'reliability based maintenance'.

Reliability Based Maintenance = Maintenance Engineering

An examination of the ingredients of maintenance engineering (Reliability Based Maintenance) is useful in understanding and applying them:

- Preventive maintenance seeks to avoid premature equipment failures with inspection, testing and calibration. It continues with lubrication, cleaning adjustment and minor component replacement (like belts and filters) to extend equipment life.
- Predictive maintenance applies a range of technologies like vibration analysis or infra-red scanning to assess equipment condition, identify the current level of deterioration and predict when corrective actions must be taken.
- Pro active maintenance is the application of investigative and corrective technologies to extend equipment life.

Pro active maintenance translates into simple, logical maintenance engineering steps that have been in place for many years. Namely:

- If equipment fails through poor design, modify it.
- If it wasn't installed correctly, do it right.
- If a rebuilt component fails, make sure others are rebuilt correctly.
- If chronic problems are causing equipment failures, analyze repair history to find and eliminate them.

Each of these actions is related to an element of the maintenance program that should exist if that program is complete and effective. For example, a maintenance engineer constantly monitors the relative cost of maintaining similar units of equipment like the same model haulage truck in a fleet. The minute he sees dramatic differences, he reviews the comparative repair history to spot chronic, repetitive problems or failure trends exhibited for one unit but not the other. With these clues he takes to the field to examine the offending truck in person. He talks with operators, mechanics and supervisors to gather other opinions or observations. Soon he is in a position to prescribe a permanent corrective action. Once the corrective action is taken, he continues to monitor cost and repair history to verify the validity of the corrective action. Once proven, he may incorporate a modification in all units of the fleet to preclude a repeat of the problem at a future time.

Maintenance Engineering Correctly Applied

Generally, every maintenance work force with 50 or more crafts personnel should seriously consider a formal maintenance engineering effort. Larger work forces should consider a maintenance engineering staff. The formal effort must be directed toward taking those steps that ensure equipment reliability and maintainability. Specifically:

- Equipment reliability is assured with an adequate PM program, the proper application of predictive techniques and the correct, timely replacement of components.
- Equipment maintainability is aided by doing things right whether they be the installation of a new conveyor, an overhaul or the replacement of a major component.

These steps are logical. But, the failure is often one of applying them effectively. In a smaller work force, the steps can be delegated to supervisors, planners and craftsmen. Either way, an overall control mechanism is necessary.

Conducting Essential Maintenance Engineering

Maintenance engineering, whether done with a dedicated staff or delegated to personnel within the maintenance organization, must accomplish the following tasks effectively.

Ensure Equipment Reliability

Preventive Maintenance - The preventive maintenance program must be verified to ensure that:

- Essential equipment is included in the program.
- Necessary actions including inspection, testing, calibration, lubrication, cleaning, adjusting and minor component replacement are being carried out.
- Frequencies of services are correct.
- Checklists spell out what is to be done and how.
- Procedures exist to schedule services at proper times.
- Equipment deficiencies are being converted into corrective actions.
- New equipment is added to the program promptly.
- Completion of PM services is verified.
- Manpower to carry out the program is known.
- PM performance is checked (has it reduced emergencies ?)

Predictive Maintenance - The adequacy of the predictive maintenance effort should be assessed to determine whether:

- All necessary technologies are being applied.
- The technology is being applied correctly (vibration analysis versus shock pulse or periodic measurements versus continuous monitoring).
- New technologies are being actively sought and applied.
- Plant personnel are being trained to use technologies versus being dependent on vendors who may not be motivated to report problems promptly.
- Readings are properly analyzed (some mines are not capable of using new on board computer performance monitors thus, this advantage is lost to them).

Standards - Maintenance engineers develop the most effective and up to date methods of carrying out repairs and performing major jobs like overhauling a unit or replacing a major component. This information is imparted to the work force in the form of standards. Standards prescribe the exact way a job is to be performed including a task list (what to do and how, step by step), a bill of materials (the purchased and stock materials required) and tool list (the tools required and how to use them). As standards are developed, supervisors are briefed and craftsmen trained in their use.

Training - Maintenance engineers monitor on-going work to observe whether craft personnel are doing the work correctly. In addition, they may trace a particular failure to an inadequately performed job and identify a poorly trained or indifferent worker. In either case, training is necessary and the maintenance engineer prescribes it and arranges for it to be provided. Training may be performed in a variety of ways from on site classes by technical personnel from a manufacturer of equipment to classes at the manufacturer's location. The result must be more competent work. Similarly, the maintenance engineer might update basic skills after noting that older craftsmen are in need of refresher training.

Ensure Equipment Maintainability

New Installations - How many times have we heard the horror stories of contractors who install new equipment poorly only to cause maintenance go to extraordinary lengths to keep it in operation much less maintain it. Again, it is the maintenance engineer who draws up the new equipment commissioning standards that preclude this from happening. The same standards apply when maintenance does the installation.

Component Rebuilding - Not all component rebuild vendors are created equal. Some are outstanding while some aren't. If a rebuilt component fails prematurely, it is the maintenance engineer who is already tracking comparative component life to identify the cause. Once known, the vendor is notified of inadequate component performance, given a grace period to improve quality or his services are discontinued.

Equipment Modification - How many operating supervisors have come up with a 'good idea' for improving equipment performance that is ultimately proven unnecessary, not feasible and unable to be cost justified. Along the same lines, how many of these projects are paid for by expensing them in increments rather that funding them properly. We then wonder why the maintenance budget is unfavorable. Often, it is the maintenance engineer who prescribes the rules to establish the necessity, feasibility and correct funding of these projects. In doing so, maintenance remains focused on what they should be doing rather than 'building stairways to nowhere'.

ISO 9000 - A major manufacturer of underground mining equipment overhauled each unit differently, adding different components, making modifications and updating control mechanisms.

Conducting Essential Maintenance Engineering 181

When restored to service, the newly overhauled equipment often did not have the proper parts in stock and mechanics were confused with modifications made. An alert maintenance engineer promptly reminded the manufacturer who the customer was, then demanded and got computer discs with proper parts listings as well as documented modifications before the overhaul invoice was paid. He applied the quality standards of ISO 9000.

Adequacy of Information- Successfully ensuring equipment reliability and maintainability is rooted in good information. Effective maintenance engineering depends on good information. Therefore it is necessary that the maintenance engineer review the adequacy of information.

Specific information is required for the effective conduct of maintenance engineering tasks. This information includes:

- Detailed cost information by unit and component pinpointing excessive, different costs between similar units
- Repair history information to track chronic, repetitive problems for immediate correction
- Repair history information showing failure trends that reveal excessive instances of improper equipment operation, for example, necessitating a need for more operator training
- The life span of major components (in repair history) permitting the development of an accurate component replacement program
- Indices like maintenance cost per operating hour to help spot the offending unit quickly
- Downtime waiting for materials to better explain why the material control department needs to provide more effective service to maintenance

In addition, the maintenance engineer may assess other factors related to the ability of maintenance to perform its mission effectively. For example:

- Does maintenance have the right number of personnel of the proper craft to carry out the workload of essential work ?
- Should maintenance perform non maintenance work or should they stick with the basic maintenance program leaving the projects to contractors ?
- Is the maintenance work order - information system adequate to the needs of maintenance or is it a warmed over accounting system ?

The Future of Maintenance Engineering

Current actions in downsizing maintenance organizations prove that plant managers understand that the cost won't go down unless it is done with fewer people, less often. Skillful downsizing will assure fewer personnel. But, solid maintenance engineering is the only way to assure more effective maintenance that ensures repairs are done less often.

Therefore, the prudent industrial will verify the adequacy of its maintenance engineering effort and, if there is none, act promptly to establish one.

Summary - Industrial maintenance is being forced into an unusual set of circumstances given the need to reduce its operating costs. They are being required to operate more effectively with fewer people. Maintenance engineering holds the key to this dilemma. Therefore, maintenance should pause and determine whether they are performing the maintenance engineering functions adequately. For plant managers, the absence of adequate maintenance engineering means that once you downsize you may be in real trouble unless you have added effective maintenance engineering to create more effective maintenance. Fewer maintenance personnel mean less labor cost but not more effective maintenance unless attention is given to implementing maintenance engineering techniques to compensate for necessary downsizing.

26

Developing and Utilizing Standards

Maintenance Quality Assurance and More

The use of standards by maintenance provides a means by which they set objectives, work toward them, measure progress and establish a level of performance. Based on actual performance against standards, corrective actions can then be taken to improve performance. See Figure 26-1.

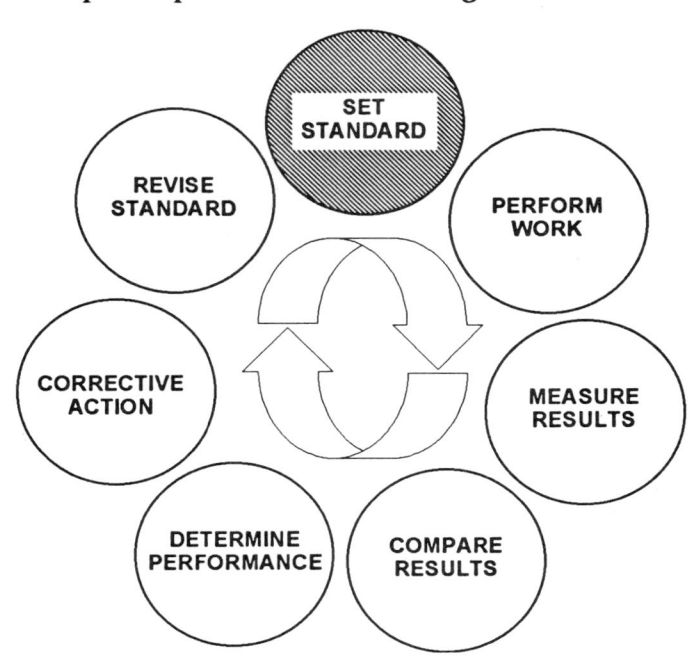

Figure 26-1. Standards are set, tried, measured and then revised.

Types of Standards - Standards include quality and quantity standards. Quality standards prescribe the manner in which the work will be done. The elements of the quality standard describe what is to be done, the scope of the task, how it is to be accomplished, the degree of quality needed, the level of service and the timing or frequency of the action. Quantity standards are time standards. They help determine the amount of time qualified craftsmen require to perform a specific amount of work under normal working conditions using reasonable effort and the elapsed time to complete the job. A quality standard for overhauling a wheeled front end loader is illustrated:

1 - Steam clean entire machine.
2.- Remove bucket.
3 - Renovate bucket lip.
4 - Repaint bucket.
5 - Strip entire unit to frame.
6 - Clean individual components as they are removed.
7 - Inspect, repair and reinforce frame as required.
8 - Replace drive train and power train components using new or rebuilt parts when operating hour data or condition requires. Accepted procedures, compression tests, etc., shall be followed before installing any used components on the unit.
9 - Install hydraulic systems using only new or rebuilt pumps. Inspect all hydraulic hoses for signs of wear or fatigue and replace when required. Locate hoses in the best possible arrangement to allow for speedy replacement and ease of access to other components.
10 - Install air and brake systems using the same hose guidelines as above. Wheel brakes shall be air over hydraulic with separate systems for front and rear axles using only 'fail safe' actuators. Drive line parking brake shall also be 'fail safe'.
11 - Install electrical and instrumentation systems. All wiring shall be enclosed in rubber hose conduit along the main frame when possible. Each unit will be equipped with the following instruments:
    (a) engine temp
    (b) oil pressure,
    (c) converter pressure
    (d) converter temp
    (e) ammeter
    (f) preheat indicator

The starter switch shall be protected to prevent it from being accidentally being depressed while the engine is running. Lights shall be wired to allow front and back to be operated individually.

Developing and Utilizing Standards

Hour meters shall be installed in conjunction with the engine kill linkage so that the meter shall run only when the engine runs.
12 - Complete body and chassis work.
13 - Remount bucket.
14 - Road test and repair or adjust any malfunction.
12 - Paint entire unit.
13 - Return to service.

A quantity standard is illustrated below. See Figure 26-2.

Occupation Code	Manhours	Elapsed Time
Mechanic	635	26 shifts
Welder - Mechanic	215	
Mechanic Helper	125	
Electrician	40	

Figure 26-2. The quantify standard prescribes man-hours by craft and elapsed time.

Developing Standards - The quality standard should be developed first to establish the job definition, scope and quality objective. Often, these documents take the form of a task list (what to do and how); a bill of materials and a tool list. It is important that operations be given an opportunity to participate in the development of the quality standard since they operate the equipment and will be the principal benefactor of the work done. Next, the quantity standard is developed to give dimension to the job in terms of manpower resources and elapsed time.

Plant personnel should be advised of the use of standards including:

- The types of activities on which standards will be used
- The type of standards to be used
- Procedures for developing and administering standards
- How the work order - information system can support them
- Procedures for measuring performance against standards

The simplest standard is the estimate. This is the time thought necessary to perform a job, based on the estimator's experience.

The estimate requires that there be a work order on which to record the estimated time and a means of comparing actual and estimated times on job completion.

Historical data are the basis for standards as well. A significant, repetitive-type activity is a good candidate for the use of historical standards, Perhaps the best uses of historical standards are preventive maintenance actions. For example, a PM inspection route that is performed every few weeks tends to set a trend for completion time. After the route has been completed several times, an average time can be determined. Providing there are no wide deviations in the results, a tentative average time can be developed. Subsequent measurements should be made to verify the validity of historical standards.

Historical standards are often the best starting point for a standards program because they emphasize the development of the procedures for preparing standards and administering them. Usually, a maintenance department that has made a commitment to using standards will do well to spend a reasonable period developing and using historical standards. This action aligns all of the key activities of the standards program effectively before adding the detail of more comprehensive standard setting techniques.

Work sampling can also be used as the basis for setting standards. Observations made on travel time, preparatory time, administrative time or actual working time are used to establish task times related to the actual job. Work sampling is an augmentation to standard setting rather than a means of standard setting itself. For example: a PM inspection route averaging 4.6 hours is actually made up of travel time, inspection time, time spent making adjustments, administrative time, etc. Work sampling would be used to help determine what percent of each of these elements make up the route time. Presuming the PM inspector had been instructed to inspect but not repair, the work sampling could be used to ensure these instructions were complied with. Thereafter, adjustments would be made on the base time.

The most comprehensive standards are engineered time standards. The basis for engineered time standards is the application of time measurements to various task elements and preparatory steps in order to develop an overall standard for an entire job. Repetitive tasks lend themselves more easily to the setting of standards because there is more opportunity to standardize and measure the task. Thereafter, the more the standards are used, the greater the opportunity to verify them.

Developing and Utilizing Standards

Non repetitive maintenance work covers a significant number of maintenance activities and it represents the area where standards use can be most cost-effective. Yet, the development of standards can be a tremendously expensive task if they are not applied where needed most. Therefore, three basic assumptions should be made when developing maintenance standards:

- Only major jobs should be considered for using predetermined time standards
- A qualified craftsman working under reasonable conditions will complete most jobs within a range of time. The use of specific range of time estimates is called slotting
- Certain types of jobs are representative of many others and can therefore be used as a general model (benchmark jobs)

The range of time or slotting technique acknowledges that the majority of maintenance jobs are two hours or less in duration. Thus, a whole group of jobs are placed in a slot anticipating the majority of them will be completed within a certain span of time. The planner who has access to this data merely selects the type of job and 'slots' it into a certain range. This becomes his estimate. See Figure 26-3.

TABLE - STATISTICAL SLOTTING STANDARDS		
Group	Manhours	Average time
A	0 - 2	.95
B	2 - 4	3.20
C	4 - 8	7.10
D	8 - 16	12.50
E	16 - 32	26.30
F	32 - 64	50.80
G	64 - 128	75.10
H	> 128	153.00

Figure 26-3. The availability of predetermined times for specific activities allows planners to visualize a major task, break it down into specific activities, assign predetermined times and adjust the situation for travel, make ready, etc., to the local situation and construct a realistic overall standard.

Benchmark jobs are selected typical jobs. They usually fall into some specific time range or slot. Benchmarks apply only to the job itself. No allowance is made for preparation, travel, clean up, etc. When standard times for these actions (travel, clean up, etc.) are added to benchmark times, the planner can develop a standard for a complete job.

Assume that the planner is able to determine from records the time necessary to perform specific jobs. By calculating the average time necessary to perform a group of jobs, the jobs can be placed into a slot. For example, assume that a series of pipe fitting jobs average at 0.95 hours. The planner would place all such jobs in Group A. When similar jobs occur in the future, the planner would research the job to determine what group it is in and then assign the correct time value as the estimate.

Using Engineered Performance Standards

Engineered performance standards are average times required for a qualified craftsman to perform a specific amount of work while working under reasonable conditions. They are provided so that the planner can estimate a wide range of jobs in less time by referring to a reliable data base for performing similar jobs. See Figure 26-4.

TABLE - Example of time standards for cleaning cooling tower condensers	
Element Description	**Manhours**
Walk 50 paces	.0075
Valve, small, turn on or off	.0032
Switch, turn on or off	.0013
Scrape, per square foot	.0045
Wash, per square foot, with water hose	.0006
Ladder, climb up or down	.0034
Tool, small, obtain and set aside	.0027
Brush, wire, per square foot	.0048
Wipe, per square foot	.0010
Tower, cooling, fill with water, per gallon	.0027
Tower, cooling, drain water, per gallon	.0049

Figure 26-4. Standards are based on elements, the smallest unit of work.

The various elements of a job, when arranged in a logical sequence, combine to form a task. Each task defines a specific amount of work to be accomplished by a single craft and it prescribes the time within which the task should be completed. This task time is called leveled time (time required for a qualified craftsman to complete a defined amount of work). A single operation becomes a task time standard.

The planner also uses a spread sheet to compare the work content of a specific job at hand with data presented on the spread sheet.

Developing and Utilizing Standards

A spread-sheet depicts (pictorially) a number of typical, related benchmark tasks showing them in the sequence of varying work content. The planner compares tasks with the job he is estimating. If the job is similar in work content, it can be completed in about the same elapsed time. The task time (benchmark) determines which slot the job belongs in. See Figure 26-5.

Task Area: Systems and Components, air conditioning (Absorption and centrifugal types). Clean, lubricate, adjust, and inspect..

Group K Allow 12.5 hours - Clean 370 tubes in 300 ton generator absorption system. Drain and fill, remove and reinstall end plates, punch tubes, and fabricate and install two gaskets. Use ladder or scaffold. Task time 9.98

Group L Allow 16.0 hours - Clean 355 tubes in 800 ton shell and tube condenser. Drain and fill, remove and reinstall end plates, punch tubes, dry, and fabricate and install two gaskets. Use ladder or scaffold. Task time 15.09

Group M Allow 20.0 hours - Clean 500 tubes in 600 ton evaporator, chilled water system. Drain and fill, remove and reinstall end plates, punch tubes, dry, and fabricate and install two gaskets. Use ladder or scaffold. Task time 19.52

Figure 26-5. In the task area shown, a benchmark time of 15.09 hours (task time) slots the task in group L. Therefore, 16.0 hours would be allowed for this task.

Benchmark jobs do not include preparation time. Therefore, an allowance must be made for these to determine the estimated time for the entire job. The type of work also reflects job preparation time and the % allowance for one or two workers. These data are built into the overall job estimate. See Figure 26-6.

Type of Work	Job Prep Hrs/man/day	% Allowance One	Two
Boiler work - shop	.2	23	33
Carpentry - general	.3	16	33
Electrical - electronic	.3	19	21
Cooling, ventilaion	.3	19	21
Machine shop	.3	23	24
Machine rpr (> 16 hrs)	.5	28	36

Figure 26-6. General data shows time values for job preparation and craft delays.

In applying these pre determined time standards, it is important to recognize that the use of standards by maintenance results in variations in times. Unlike production standards, there should not be an attempt to pin point maintenance standard times. Rather, deviations should be expected especially for long duration, non repetitive tasks with variable work content left to the craftsman's judgment. See Figure 26-7.

Job Duration versus Percent Deviation in Actual and Estimated time	
Duration (Hours)	Deviation (+/-)
120	4 %
100	5 %
10	10 %
4	20 %
2	25 %

Figure 26-7. Jobs will yield various deviations according to job duration.

The use of these standards requires an extensive training program. However, once a planner becomes familiar with them, they are a ready, more rapid source of estimating data. With their continued use, standards can be used more effectively.

Assembling The Elements

Each job is a building block of times. These times are broken down as follows:

- Total Allowed time for the job = Allowed Time + Travel Time
- Allowed Time = Task Times + Craft Allowance Time + Job Preparation Time
- Task Times = Task Time 1 + Task Time 2 + Task Time 3 (See Task Areas)
- Craft Allowance Times = Selected Task Area and time allowed
- Job Preparation Time = Time allowed for preparation based on type of work and number of workers (hours per worker per day)

Using a nomograph, the planner relates the craft data and the general data to arrive at the allowed time for the job. This considers travel zones in the plant and allowances for job preparation. See Figure 26-8.

Developing and Utilizing Standards

Travel time	
Craft allowance time	
Job prepartation time	
Task time 4	
Task time 3	Leveled time → Allowed time → Total allowed time
Task time 2	
Task time 1	

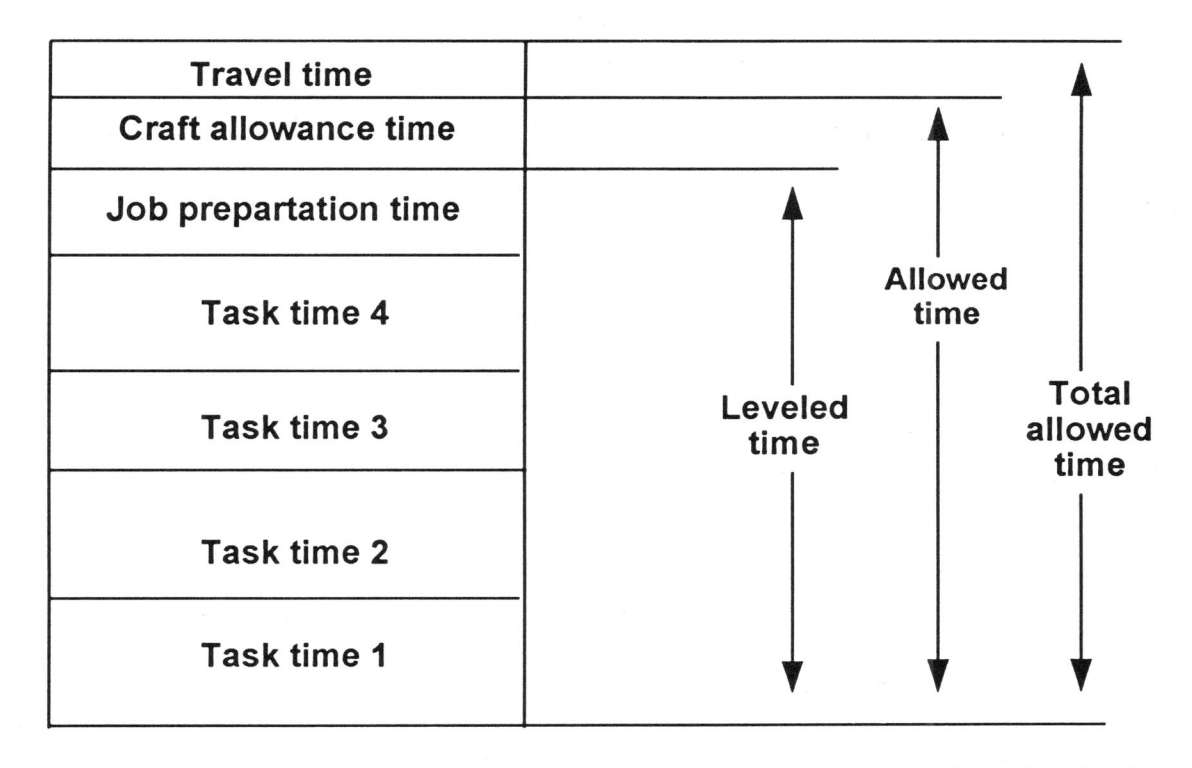

Figure 26-8. The total of all individual tasks times plus job preparation time is the leveled time. Craft allowance times are added to provide the allowed time. To this is added the travel time to determine the total allowed time.

While engineered performance standards define the average time required by a qualified craftsman to perform a specific amount of work using average skill levels and reasonable effort, other standards are used as well:

- Universal Maintenance Standards establish time standards for various jobs thus, making it possible for a planner to assemble applicable sub jobs to compose an overall standard for a major job.
- Work Factor bases its standards on distance, specific actions of the body and degree of control over work.
- Methods Time Measurement bases its standards on a predetermined standard time for each motion involved.
- Motion Time Analysis bases its standards on differences in various motions and the effect of common usage of physical body functions and movements.
- Maynard Operation Sequence Technique bases its standards on pre determined times linked together in common sequences such as preparation, tool use, material handling, etc.

Summary - The use of standards provides a reliable means of estimating manpower required for specific maintenance tasks. The most significant result of the effective use of standards is better use of the labor through tighter scheduling and improved job completion rates. The use of standards requires a commitment by maintenance, the talent to install them and the ability to administer them. The effective application of standards is a positive means of tightening labor control and helping to reduce overall maintenance costs.

27

Conducting Shop Operations and Support Services

Essential Maintenance Back Up

Shop Operation - Shop facilities such a machine shop, fabrication or welding shop, carpentry shop, paint shop, plumbing shop etc., provide support for field maintenance as well as for non maintenance work like equipment installation or construction. In effect, they are a support service to field maintenance in much the same way as are purchasing or the warehouse. Because they support an entire maintenance department, individual supervisors, planners or maintenance engineers who require their services must follow specific work control procedures and allow sufficient lead time to get work done.

Similarly, shops must define how their services are obtained in much the same way as maintenance must define its program. When customers are aware of these procedures, they can follow them properly.

Support Services - Transportation to move personnel, materials and supplies, rigging, the provision of equipment like mobile cranes and the operations of heating or compressed air systems are types of support services provided by maintenance. In addition, buildings and grounds work or custodial services are commonly carried out by maintenance. Any support service required at a certain time should be scheduled and its use properly coordinated. Therefore, the procedures for arranging for their use must announced in the same way shop support is requested and obtained. In both instances, the field planner is the liaison with the shop or service department. If there is a shop or services planner, he works through them. Alternately, needs are coordinated through shop or service supervisors.

Procedures for the control of work associated with obtaining shop work and support services must be documented and publicized. This type of support, in turn, is coordinated with the weekly maintenance scheduling meeting held with operations. As the field planner prepares for this meeting, he must have the up to date status of shop work before maintenance field personnel can be committed to the jobs. However, after operations has committed to making equipment available, services like transportation and rigging can be coordinated. Once the weekly schedule has been approved, daily coordination meetings are held to adjust the timing of support services, like transportation to accommodate operational delays. Daily coordination meetings should include shop and service supervisors as well as planners. See Figure 27-1.

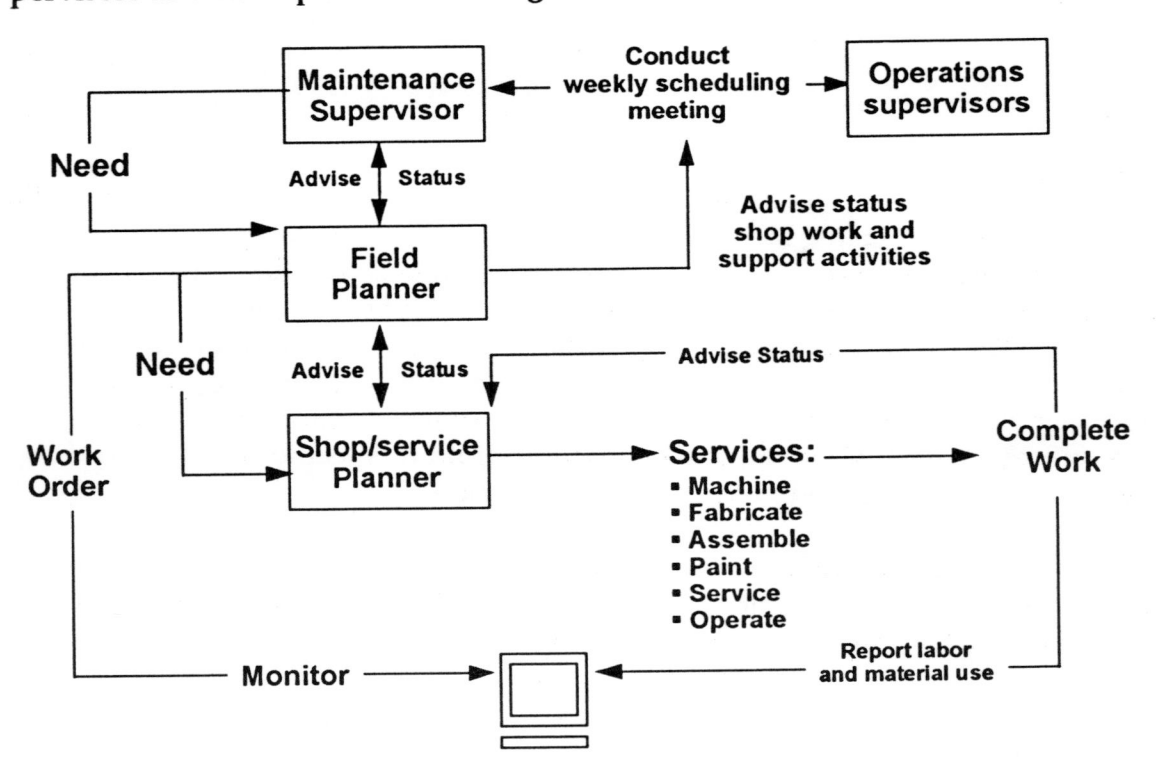

Figure 27-1. A well defined procedure for obtaining shop work and support services must be provided for maintenance customers.

    Shop Control - Effective control of internal shop work requires a balance between the planned and scheduled work it performs, the occasional emergency and a need to provide some crafts for direct support of field projects. Generally, the majority of the work performed in a shop can be planned, scheduled and marshaled through the shop in a deliberate fashion. Therefore, an efficiently run shop must have specific control measures in place. Among them:

Conducting Shop Operations and Support Services

- Shop activities like welding, assembly, painting and pick up areas must be arranged to ensure an orderly flow of work through the shop.
- Planning and control of work should be related to work stations to ensure effective work control so the status of jobs can be easily determined.
- The work order system must be used to control all work so that costs and job status can be monitored.
- Information on job status must be provided to shop customers who need to know progress to better schedule related field work.
- There must be procedures that preclude emergency work from hindering shop work progress.
- Shop drawings must be complete and well organized especially if the shop is required to rebuild components or manufacture stocked parts.
- Procedures for requesting work must be explained to customers.
- Work requesting procedures must be followed to ensure all customers have equal opportunity to obtain services.
- Shop performance and productivity should be measured to help spot problems of control or poor resource utilization to be able to correct them.
- An equitable priority system must be used to allocate shop resources and help control the flow of essential work.
- When required, shop personnel must be able to be assigned to perform project work in the field.
- Shop materials should be properly stored and organized so they are readily available for use.
- Job costs should be provided and linked with related work orders.
- Tool control must make the right tools available and keep them in good repair.
- The skills of shop personnel must be kept up to date as new machines and controls are introduced.
- All shop personnel must be aware of internal control procedures.
- Planners should be knowledgeable of all phases of shop operation and be especially effective in liaison with field planners and supervisors 1.

Service Activities - Support services cover a broad range of activities across the plant. While some services like utility operation or buildings and grounds work are carried out continuously, many others must be carefully coordinated with the maintenance effort to support operations. Normally, a high level of service is expected with activities such as:

- Buildings and grounds maintenance
- Collection and disposal of debris and scrap
- Road maintenance

However, the activities directly affecting maintenance support of operations, thus profits, must be properly anticipated, planned scheduled and coordinated. These are especially acute with:

- Transportation for people and materials
- Rigging
- Scaffolding
- Support equipment like mobile cranes, loaders and derricks
- The use of overhead cranes
- Field carpentry support
- Operation of critical utilities
- Operation and maintenance of elevators, hoists and cages 2.

Summary - Shop operations and support services have the same relationship to maintenance as does maintenance to operations. Therefore, just as maintenance must organize properly and carry out its program diligently, shops and services have the same requirement. But, there is a two way street. Shops and services have established procedures that must be honored by maintenance in the same way that maintenance expects operations to support its program. Through a collective, cooperative effort maintenance and its shops can provide effective operations support to better assure plant profitability.

1 - 2 - Images, copyright New Vision Technologies Inc.

28

Achieving Effective Material Control

Half of Maintenance Cost is Material

Maintenance organizations require that the right materials, in the proper quantity be available at the right time. If these basic conditions are not met, maintenance will be ineffective. In response to these needs, a material control department must provide the means to identify stock and purchased materials in proper quantities and make them available to maintenance as requested. To take advantage of this, maintenance must determine what it needs and specify the quantity and the time they are required.

All of this sounds very simple but, it is the biggest problem most maintenance organizations have. Some, in desperation, try to take control of the warehouse. They only manage to split the material control function while alienating the accounting department. Others disregard procedures, enter the warehouse and remove parts in quantities far in excess of current needs. They then establish their own 'treasure islands' where bounteous, unclassified parts are made handier in a 'junk yard' arrangement. Having successfully violated every sensible rule of accountability, maintenance costs escalate with all watching in anguish, unable to do anything about it. But, something must be done because materials are half the cost of maintenance.

Planning and Material Control - Corrective actions begin within maintenance. A good place to start is the planning effort because, if requirements are not stated in advance, there is little chance of tracking down the underlying causes. A principal task of maintenance is the installation of materials. But, the only way maintenance can directly control the cost of the work they do is the efficiency with which they install the materials.

Thus, a maintenance organization doing emergency repairs will not install materials efficiently while an organization that plans most work will. Whether a job is well planned or not, the cost of the materials to complete the job will be about the same. The difference lies in the efficiency of their installation. Cost wise the difference between two similar major jobs is about 15 % less labor, materials will be the same. Thus, good planning affects labor productivity not material consumption. Obviously, the maintenance organization that experiences too many emergencies consumes materials quickly as they are install hurriedly often without proper lubrication or alignment.

A major task for planners is the identification and procurement of materials. Generally, about 40 % of the planners time is spent on some form of material control. Similarly, maintenance supervisors must help in the identification of materials for work done by their crews. Procedures to ease this task are currently available in most computerized maintenance management systems that include inventory control and purchasing links. However, it is useful to first establish who obtains materials for what type of work. Generally, the planner obtains only the materials required for planned work. It is helpful to have criteria that spell out which work will be planned. This will avoid confusion. The supervisor and his crew members should obtain materials for all other types of work: unscheduled repairs, emergency repairs, PM services and routine maintenance. The difference lies in whether the supervisor or the crew members actually obtain the materials. When there is difficulty, it is invariably due to a problem of material identification. As a general rule, craftsmen should be able to identify the correct materials with the same skill as they select the correct tool. If the difficulty is not corrected:

- Craftsmen depend on supervisors to get materials
- In turn, supervisors are getting materials rather than supervising
- Planners then get caught up in getting parts rather than planning jobs
- Even superintendents get involved when they should be managing the whole department

It follows that there must be easy to follow procedures that will allow craftsmen to obtain materials easily for on going jobs. Similarly, the supervisor should help to identify materials for future unplanned work. He may then make the pick up of the materials part of the crew members job assignment. However, most material control problems start at the time the materials needed must be identified. If this one aspect is improved it has a very beneficial effect on the balance of the material control chain. Consider the following points in easing the material identification and procurement situation:

Achieving Effective Material Control

- Make sure craftsmen can identify and obtain materials using established procedures
- If procedures are inadequate, get them corrected
- Give craftsmen direct access to the computer for both parts and equipment specifications (assembly drawings etc.,)
- Set up a periodic maintenance program that prescribes standard bills of materials as often as possible to reduce the need to identify materials for the same type of job over and over
- Verify the adequacy of the linkage of the work order system with inventory control and purchasing to ensure that material data is available and correct
- Give all personnel adequate training on material control techniques
- Pin down specific material control responsibilities such as what the planner, supervisor or crew members do

Beware of impractical administrative procedures: Don't require a craftsman to get his supervisor's OK to withdraw a $ 7.00 item and then expect him to assemble all the materials to repair a $ 500, 000.00 crane, for example..

Using the Maintenance Program to Solve Material Control Problems

One of the most useful elements of the maintenance program that can contribute to the more effective control of materials is the forecast. By identifying the future time when major components might be replaced, it signals an advance need for materials. See Figure 28-1.

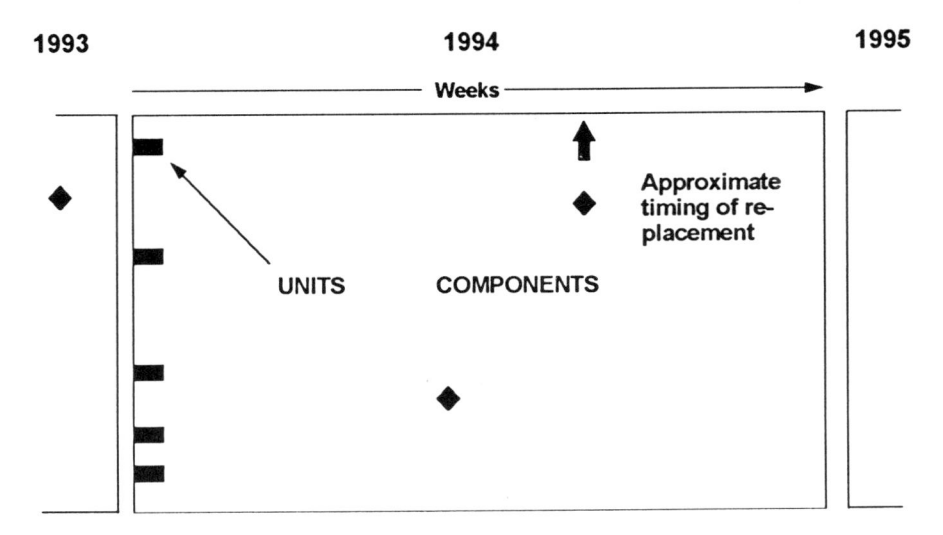

Figure 28-1. The forecast signals the future timing of a major component replacement and a requirement for materials to support the work.

Each component replacement action contains useful information that can help in controlling materials better, including:

- A task list explaining the job and how to do it
- A bill of materials listing the major components required and required stocked parts
- A tool list describing the special tools required to perform the job

If the bill of materials were shared with the purchasing agent, for example, it would help him to meet current component replacement needs. In addition, he could better anticipate requirements for several months. Similarly, the PM program might assist the warehouse in pre packaging kits of belts and filters required for PM services. In these ways, maintenance program elements can help to improve material control, providing they are shared.

Material Control Information - The most important material control information is the linkage between the work order system and the inventory control program. This link permits maintenance personnel to identify an order stock materials easily. When this can happen, craftsmen are more actively involved in material identification. This is a very positive objective as it frees the supervisor for more overall control of crew activities as opposed to dealing with the details of parts identification. See Figure 28-2.

```
CC  TP  UNIT  CP        STOCKED PARTS FOR COMPONENT

00  01  000   05        141576  Hydraulic hose 1/2"
                        141588  Fitting 1/2"
                        141592  Adapter 1/2"
                        141598  Sleeve clamp assy
                        141599  Sleeve clamp bracket
                        141602  ...
```

Figure 28-2. The user selects the equipment type (TP) and the component (CP). The computer responds with all of the stocked parts to fit the component for the type of equipment specified.

To order selected materials, the user highlights parts needed, adds the quantity of each plus the work order number and transmits the order to the warehouse. Later, he picks up the parts or has them delivered to the work site.

Achieving Effective Material Control 201

The availability of computer graphics helps as well. Both maintenance and material control personnel are able to quickly identify parts as well as see exactly how a component is re assembled. Moreover, these graphics can be printed and taken back to the job site by craftsmen. See Figure 28-3.

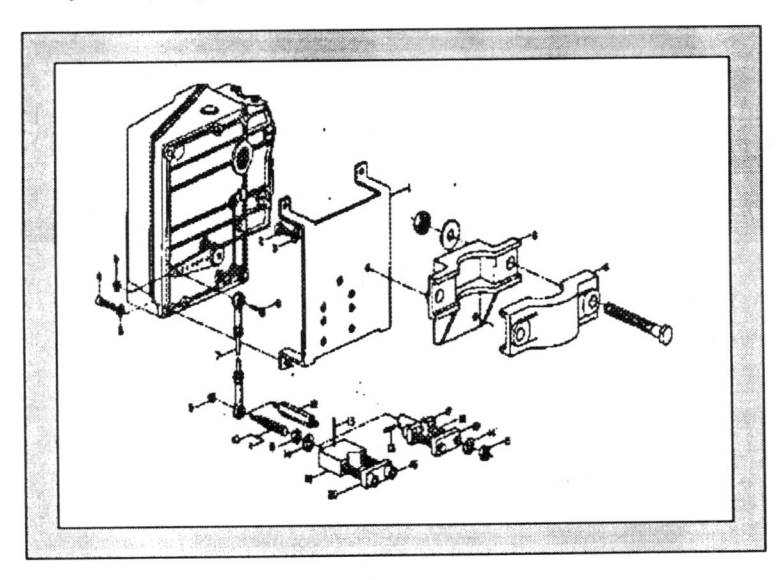

Figure 28-3. The use of computer graphics also improves material control.

A major problem of material control is the flow of damaged components back into the classification and rebuild procedure. See Figure 28-4.

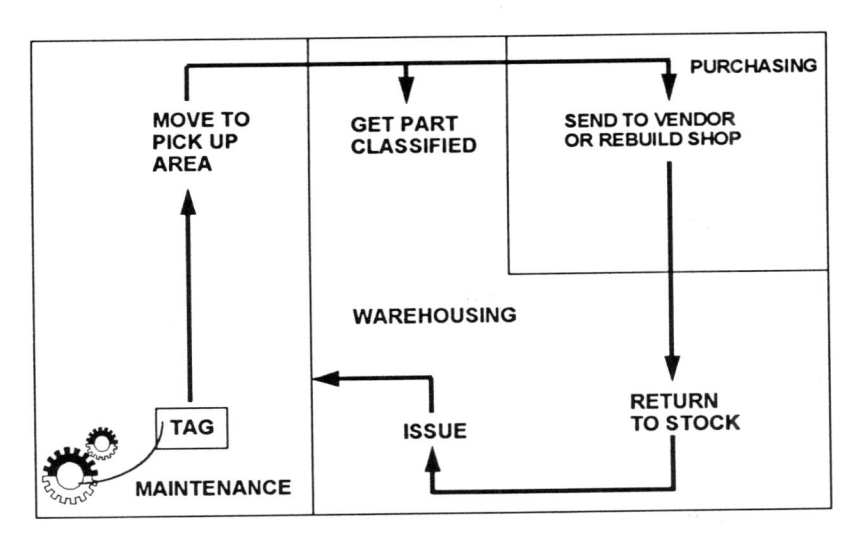

Figure 28-4. Clearly tagged, damaged components are moved to a pick up area. Then they are removed to a classification point. If rebuildable, they are sent to the vendor or a local shop for rebuilding. Once rebuilt, they are returned to stock for subsequent issue.

What Maintenance Expects From Material Control

Maintenance expects that the material control department will provide a high level of service with special attention to:

- Ensuring that material control procedures are well explained, easy to follow and effective
- Providing an easy to access critical spare parts lists for all important equipment
- Making sure that the items and quantities maintenance specifies are properly stocked
- Promptly replenishing out of stock materials
- Keeping the stock room properly staffed
- Making the stock withdrawal procedure efficient
- Providing an easy to use procedure for returning unused stock parts
- Ensuring proper accountability for stocked parts
- Making stock materials easy to identify
- Providing well documented parts interchangeability
- Providing a procedure for reserving stock parts in advance
- Helping to develop standard bills of materials for major repetitive jobs
- Making stock issues on presentation of approved work orders
- Issuing stock material against either work orders or equipment numbers
- Summarizing material costs by equipment and component
- Providing material costs job by job
- Allowing purchase order status to be tracked
- Providing drawings for the manufacture of spare parts
- Delivering materials to job sites when requested

Of these, there are special concerns of individual craftsmen, supervisors and planners. If teams are formed the concerns of craftsmen and supervisors are re-inforced.

The craftsman is most concerned with a need to obtain materials while the job is on going. Even though the planner did a good job on planning a major job and the supervisor tried to anticipate needs for an unscheduled job, the actual work often reveals a need for additional, supplemental stock parts. Getting them is seen by the craftsman as his responsibility. As a consequence, his most important need is to be able to identify the needed parts quickly and easily. He is also concerned with the easy availability of free issue items, the nuts and bolts that allow him to get on with the job. Most craftsmen are well qualified to help specify what these items should be and how they should be stored.

Achieving Effective Material Control

In addition, craftsmen are concerned with the availability of large crew and power tools, often controlled through the warehouse tool room. These are often in short supply and an effective tool control procedure is necessary to ensure they are able to be reserved and returned on time, in good condition. Many mines, acknowledging the need to downsize the warehouse staff are assigning the responsibility of tool repair to certain craftsmen with considerable success.

Immediately behind the need for power tools is the requirement that the steel rack be kept stocked. If warehouse personnel do nothing else well this task must be done effectively if good will between craftsmen and warehouse personnel is important. Craftsmen can easily scrounge a nut or bolt but, they can't easily get needed steel.

Next, ladders and rigging slings and cables are required.

Supervisors see things a little differently in the sense that their material procurement focus occurs before the task is started rather than once it is underway. Therefore, their interest is in identification as well as the ability to reserve parts. Thus, as daily assignments are made to crew members, they can be advised on the availability of needed materials. Supervisors are also concerned with getting parts quickly but, primarily when emergency work is at hand. Generally, the warehouse rather than purchasing should be the primary source of the supervisors material needs. If he finds himself constantly preparing purchase requisitions, there may not be enough of the right parts in stock.

Planners are intended to organize major jobs as far in advance as possible. Their sources of materials include purchasing, local and vendor shops and the warehouse. They must be careful to allow sufficient lead time for shops and purchasing to obtain materials and be able to reserve stock parts. Similarly. because the major planned jobs require crew tools, they must provide the same lead time to the tool room. Generally, the planners material needs are divided into two phases. During planning he would identify and order purchased materials and shop work. Once operations approves the schedule for the major jobs, stocked parts and tools are reserved and rigging plus transportation services are lined up.

What Material Control Expects of Maintenance

Material control expects maintenance to follow the basic rules of accountability, even 'during the heat of battle'. They feel that if maintenance is having trouble doing so, they should say so not continue to violate them.

Material control personnel also want to know what's going on. Thus, if maintenance has a forecast, they must share it with purchasing. Further, material control personnel cannot be omitted from weekly operations maintenance scheduling meetings. Maintenance must know as much about the status of materials as it does about the other half of its costs, labor. By not having a material control representative at every scheduling and coordination meeting, maintenance may be trying to control its work in ignorance.

Other concerns permeate the serious material control department over which they may have little control yet, they are important. One of the most significant ones is the tendency to reduce the warehouse staff to the point that the warehouse must be operated as a maintenance self help depot. This will only work if all maintenance personnel understand accountability and follow its principles.

Summary - In this era of downsizing industrial operations, the team spirit is essential across the plant. The old operations bromide of "we break it, you fix it" is out with maintenance. Similarly, the carelessness of maintenance in disregarding material accountability must end as well. Certainly, the availability of better data and refined control measures will help. However, maintenance and material control personnel have a joint responsibility to ensure equipment reliability and plant profitability through the services they provide. They are compelled to work together.

29

Conducting Non maintenance Project Work

If so, how Much ?

Maintenance departments are often called upon to perform construction, equipment installation or modification and equipment relocation. While the same craft skills are used, limits should be observed to ensure this work does not displace regular maintenance work. The most important question concerning non maintenance work is how much. Non maintenance work must not be allowed to interfere with the conduct of the maintenance program. This is often a major problem for maintenance. Consider:

Maintenance personnel are often perceived as co equal in getting things running again (maintenance), building or changing things (construction or modification). Therefore, they are expected to get the 'work' done no matter what it is.

Although crafts are thought interchangeable, a construction craftsman often lacks the diagnostic skills that distinguish a good maintenance craftsman. A construction type relishes 'building a stairway to nowhere' while the maintenance type wants to get on with fixing things.

Operations personnel often do not distinguish non maintenance from maintenance and levy requirements on maintenance for equipment modifications just as if they were regular maintenance. Often, requirements for modifications are found to be unnecessary and not feasible. Generally, there is little appreciation for the fact that they must be capitalized. For all of these reasons logical ground rules must exist.

Developing the Ground Rules

A plant should start developing its ground rules for handling non maintenance by considering the definition of maintenance:

Maintenance is the repair and upkeep of existing equipment, facilities, buildings or areas in accordance with current design specifications to keep them in a safe, effective condition while meeting their intended purposes.

The word 'existing' and the phrase 'current design specifications' suggest that construction, new equipment installation and modification be excluded from the definition of maintenance.

This does not mean that maintenance personnel would not do this work. Obviously, they must be involved in the construction and installation so they can maintain the new facilities and equipment properly. Likewise, they must understand how the modifications were made.

But the word involvement also means balance.

If the work force size is limited, there must be boundaries on how much non maintenance work they can perform and still be able to carry out the basic maintenance program. Therefore, there is a requirement for the plant to know how much manpower is required for a successful maintenance program. If the amount of manpower required for maintenance is less than the total available, there are few problems. However, if it is equal or less, non maintenance work can only be done during periods of low maintenance work demand. Even then, limitations must be imposed through the use of effective priority schemes.

Even more important are steps to qualify the non maintenance work for further consideration. For example, the plant must determine:

- Whether it is necessary (some work is proposed only to find that the equipment is to be scrapped)
- The feasibility of the work (before it is determined that one modification fixes one unit but undermines six others)
- How it will be funded (expensing obviously capital intensive non maintenance work is the best way to ruin the maintenance budget)

Conducting Non Maintenance Project Work

Generally, all non maintenance work should be planned. Thus, a deliberate examination of the scope and nature of the work will be made. Working within guidelines, the planner would obtain the view of the maintenance engineer before proceeding with the planning. It is the maintenance engineer who would assess the necessity and feasibility of the proposed work.

Non maintenance work should be subjected to the same planning steps as is regular planned maintenance. It should be awarded a priority so that it can compete fairly with other major tasks for the allocation of labor. During the weekly operation maintenance scheduling meeting non maintenance jobs should be considered on their merits but, within the position dictated by their priorities. Like all other jobs, if their priorities are not sufficiently high, they should be deferred.

The worst thing is to allow a maintenance supervisor and his crew to simply do the work. In their eagerness to please, a modification may not only undermine a process circuit but, they may fail to document the work and account for the cost. As a result, the 'mystery project' is undocumented and free. While this may please the initiator, it does great harm to maintenance.

Preliminary ground rules can now be identified. Non maintenance projects:

- Should be examined to ensure they are necessary and feasible
- Must meet an approval and funding criteria
- All non maintenance work should be planned
- On scheduling, resources should be allocated by priority

Project Control - Often overlooked, but necessary, is the adequacy of the work order system for the control of non maintenance work. Since work may be done by either a contractor or the maintenance department, the work order system must provide a means for adequate control of this work either way.

Usually this is accomplished with an engineering work order. Generally, the contractor's work is authorized by a purchase order, while a maintenance work order is used when maintenance does this type of work. Yet, both must tie into the accounting system. When resource use is reported, the contractor uses an invoice referring to the purchase order while maintenance uses a time card and a stock issue document. Therefore, the ground rules should now be expanded to specify the correct use of the work order system.

Logically following are considerations to ensure the non maintenance project goes well. Typically, maintenance crews should:

- Have adequate job instructions to ensure they can carry out the work properly (don't make them improvise)
- Get properly coordinated on site support like materials delivery or the use of cranes to ensure the project stays on schedule

As the project is being conducted, information must be provided on the cost, project status and performance of the work. Then. at the completion of the work, installations should be checked by the maintenance engineer to ensure they are maintainable.

Summary - There is little question that a plant maintenance department will perform non maintenance work. However, steps must be taken to ensure that it is done with proper consideration that it not interfere with accomplishing the basic maintenance program.

30

Establishing Information to Support Maintenance

Who Needs to Know What and How do they Get it ?

Function of Information - It is the function of a maintenance management information system to convert field data into useful information so that decisions affecting the efficiency of maintenance can be made. Since the success of a maintenance program depends on key people making correct decisions, their information must be timely, accurate and complete. To ensure this takes place:

- Essential information for the management of maintenance must be correctly identified.
- Field data necessary to produce this information must be properly collected, complete, accurately and timely.
- A practical work order system must exist to focus this field data into an efficient data processing scheme which, in turn, produces the required information.
- The information, once obtained, must be presented in useful, easy to interpret formats.
- Those receiving the information must utilize it effectively to make correct, necessary decisions.

Leadership in Development - An important element in developing an information system successfully is the leadership maintenance must take in establishing its information needs. Thereafter they must influence the development of the program based on what they know they must have. An effective maintenance management information system must build on maintenance expertise. It cannot be developed effectively by a source unfamiliar with plant maintenance and the environment in which it operates.

Four Phases to Establish System - A Maintenance information system is the means by which field data are converted into useful information so that maintenance can determine work needed, control the work and measure the effectiveness of the work done. Four phases are necessary to establish an effective maintenance management information system, put it into operation and attain improvements as a result of its use. They include:

- Conceptualization - The identification of needs and factors which influence development of the system plus an assessment of support for the system
- Development - A plan of action for bringing the conceptualized information system to life considering the practical aspects of collecting field data, processing it and converting it into useful information
- Implementation - Steps by which the newly developed system is integrated into the organization and put in the hands of its personnel so that value can be derived from its use
- Utilization - The means by which personnel utilize the information to accomplish maintenance improvement objectives

Implementation gets the system working but, support from management, operations and staff departments will be essential to gain improved performance. As maintenance uses the system, information must be carefully integrated into the overall plant information picture as it is used by other departments. When these steps are carried out properly, information will contribute significantly to improving maintenance performance and productivity and, subsequently, helping plant profitability. Maintenance has responsibility, as the system user, to take the initiative in acquiring the information system best suited to their needs. This responsibility cannot be delegated. While they can obtain technical assistance from data processing for instance, they remain ultimately responsible. Similarly, maintenance may elect to purchase a system but, first they must determine what they need and then evaluate candidate systems against these needs. The same is true for the use of professional consulting assistance.

System Objectives - The information system is the communications network of the maintenance program. Information is the ingredient that holds the maintenance program together. It is the basis for actions to ensure effective work control, good planning, proper resource use and the meeting of improvement goals. Maintenance information has the primary purpose of gaining control of maintenance activities and every element of an effective maintenance management information system must satisfies one or more of three objectives. These objectives include:

Establishing Information to Support Maintenance 211

Determine what must be done. For example:

- Study repair history data to learn the pattern of failure and observe the lifespan of specific components to guide planners in scheduling their subsequent replacement.
- Analyze the pattern of equipment failures to determine the basis for corrective actions that ensure the failures are not repeated.

Justify actions. Typical:

- Use the cost information to determine the highest equipment repair costs and investigate specific problems through repair history in order to make repair decisions.
- Observe unit costs and, based on the magnitude of these costs, establish priorities to determine which units will receive attention first.

Confirm or measure effectiveness. For instance:

- Assess the use of manpower and materials to confirm how effectively work was done and resources controlled.
- Observe the actual costs against the budget and determine the effectiveness of cost-control measures.

Control Through Information

Control Overview - Maintenance work is carried out using the work order system in combination with the information system and its interface with the accounting system. In turn, control is provided by the actions of key personnel who are guided by the information they receive. It is important in establishing the overall picture of the information system to identify the relationship of the work order system elements with the accounting system. Then, to ensure the system fits maintenance needs, diagram the flow of each type of work, including non maintenance work, if applicable.

Next, show the relationship of key personnel like the supervisor, planner or crew members to the work order system elements they use and the work control information they portray. Be certain to include a picture of the handling of non maintenance work and, as appropriate, show the coordination between engineering and maintenance planning as the non maintenance and maintenance work are integrated on the weekly schedule. See Figure 30-1.

To illustrate:

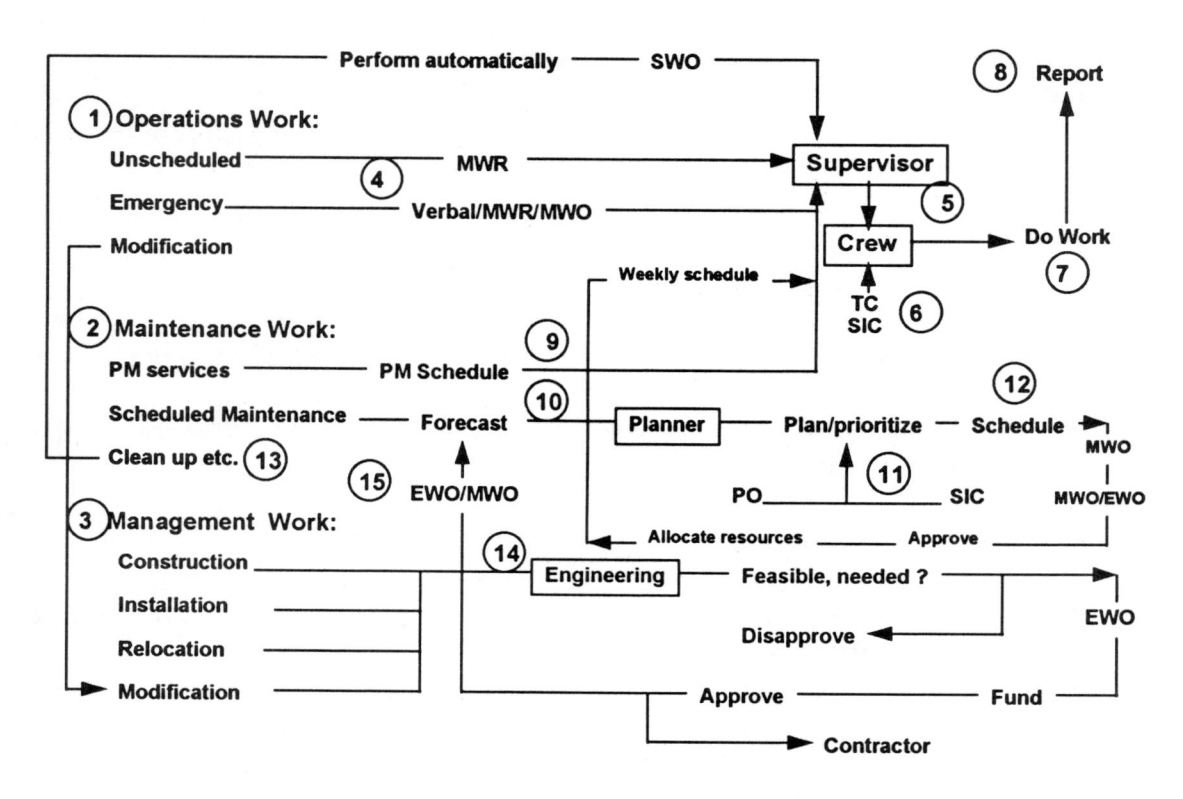

Figure 30-1. The Maintenance Work Control Overview illustrates how key personnel are guided in the actions they take by the information they receive.

Legend - Maintenance needs are generated by operations, maintenance and management. They are converted into actions using the information system:

1. Operations generated work includes unscheduled and emergency repairs plus non maintenance work like modifications.
2. Maintenance work includes PM services, scheduled maintenance and routine activities like clean up.
3. Management generated work includes: construction, installation, relocation or modification.
4. Unscheduled and emergency repairs are communicated directly to the maintenance supervisor.
5. The maintenance supervisor assigns work to his crew.
6. Crew members use the time card (TC) to report labor and the stock issue card (SIC) to report material use.

Establishing Information to Support Maintenance

7. Crew members do the work and keep the supervisor informed of work status.
8. Reports provide information for decision making.
9. The PM schedule specifies services due.
10. Scheduled maintenance is forecasted to help determine its timing, along with major non maintenance projects.
11. The planner uses the purchase order (PO) and stock issue card (SIC) to obtain materials for planned jobs.
12. Maintenance negotiates with operations to schedule equipment shutdown once jobs are planned and resources identified.
13. Maintenance controls routine activities (like clean up) using the standing work order (SWO).
14. Work generated by management is reviewed by engineering to determine feasibility, need and funding requirements.
15. Non maintenance work follows the standard pattern for planning and scheduling and uses an engineering work order (EWO).

The Impact of Information on the Maintenance Organization

The Situation - Not many years ago, when the master mechanic said "shut it down we're going to fix it ", the equipment was shut down and it was fixed. Today, the master mechanic still exists but, he is a less powerful figure. 'Shut downs' must now be negotiated with operations. Production efficiency yields high product output. In turn, maximum product output, on time, at least cost with quality product assures profitability. Minimum maintenance downtime contributes to optimum equipment availability. Thus, avoidance of downtime is a major, valid objective of maintenance. Downtime is the converse of production operating time and it has become a principal index of maintenance performance. Therefore, an important objective of information is to help maintenance avoid downtime.

Today, a maintenance organization cannot be judged by the actions of one person as in the days of the master mechanic. It acts and must be judged as a total organization. Similarly, maintenance performance cannot be judged solely on the basis of downtime. A broad range of information is available by which maintenance performance can be judged precisely as well as fairly. Thus, as the maintenance organization has changed, so has the quality of information available.

The Changing Maintenance Organization - Maintenance has been a close-knit group. Those not 'graduates of the school of hard-knocks' simply weren't allowed in maintenance. Until recently, industrial maintenance organizations were populated primarily with people who had always been in maintenance. Over 75 % of supervisors were once hourly workers. Of these, many became maintenance managers without ever holding a job outside of maintenance. This 'inbreeding' orientation permeated the ranks of plant managers who placed a premium on the singular talent of 'being able to make equipment run again' as a primary qualification for leading maintenance.

Thus, the need for managerial skills among supervisors was overlooked in favor of technical repair skills as a requirement for maintenance supervisors and managers. A good maintenance supervisor, general supervisor or superintendent was often elevated to the status of 'folk hero' because he could bail the plant out even after 'maintenance neglect' and a 'meet production targets at all cost' mentality had reduced equipment to 'junk'. These circumstances raise the question of whether plant management intended to elevate 'indispensable' maintenance personnel to management positions.

An assignment in maintenance for someone who could go elsewhere in the organization used to be looked upon with disdain. If assigned there, those aspiring to management positions contrived to escape as rapidly as possible and often did not place their maintenance assignment on their resumes. The old timers, the practical maintenance people who could make things run again and enjoyed the challenge of tinkering, diagnosing and hearing a machine 'hum smoothly' retired. Machinery was simpler in their day. Equipment is much more complex now and more costly because of it. More can go wrong and it usually does.

As a result, equipment condition must be carefully monitored. New testing techniques help diagnose equipment condition and pinpoint problems to ensure timely, correct repairs. Repair techniques are more complex. Maintenance cannot afford to field an 'army of parts changers'. Its personnel must be able identify and make correct, effective repairs rapidly. Prevention is more applicable than ever and it pays dividends in the avoidance of equipment failure and unnecessary downtime. As the technical complexity of equipment has increased so has the complexity of the maintenance organization. As a consequence, the personnel performing maintenance have had to change as well. Typical is the maintenance engineer. His task is to ensure the reliability of equipment. He accomplishes this, in part, by using modern equipment condition monitoring techniques.

Establishing Information to Support Maintenance

His performance is measured in successfully predicting wear out before failure. As he goes about his task, he uses repair history and cost information to ensure reliability. Thus, his contribution has been made more effective through the use of information.

Similarly, planners can monitor the status of a major job. One planner put it this way: "It used to be a disaster to find out that 4 weeks into a 10 week overhaul we were already 375% over budget. With the information we can get now, this can't happen." Nor is it unusual to see a supervisor and his crew gathered around a computer scanning the previous 6 months of repair history on a unit about to undergo a major repair. A maintenance manager is able to determine overall performance by watching labor use trends. He can anticipate which units are due for overhaul or replacement with cost and performance information easily available to him. The future maintenance organization will be more effective as a result of better, more complete and up to date information. Its people will have changed. Much of that change will have been made possible because information warns them of problems, helps them predict when actions must be taken, points out failure patterns, excessive costs and poorly performing units. Information will also enhance internal communications and help control labor, material, time and tools more effectively. With better information available, those using it are moving ahead. They are the future of maintenance.

The advent of better information is attracting different people to maintenance. The number of degreed engineers seeking maintenance assignments is increasing. In part, the more enlightened approach to maintenance made possible by better information is responsible. Junior engineers are no longer shunning a maintenance assignment, they are welcoming it. Craftsmen's technical skills are enhanced with the availability of better information.

A Preliminary Checklist

In preparation for the conceptualization, development, implementation and utilization of its future information system, questions should be prepared to guide development. The answers to these questions will then help to determine the adequacy of the newly developed information system before its implementation. Typical questions to be asked include:

1. Has all information necessary to manage maintenance been identified?
2. Has the content and use of the information been properly explained?
3. Have sufficient meaningful performance indices been provided?
4. Have we asked operations about reviewing maintenance cost?

5. Have we asked management about reviewing maintenance cost ?
6. Who will get what information ?
7. How will we train recipients to use information properly ?
8. When will we compare actual versus estimated man hours ?
9. How will we show labor use on PM, emergencies or scheduled work ?
10. Have we made provision for information on absenteeism ?
11. Will information on overtime be available ?
12. How will the backlog show whether we keep up with new work ?
13. Will backlog information be good enough to change the work force size ?
14. Will we have an open work order list to show all open jobs ?
15. Who gets cost and performance on major jobs ?
16. Will we summarize the cost and performance major jobs ?
17. Will accounting summarize equipment costs month and year to date ?
18. Should we make costs available on components like drive motors ?
19. Should we bother with costs on functions like grass cutting ?
20. In what format should we present cost information ?
21. Should the actual costs be compared with budgeted costs ?
22. Will repair history adequate to trace both repairs and failure patterns ?
23. Do we want repair history to include the life span of critical components ?
24. Which reports should be display formats versus printed reports ?
25. How will we verify that report formats are clear and easy to interpret ?
26. How can we ensure the information system satisfies genuine needs ?
27. How will we explain that the information system is the primary communications system for control of work and cost ?

Summary - Because the information system is the mines total communications and management system, it is often developed with a strong accounting flavor and an emphasis on operations control and statistics. Maintenance sometimes has to wedge itself in to get what it needs. Therefore, a well thought out plan of action to ensure the information system will support maintenance properly is necessary. This chapter outlines the development of the plan of action. Chapters 32 - 36 on Maintenance Information Systems cover the conceptualization, development, implementation and utilization of information and work order systems.

31

Conducting Mobile Equipment Maintenance

Characteristics of a Successful Mobile Equipment Maintenance Program

A successful mobile equipment maintenance program emphasizes preventive maintenance. Services are performed at prescribed intervals allowing equipment to operate reliably until the next service. The successful program also has a component replacement element in which key components are identified with a reliable forecasting procedure for possible replacement. Replacement timing is always verified by the actual condition indicated by the latest PM inspection. Both the PM services and the component replacement actions are documented with, respectively, checklists and standards. The standards of the component replacement program include a task list (what to do and how), a standard bill of materials and a tool list. Successful programs always document actions that are performed repetitively like PM services and major component replacements. While scheduled PM and major maintenance are performed in a garage location or in a field location, a successful program also make provision for the conduct of unscheduled and emergency wherever the need arises.

Prior to any major service or maintenance carried out in the garage, the unit will be cleaned to ensure all components can be inspected, serviced or repaired properly. ₁. Generally, field repairs are controlled by operations supervisors by radio via the dispatcher. Maintenance field repair crews may be dispatched from the garage or may operate exclusively in the field depending on the size of the operation.

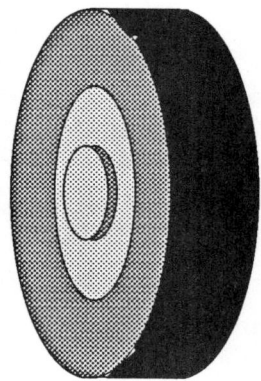

Tires often represent a significant portion of total maintenance cost for most large mobile equipment fleets. Therefore, a successful mobile equipment maintenance program should include a tire program to monitor the repair history of individual tires. It records repair details and monitors wear performance. This program should be integrated with the regular work order system and used to produce diagnostic details establishing relative performance of each tire by type, brand and mounted location on the unit.

A successful mobile equipment maintenance program will also schedule the majority of its work including:

- PM services
- Major Component replacements
- Other unique planned major jobs
- Overhauls

They will also use a work order system that provides separate control elements for each type of work and allows the use of verbal orders without loss of either control of work or information.

Their information system will emphasize the control of labor, permitting them to track each major job from inception to completion and to summarize cost and performance on completion. They will also be able to obtain the cost of labor and material by job, by unit and component, with fleet summaries and comparisons. They will also have indices to show relative performance unit by unit. Their repair history will show a chronological listing of chronic, repetitive problems and failure trends allowing them to quickly identify and correct problems. They will also be able to record the life span of every critical component to enable them to accurately forecast future replacements. Their backlog data will permit them to determine how well they are keeping up with the generation of new work and, changes in the backlog will permit them to adjust work size and craft composition as the work load changes.

Conducting Mobile Equipment Maintenance

By defining their program effectively and educating their own personnel as well as operations and staff departments, maintenance will gain their support for and cooperation with the program. For example:

- Equipment operators will take care of the equipment, perform simple repairs and adjustments and report problems as soon as they are encountered.
- Key operations people will meet with maintenance weekly to negotiate the best timing for PM services and major repairs.
- There will be joint coordination meetings to adjust the timing of major jobs in the event of operating delays.
- Material control personnel will come to the weekly scheduling meeting to report the status of materials they have been asked to assemble for up coming major jobs.
- Warehouse personnel would attend the daily coordination meeting to get instructions for the on site delivery of materials.
- Operations managers would be aware of maintenance costs and trouble spots and work with maintenance to solve problems.

The successful program will be carefully monitored by the maintenance engineer who has set up procedures to help guarantee the reliability and maintainability of the equipment. He is also aware of the latest techniques that will contribute to better maintenance. Currently, for example, he is studying the best way to download the data from the on board computers so it can be diagnosed quickly and used to guide repairs.

The Organization Must Match the Program

It may be useful to speculate on the future maintenance organization. Certainly, there are many organization arrangements being tried out. Many are being successful simply because they break away from tradition and are willing to try new approaches. Others are successful because they allow the natural talent of individuals to grapple with problems and empower them to make their own decisions. In all instances, the information revolution permits everyone to be better informed and able to exert initiatives previously impeded by the uncertainty of poor or missing information. 2.

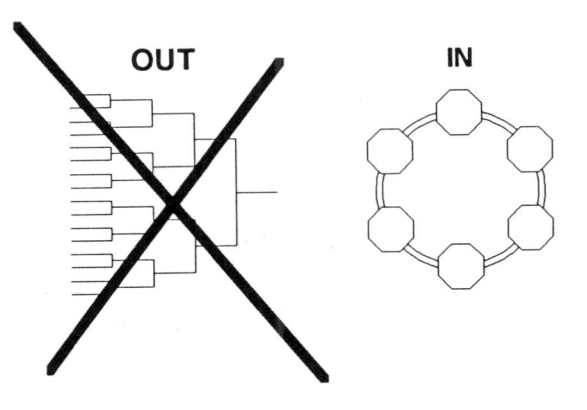

To support a more effective maintenance program, organizations must change. Therefore, let's speculate on the future maintenance organization.

In the future maintenance organization, there may not be a master mechanic or even a maintenance superintendent. Instead, there will be a manager who controls both operations and maintenance. The master mechanic will be more of a maintenance engineer than a line supervisor. He will not be a mechanical or electrical person but, rather, one who monitors the overall maintenance program to ensure it is followed closely. He would be an advisor to an area manager, who controls both production and maintenance. In a large operation, the maintenance engineer will have a staff of technicians who branch out into various critical areas of the program. One, for example, might be expert in diagnosing the over 180 items of data that reveal the precise behavior of each component on selected equipment. Another might oversee the predictive program with emphasis on reliability engineering (find the problem quickly and eliminate it). Little would escape notice on the computer network as all work assignments are 'broadcast' even before work starts. Their status is reported immediately on work completion. Production data are assimilated into the maintenance performance picture instantly to yield a picture of maintenance cost versus operating hour or ton, by unit, by component and by fleet. When downtime occurs it is immediately pinpointed by component, cause and duration. Human beings are seldom reporting data anymore as on board computers now transmit data as the equipment works. 3.

There are still maintenance planners but, they are no longer craft oriented. They are job oriented. Data from previous jobs is instantly available to them so they can estimate time virtually to the minute. Historical material use is also available, cross referenced to jobs by component, by unit and with fleet averages. Instant communication with the inventory control program informs of parts availability and order status. Purchase order status is tracked instantly providing a 'just in time' arrival of materials to coincide with the scheduling of jobs. The planner is probably part of the maintenance engineering group with a clear focus on planning only those jobs that require planning. 4.

Conducting Mobile Equipment Maintenance

While most jobs, regardless of duration or size, will get some degree of planning, it will be the big jobs that meet a planning criteria that planners will focus on. Crews will now appreciate the planners efforts more since:

- Equipment is shut down according to the schedule
- Areas in and around the equipment are cleaned up by attentive operators
- Materials needed are delivered to the work site before work starts
- Mobile cranes are at hand when needed
- Tools have been reserved
- Rigging is complete before the crew arrives on the job site

Impact of Teams - Maintenance craftsmen are all well trained and receive periodic updating of existing skills. When new equipment is added to the fleet, training is provided promptly. But, the most outstanding characteristic of the craft groups is that there is no supervisor. Nor do we see craft groups. All of the skills required are in the same crew. The crew works harmoniously with everyone pulling his own load, contributing ideas and quickly saying "let's do it ". There seems to be total familiarity of the crew with the maintenance program, the computer and the help available from maintenance engineering. Crew members seem interchangeable, while some do jobs, others inspect, coordinate or line up materials using an interactive work order, material control network. Others check job sites or scan the repair history of a troublesome unit about to undergo a major repair. Some are conferring with operators who now know as much about equipment maintenance needs as some of the craftsmen do. 5.

On job completion, each crew member reports what happened whether it be a matter of time required or an item needed for repair history. Most weeks, other craftsmen join the group to help perform major jobs. These other craftsmen are from a pool and are used to reinforce each crew during peak work loads such as an overhaul or a major component replacement. The regular crew carries out day to day activities including PM services, unscheduled and emergency repairs (in those few instances that they now occur). 6.

When pool craftsmen are needed, it is usually the planner who has arranged for them after consulting with the crew and getting their input on how the upcoming major job should be done. As the pool craftsmen join the regular crew, they become part of the 'team' for the time they are there. Agreement is reached on who might do what between the crew an the pool. After all, every major job has a detailed quality standard explaining what is required, how to do it etc. This helps to make personnel interchangeable. Operations has recently remarked that overhauls are now being completed in 30 % of the time it used to take when serious work only happened on the day shift. Now, thanks to standards and better training the interchangeability extends beyond the crew and into the pool. Work can go on around the clock and who knows, the pool people might like a day in the fresh air in the field doing other types of work. Certainly they are qualified.

In another part of the operation, we see another type of maintenance organization also with mechanics, electricians and other crafts on the same crew but, with someone who seems to be calling the shots. This crew has not been out of its former craft organization as long as the previously described crew and is using a rotating coordinator with its team. Every one is contributing ideas in the same way as the other group. But, this group is working through its 'group dynamics' to settle in on how they will operate when they take the next organizational step into a self directed work team.

The people who used to be maintenance supervisors are still around lending their experience and savvy in other ways. Some are operations - maintenance coordinators and some are planners. Most electrical and instrument supervisors are now on the maintenance engineering staff because they always preferred solving technical problems rather than trying to control people. Most are happy and productive because they are doing what they prefer to do in the way they do it best. 7.

The most dramatic changes are in the new work control procedures. Anyone can enter the network. Operators who see problems highlight the on screen equipment lay out diagram and touch the code bar to identify the problem and its priority. Soon the work order is on the network and because the program already knows where every individual is working, signals the closest craftsman to help the operator.

Conducting Mobile Equipment Maintenance

Planned jobs and PM services requiring equipment shutdown emerge from the weekly scheduling meeting in which operations states when the equipment will be shut down. The computer scans the status of all pending jobs on the same equipment, noting their ready status (like materials OK) and appends them to the scheduled jobs. Scanning the historical use of manpower and job duration plus priority the program presents the crew with a workable daily schedule. If OK, the crew accepts it and the network now sends out messages telling operations when to shut down equipment and the warehouse when to deliver materials. 8.

Within the network, the computer is also following the production plan and it verifies the mix of equipment necessary to meet production targets and still have time to shut down for maintenance. The electrical - mechanical team in the field requires several parts and they enter the linked on board computer to order the stocked parts which are soon enroute to their pit location. In the warehouse, parts are retrieved by robotic trolleys and packaged onto bar coded containers for delivery to the right field location. Close by, a warehouse truck is making its way between the loading platform and the plant without a driver, it has been programmed along a certain path and it sensors provide signals to guide it as obstacles or other vehicles approach. Meanwhile, its on board computers are busy monitoring its every action, alert to the abnormal and recording all for continuous analysis.

When a craftsman wants stocked parts he selects them from a computer graphics display showing an exploded view of the total component. He highlights what he wants, adds the quantity needed plus the work order. Since the work order contains the equipment and component being worked on, material costs are correctly charged and the location for work site parts delivery is acknowledged. While he is at it the craftsman prints the diagram to use as he reassembles the component. He can study it in the 15 minutes it will take to get the parts he requested out to him.

Vendors share the network as well. One was surprised to find that the mobile crane recently overhauled at the factory lacked the computer disc showing its parts by component in the format to fit the plant's inventory control program. Another vendor entered the inventory control program and identified all of the parts that no longer were used on all retrofitted dozers. At about this time another vendor was being warned electronically that electric drive motors they were rebuilding were not lasting as long as those of their competitors and steps were suggested for improving work quality with 90 days.

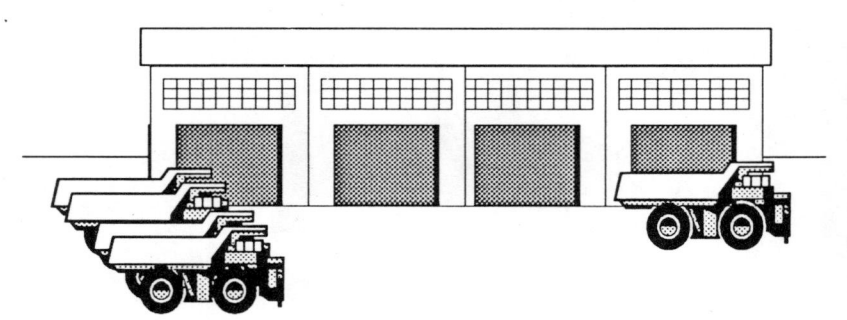

Meantime, back at the garage, the trolley that is now pulling equipment through the newly reengineered, unattended wash facility is cleaning equipment more thoroughly than ever before. Similarly, the garage itself was different. Not only is there a well organized area for units awaiting work but, the amount of time they are there is recorded and analyzed. Within the garage, PM services are carried out in designated areas and inspection deficiencies are reviewed to determine further repair actions. Major repairs are not done until PM services are completed. The garage is divided up into efficient working areas with bays designated for long-term repairs (like overhauls) or quick fixes of two hours or less. The garage is organized to maintain all types of equipment. Warehouse facilities are properly located and well run. Utilities like compressed air lines are convenient and areas for welding or tire repair are properly located. Generally, the garage configuration enhances maintenance activities and internal communications operate very effectively. Noteworthy is the fact that completed work is picked up promptly by operations because, the amount of time on the ready line is recorded and analyzed. 9.

Conducting Mobile Equipment Maintenance

It not unusual to see a mechanic and an operator going off for a road test to add their own quality control as well as to discuss the merits of a new modification prescribed by maintenance engineering.

Field repair is done efficiently as well. Not only are the craftsmen tied in by radio to the pit operations manager, the dispatcher and every unit but, they also have notebook computers that link them by modem with maintenance, inventory control etc. They are particularly fond of the new bar coding wands that permit real time recording of events virtually eliminating any keyboard entries. The installation of suspended coaxial cable to transmit telephone signals has revolutionized communications between the garage and the field. Parts can be hustled between locations using overhead conveying systems that are quickly suspended from the roof and easily bend around corners. As field operations move, the conveyors follow them. 10.

In the average crew, it is difficult to distinguish the operators from the maintenance people. The operations-maintenance team effort is starting to show.

New Challenges

Whether it is the garage or the field, there are numerous ideas to be tried out, challenges to be met and probably lots of people eager to try them. But, the economics of industry are requiring more efficient organizations. Downsizing is common but, it must be a process that results in a smaller, more efficient organization. Old organizations that were not very productive are giving way to new, smaller organizations that are more flexible and make much better use of personnel. But, the organization is not the only thing that is changing. Maintenance programs are now oriented toward better maintenance engineering to better assure equipment reliability. Preventive and predictive techniques are being enhanced. The objective is changing from fix the problem to eliminate it. Information is dramatically better and more widely available. Cost reduction needs and better technology are pushing organizations to perform better than they ever have before. Most acknowledge the challenge.

Summary - Whether the mobile equipment maintenance program is being carried out in a distribution warehouse, a mining operation, in support of a road construction project or a logging operation in the 'bush', the effective mobile equipment maintenance program must:

- Have a superbly defined program
- Be capable of performing garage and field repairs equally well
- Implement a solid operations-maintenance team
- Take full advantage of repair technology
- Establish the most effective communications

Because the equipment operator is with his unit continuously, he will be a critical element in the successful program. This suggests that these objectives can be met through a team effort between operations and maintenance.

1. - 10. Images, copyright, New Vision Technologies, Inc.

IV

Maintenance Management Information Systems

32

Conceptualizing Information Needs

The Vital Maintenance Information

Maintenance information is divided into decision making information and administrative information:

- Decision making information is necessary to control day to day maintenance and determine current and long term cost and performance trends. This information is the basis for management decisions
- Administrative information is used to communicate within maintenance and operate the maintenance information system

Decision making information includes:

- Control of labor
- Backlog
- Status of Major Jobs
- Repair History
- Cost
- Selected Performance Indices

Control of Labor - This information should ensure that:

- The workload has been measured
- Labor is being controlled properly.
- The work force is being utilized effectively.
- Absenteeism is being controlled.
- Overtime is being used properly.

The information system designer must appreciate the importance of labor control because it is the principal way in which maintenance can significantly control the cost of its work. The cost of maintenance is a function of the number of units of equipment maintained and the speed at which their repair consumes materials. While PM inspections slow the rate of material consumption by reducing the incidence of emergency repairs, they also allow more work to be planned by finding problems sooner. Thus, more lead time is available for planning.

In turn, planned and scheduled work allows more control over costs because personnel work more productively. This increases the efficiency with which they install materials. Labor control, therefore is a vital and necessary maintenance function to which the system developer must provide quality information.

Since field labor data is the basis of this essential labor control information, it must be reported accurately, completely and on time. Once reported, it must be processed and presented in a meaningful way. Consider a graphic presentation that portrays the week to week use of labor. This is the most important of all labor control information. See Figure 32-1.

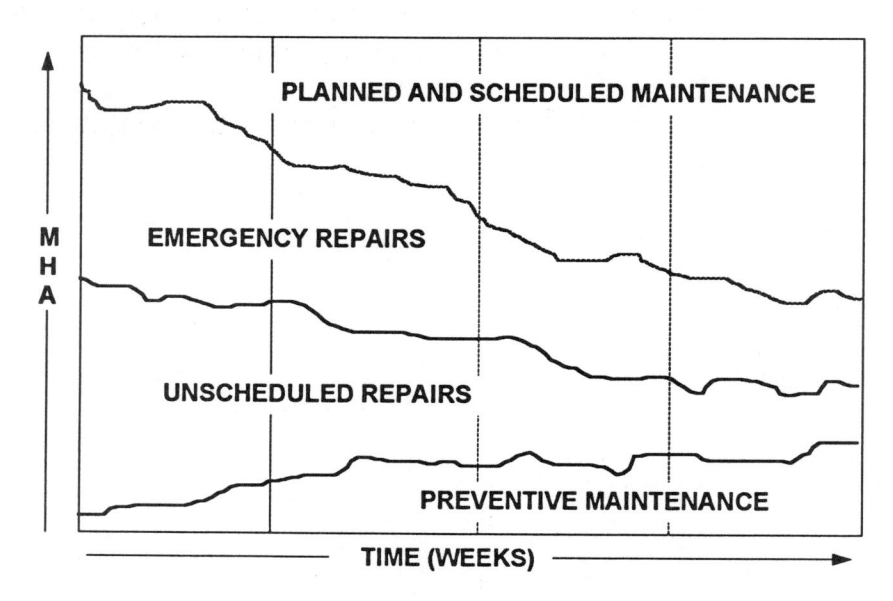

Figure 32-1. When PM is conducted regularly, it will reduce the amount of manpower used for emergency and unscheduled repairs. This manpower can be shifted to more productive planned and scheduled maintenance.

Conceptualizing Information Needs

Backlog - The backlog is the total number of estimated man-hours, by craft, required to perform all identified, but incomplete planned and scheduled work:

- It illustrates the degree to which maintenance is keeping up with the generation of new planned and scheduled work
- It allows adjustment of the work force size and craft composition as work loads change

Unscheduled jobs, emergency repairs and PM services are not part of the backlog:

- Emergency work is excluded because it must be done immediately
- Unscheduled repair man-hours are seldom estimated
- The inclusion of PM service would cancel themselves as repetitive services are performed

An open work order list, which shows work orders opened but not complete, is often confused with the backlog. As the list grows it only indicates that maintenance is falling behind. It cannot help in the adjustment of the work force size because it treats a 2 hour job the same way as a 200 hour job. Thus, useful backlog information should be measured in estimated man-hours, not jobs. See Figure 32-2.

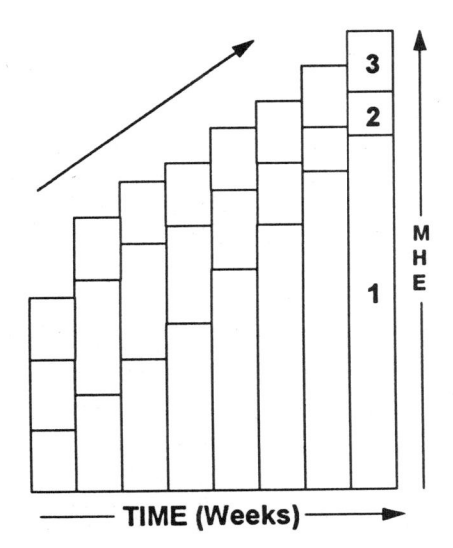

Figure 32-2. The backlog shows that there are too few of craft 1, too many of craft 2 and the correct number of craft 3 personnel. The open work order list would only reveals that the number of jobs is growing.

Work Order Status - Costly, important jobs performed on vital equipment must be managed. Information on the status of such jobs permits control of the work. Since each job is planned, targets like man-hours needed, cost and elapsed time required can be established for each work order. The work order prescribes the cost and performance targets while accounting documents, like the time card, accumulate actual cost and performance data against the job. The information system converts this data into information permitting users to observe the status of each planned and scheduled job from inception to completion. See Figure 32-3.

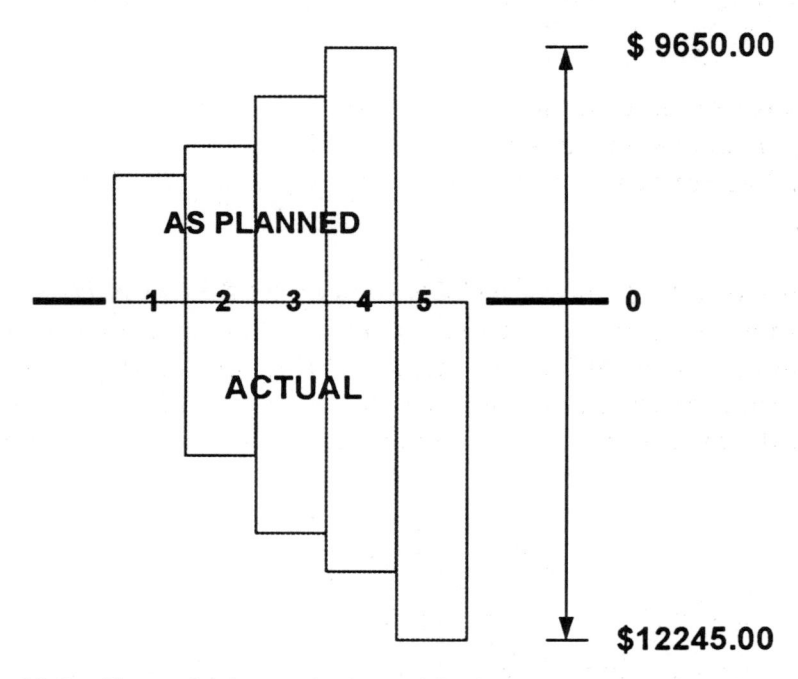

Figure 32-3. Above the heavy horizontal line are the targets for labor and material costs, manpower by craft and job duration in elapsed shifts as planned. Below, the actual costs and resources used and job duration are shown.

Repair History - This is record of significant repairs made on key equipment. Its analysis reveals patterns of chronic, repetitive problems which help identify specific corrective actions. Repair history also includes failure patterns. Component life span permits the development of a component replacement program. While cost information can identify high equipment costs, it is the repair history that pinpoints specific problems which are then confirmed by field investigation. Cost information (how much it cost) should be separated from repair history (what happened) to avoid over detailed reports.

Cost - Maintenance in many mines represents over 30% of operating costs. It is potentially the best way to judge how well maintenance is performing. Therefore, cost management should be given priority in the information system. The information system should convert field data into reliable, timely cost information so that correct decisions can be made. As the cost of labor and materials accumulate, they focus on units of equipment, facilities or functions (like building maintenance). Units of equipment have numbers so that costs can be accumulated against them. By assigning a work order number to a job, it is possible to isolate the cost and performance for that job. Subsequently, when the work is completed, costs close against equipment. Cost data is the basis for most economic decisions to overhaul equipment, replace equipment or standardize equipment based on its cost performance.

It is desirable to provide two levels of cost information:

- Cost Summary - The cost of labor and material, actual versus budget, month and year to date for each cost center, summarized by area and plant totals
- Cost Detail - The cost of labor and material for each unit of equipment and its components, with month and year to date summaries showing the average cost of components (A cost related performance index such cost per operating hour might be shown following each unit)

Performance Indices - In addition to the formal major information elements above, selected performance indices round out the decision making information picture. Typically, regularly used performance indices might include:

- Maintenance man-hours per ton of product
- Maintenance cost per ton produced
- Percentage of man-hours used on emergency repairs
- Percentage of man-hours used for PM services
- Percentage of the weekly schedule complied with
- Labor cost to install each dollar of material

In addition, unique, periodic performance measurements might include:

- Work sampling to establish worker productivity
- Percentage of first line supervisor's time spent on direct supervision
- Effectiveness of bay utilization in a mobile equipment shop
- Comparison of overall absenteeism with established guidelines

As a plant assembles this decision making information, they must organize it so that they are able to utilize the it effectively. Typically, information is arranged in a pyramid. See Figure 32-4.

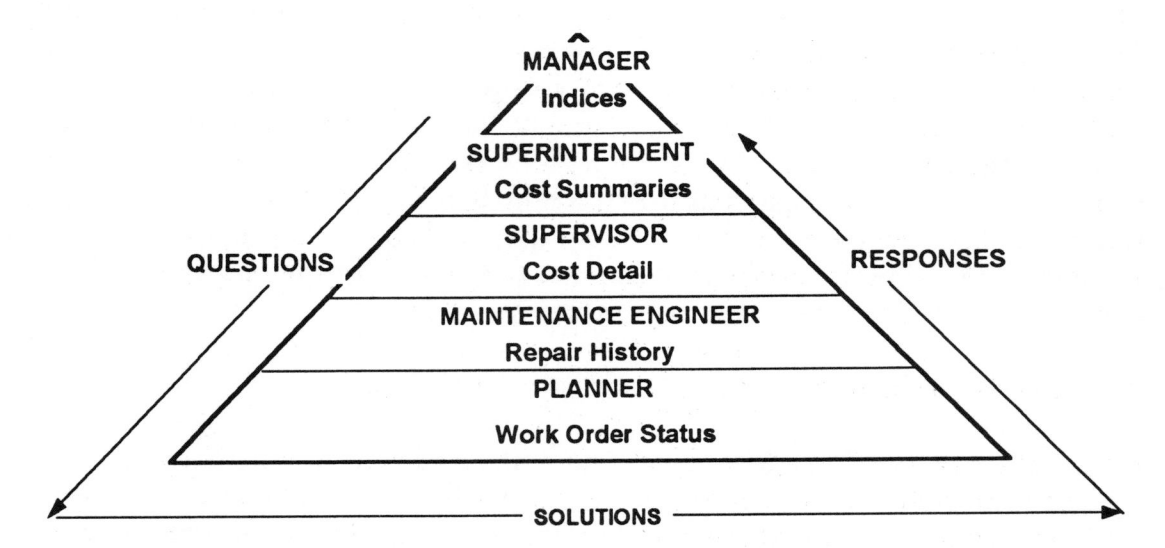

```
                    MANAGER
                    Indices
                 SUPERINTENDENT
                  Cost Summaries
   QUESTIONS       SUPERVISOR        RESPONSES
                   Cost Detail
               MAINTENANCE ENGINEER
                  Repair History
                     PLANNER
                 Work Order Status
                    SOLUTIONS
```

Figure 32-4. At the top of the information pyramid, the manager uses performance indices to note overall performance. If he has questions, the structure of the system permits rapid development of solutions and responses to his inquiries with corrective actions taken.

Administrative Information

Administrative Information is used to communicate within maintenance and operate the maintenance information system. It consists of the data necessary to run the maintenance department, meets a communication need, produces reference lists, provides exception information and, in general, is utilized to control day to day internal maintenance administration.

A major conceptualization problem for system developers is the separation of administrative information from the more important decision-making information. A good guideline to follow is to give emphasis to the development of decision making information. Then set it aside and, separately, develop administrative information requiring that it be fully justified against strict criteria.

Tough criteria and full justification for every administrative report should be established. Too often, even key maintenance personnel fail to analyze their administrative needs. This results in duplicate, unnecessary reporting.

Conceptualizing Information Needs

Unless excessive administrative information is limited, the end result will be an information system made up of numerous reports serving no real purpose. Each suggested administrative report should be measured against three criteria:

- Why is the information necessary?
- Is the information available elsewhere?
- What actions will be taken and, by whom, as a result of the information?

Adherence to these guidelines will eliminate considerable grief!

Among the typical administrative reports used by maintenance are:

- Incorrect entries made on time cards
- Incorrect data entered on work orders
- Open work orders
- Closed work orders
- Stock parts issued to crew members
- Stock parts utilized for a specific work order
- Overtime hours worked by each crew member
- Craftsmen exceeding allowable time for personal business
- Labor reported by one individual for a weekly period
- Comparison of overtime with absenteeism for one employee
- Employees by number sequence or seniority, by craft.
- Occupation codes.
- Overtime codes.
- Absentee codes.
- Failure codes.
- Categories of work.
- Standing work orders, their descriptions, crafts man-hours, etc.
- Departments and their cost centers.
- Equipment types.
- Equipment by type, unit and component by cost center.
- Components by equipment types.

These typical reports, while necessary for administration, are not the tools by which the maintenance department is managed. Every effort should be made to keep their number and purpose in perspective.

The system developer should always give priority to decision making information.

Field Data

Maintenance information is developed from field data reported by individuals as they go about or complete work. Field labor data such as man-hours are reported on time cards while stock material data like part numbers are recorded on stock issue cards and direct charge material data on purchase orders. System programming then converts the data into information used to control the execution of the maintenance program.

The quality of maintenance information will be no better than the quality of field data on which it is based. In turn, the quality of the information will be no better than the quality of the system that converts the field data into information. It follows that, unless field data are accurate, complete and timely, the resulting information will be faulty, as will the decision making resulting from it.

Maintenance field data is developed from the day to day use of accounting documents: the time card, stock issue card and purchase order in combination with elements of the work order system. The work order system focuses the field data of accounting documents into system programming where it is converted into information. See Figure 32-5.

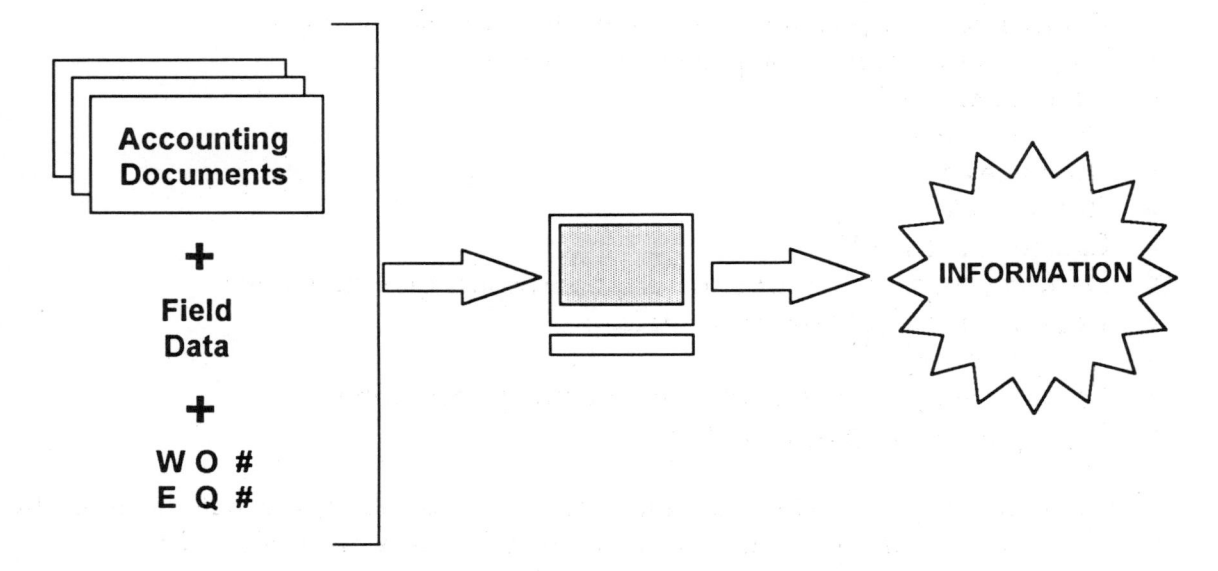

Figure 32-5. Accounting documents are used to record field data such as man-hours, parts numbers, work order numbers etc. Numbers provided by the work order system guide the data through system programming to emerge as information.

Conceptualizing Information Needs

      Labor data - Since maintenance relies so heavily on good labor data to better control the cost of their work, the data must be reliable. To obtain good labor data in the conventional mode of labor data reporting, the time card must be compatible with the work order system. Generally, labor data reported on the time card should include:

- Employees identified by craft code and employee number
- Labor data recorded against the date and shift.
- Man-hours reported as regular time or overtime (should be coded to indicate use)
- Absenteeism reported (by the supervisor) and coded to indicate cause
- Man-hours should be reported against cost centers, units and components, functions performed and work order numbers

Individual maintenance workers are the best source of information on work performed. They must report data initially and supervisors should verify it.

The reporting of labor by supervisors is seldom adequate when the supervisor has a large crew with each member performing numerous jobs during each shift. This expects too much of the supervisor and the data is often incorrect. It will yield equally incorrect information.

The best rule to follow is to treat the work order as a control document and the time card as the primary source of field labor data.

Current information systems that display crew assignments by work order on the computer screen, allow the craftsman to record hours against the work order on the screen. This procedure covers only the maintenance requirement for data. In this case, alternate means such as work place, entry and exit, bar coding are used to collect accounting labor data for payroll, absenteeism etc.

      Material data - Field data on the consumption of materials is obtained from the stock issue card and the purchase order. However, material cost data may also be developed from a combination of work orders, stock issue cards and time cards when parts and assemblies are machined or fabricated in shops. The stock issue card and the purchase order should have the capability of recording the:

- Cost center
- Equipment number and components
- Work order numbers

System Integration

One of the most important decisions to be made during maintenance system conceptualization is the degree to which the maintenance system needs to be integrated with other systems. If information required from other systems is not acquired by system interface, then maintenance must produce the information itself. Typically, duplicate labor reporting may be necessary or material data must be obtained and re entered into the maintenance system. Both situations are undesirable and the added administrative burden may influence data and information quality. Maintenance may elect to have a fully integrated information system, a partially integrated system or a stand alone system. Each option must be considered against genuine information needs.

Fully Integrated System - Advantages:

- Carefully integrated with production strategy
- Staff departments supportive
- Management committed
- Accounting information available
- Information like production statistics available from other systems
- Maintenance information shared with others
- Development speed and cost reasonable if system properly designed

Disadvantages of fully integrated system:

- Dependency on other systems
- Procedures allow no deviation
- Development speed slow and cost high if system not properly designed

Partially Integrated System - Advantages:

- Labor and material data available
- Selected operational data available
- Limited system operation and administration required
- Full control over exclusive maintenance functions
- Some cost reconciliation possible

Disadvantages of partially integrated system:

- Labor and material data incomplete for maintenance use
- Dependency on other systems

Conceptualizing Information Needs

Disadvantages of partially integrated system continued . . .

- Difficulty to obtain modification
- Limited interest by staff departments
- Limited relationship to production strategy
- Management support questionable

Stand Alone System - Advantages:

- No dependency on other systems
- Data for exclusive use available
- Full control over exclusive maintenance functions
- Low cost, fast installation

Disadvantages of stand alone system:

- Must operate and administer entire system
- Must develop own labor data
- Must develop own material data
- May have to run own stock control
- Dependent on purchased material cost
- No access to other systems
- No operating data available
- Cost reconciliation difficult
- Information picture often incomplete
- Little or no staff department support
- Danger of non compatibility with production strategy
- Implies questionable management support

Typical Maintenance Management System

The system designer should summarize his system conceptualization with a schematic diagram of a maintenance management information system to help consolidate his thoughts and present them to decision makers. The schematic should portray the sequence of maintenance actions taken as information is developed. It should also show the interface with accounting. The resulting reports are also shown. By developing a model, the conceptualization can be verified and the key personnel with whom it is discussed can offer constructive comments. This step will help to get system development underway sooner by defining the development roles of each group involved. See Figure 32-6.

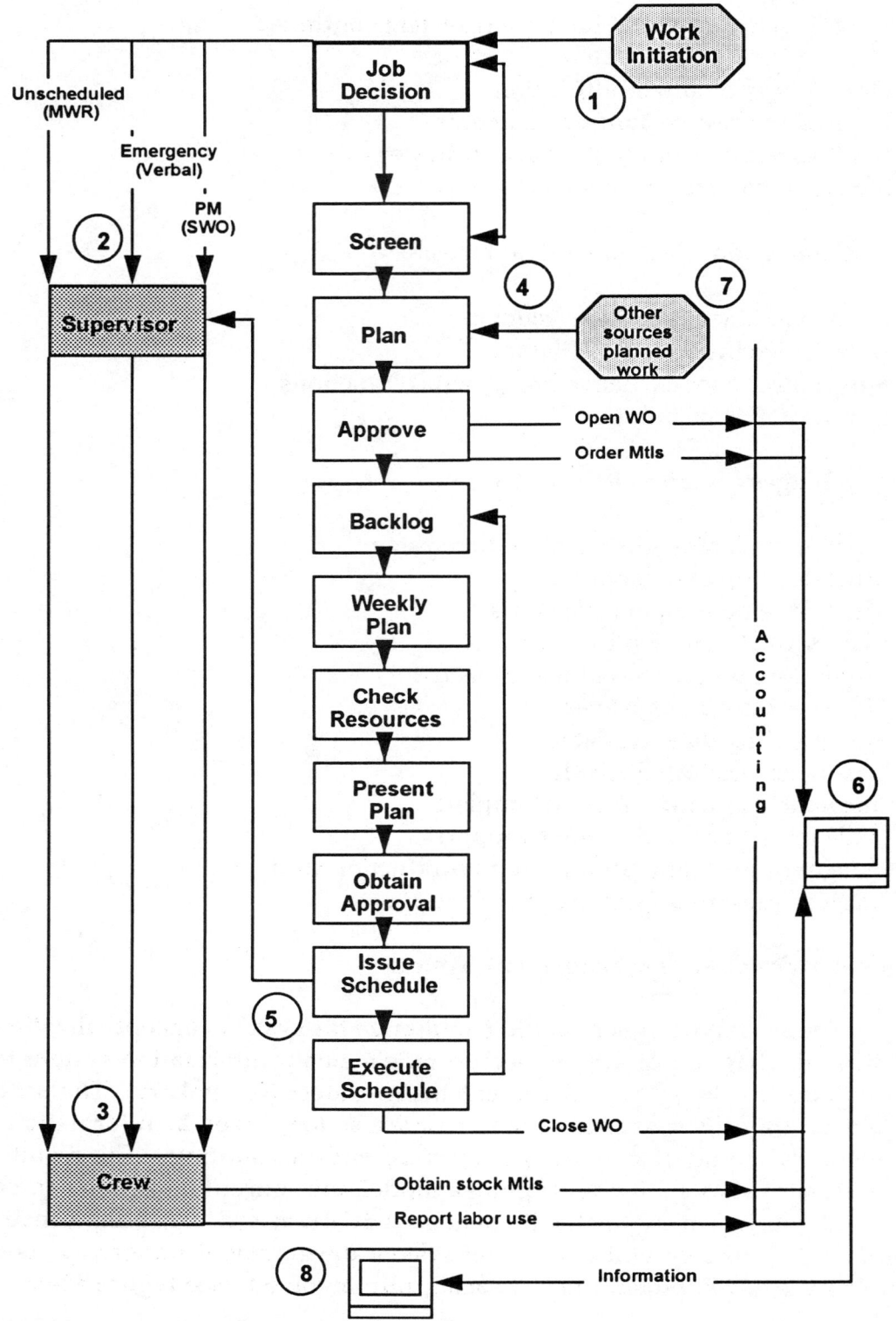

Figure 32-6. Maintenance Management System Overview.

Conceptualizing Information Needs

Refer to schematic diagram

1. As jobs are initiated, unscheduled and emergency repairs and PM services would be sent to the supervisor. Major work should be planned. Each type of job should be controlled by the appropriate element of the work order system.
2. The supervisor makes direct assignments to crew members.
3. Crew members perform work, report labor and obtain stock materials.
4. Jobs requiring planning are selected and planned.
5. The approved schedule of planned work orders is assigned to the supervisor for execution.
6. As work progresses or is completed, information is provided so that work may be controlled, evaluated or converted into additional work.
7. Other sources of planned work such as PM inspection results are sent to planning.
8. Decision making reports include:

    - Labor utilization
    - Backlog
    - MWO Status
    - EWO Status
    - Cost Reports
    - Repair History
    - Performance Indices

Ensuring System Compatibility

The plant manager expresses his objectives in terms of production targets, units of production related to time: tons per day, units per hour etc. Accompanying these targets are corresponding cost values: dollars per ton, dollars per operating hour etc. Achievement of production targets and adherence to budgetary limits are requirements that he must impose on the organization. Within these guidelines, he provides mutually supporting objectives for operating and staff departments. He accompanies these with guidelines on which departments build their coordinating day to day procedures. Taken together, the plant production targets, fiscal requirements, departmental objectives and managers guidelines constitute the plant's production strategy, the plan for achieving profitability. The day to day procedures spelled out by individual departments in response to the guidelines and departmental objectives are the responses to the plant production strategy. Accurate, timely, complete information being utilized effectively by the right people, assures quality maintenance performance.

Summary - The effective marriage of the formal information reports and complementing performance indices is a mark of management skill. Worker productivity measurements derived from informal sampling, for example, should accompany the more formal efforts to measure the workload. A criteria for approval of administrative information will help to avoid duplicate, unnecessary reports. The quality of the information on which maintenance managers will base their decisions is derived from the field data. Therefore, verify its accuracy, timeliness and completeness. Weigh the degree of system interface carefully and request that managers provide guidelines for developing information needs.

33

Developing the Information System

Development Techniques

Development Options - Once the information system has been conceptualized and necessary information identified, decisions can be made on the method of development. There are four methods by which an organization can develop a maintenance management information system:

- In house Development - Use the company data processing department for development assistance
- Package Programs - Consider the purchase and installation of a package
- Buy Another Company's System - Purchase the existing system of another company and adapt it
- Consulting Assistance - Utilize the services of a consulting organization to help develop the program

There are pro's and con's for each method of development and they should be weighed to determining the best one.

In house Development - The primary advantage of the use of the company's own data processing group is that, once agreed upon, the resulting system is usually able to be fully integrated and it will perform according to plan. There are few surprises, and because the staff departments have worked with maintenance during the conceptualization stage, development moves along smoothly. It follows that data processing will be able to maintain the system and, as necessary, modify it.

The use of in house resources for development of the information system assures:

- The system will be consistent with the production strategy
- Staff departments will be supportive
- Management will have committed to it
- Information from other systems will be available
- Maintenance information can be shared with other departments
- Development speed will be pushed by all
- Development cost will be shared by the total plant

A drawback of 'in house' development is the competing demand for data processing services from other departments to support existing needs. In these instances, development can lose its priority. Therefore, a careful assessment of 'in-house' resources should be made before committing to development within a specific, short term period. If appropriate, contract programming assistance could be used to avoid this problem while still accruing the advantages of 'in house' development. There will be:

- Dependency on other systems
- Procedures may allow few deviations
- Priorities of other commitments might slow development

Package Programs - There once was a tourist who went on a 'package' tour. While, at first, the price and content looked good, the total experience was not. The tourist went a lot of places he did not want to go, paid for some services he did not need and got a few surprises. The same thing can be said for many 'package' maintenance information systems. A 'package' is a generalization. Since the creator cannot know accounting details in advance, they require a work order for every job no matter how trivial. This is the only way the vendor can sell his package to a range of users, many of whom are in schools, office buildings etc.. These programs are strong on manipulative record keeping but weak on their ability to collect and convert field data into information. Almost overnight some form of repair history can be initiated. The problem is however, that the clerk creates it by scanning completed work orders to list what the outgoing work order directed rather than what actually happened ! As a supplement to the regular, integrated information program or as a transitional step while the full program is under development, some 'packages' could be helpful. While a 'package' will allow rapid installation, an obvious advantage, the buyer may have to adopt the system philosophy of the package developer.

Developing the Information System

Yet, there are advantages in using a 'package', among them:

- There is little dependency on other systems
- Data is available for exclusive use by maintenance
- There is full control over maintenance functions
- Fast installation is possible

However, integration with existing systems could prove difficult. Some packages depend for material data on the sale of an inventory control program as a module. If inventory control is already covered then, some interface must be programmed. In many instances, inventory control is not a maintenance function. How then does the package produce material costs ? How does it accrue material costs from direct charge purchases ? Another problem may arise if there is language incompatibility. Equipment numbering could be a problem as some packages may require renumbering to match their programs.

Labor data can be a critical decision for or against a 'package'. If it is suggested that labor be reported on the work order, this reveals little real conception of the way maintenance is carried out. Few jobs are a one man - one work order jobs. Few craftsmen or supervisors want to take on the task of matching the labor data of many crew members with various jobs when they know it means duplicate reporting and unnecessary work. The time card performs this function without the hassle. On screen reporting to a work order is also a common 'package' solution but, it requires that the data so collected must then be re distributed to other accounting requirements. This will require dual reporting or the development of communications software to up load the data.

Packages which promote a 'PM' module should be looked at cautiously. Most include almost everything maintenance does under PM. Component replacements, sometimes overhauls as well as inspection or lubrication are all classified as PM. Checklists are printed like confetti. In many instances 'events' are scheduled piecemeal. Typical, one maintenance department noticed that they visited the same unit 4 times in a 2 week period to perform services that could have been done in one trip.

Some of the disadvantages include:

- Maintenance may have to operate and administer the entire system
- Often maintenance must develop its own labor data and this may mean duplicate reporting

'Package' disadvantages continued . . .

- Maintenance may have to develop its own stock issue data and purchasing data may not be available
- There may be no access to other systems that contain needed data
- Cost reconciliation may be difficult and accounting may be critical of cost data that is at variance with theirs
- The total information picture may often be incomplete
- If maintenance bought the system for its exclusive use, expect no MIS support
- The system may not be compatible with the production strategy
- Use of the system implies questionable management support because maintenance acted alone in procuring it
- The program may require significant administrative support that the vendor will not supply
- An often too 'busy' program will confuse users who then make poor use of it

A maintenance department considering a 'package' must first determine its real information needs. Only then can the acceptable packages even be identified. Next, maintenance must be prepared to accept some compromises in order to gain the advantage of installation speed. The direct choice of a 'package' program by a maintenance department without considering the other options available may indicate that:

- Speed of installation is a factor or . . .
- Ignorance of how to proceed forces acceptance of this solution

Buy Another Company's System - There is a strong temptation to purchase a system being used by another company, particularly if they are in the same industry. Usually, such systems are comprehensive and they include most elements contained in a fully integrated system like payroll, inventory-control and purchasing. A major purchase of this nature is a matter of serious consideration and it has impact beyond maintenance. The decision should be made by a team, never by maintenance alone.

Once a decision has been made to consider such a system, the usual next step is a 'team' visit to the selling plant. Be aware that everything demonstrated will be shown under ideal circumstances in which the seller knows beforehand exactly what will be produced. Also, the selling plant has already invested considerable development costs and has now added marketing costs to them. They intend to get their investment back and you will be their financier. Thus, the purchase is likely to be expensive.

Developing the Information System

There will be some drawbacks once the system is acquired. Of those organizations who purchased systems from other organizations, many never finished modifying them to their satisfaction. They either abandoned them or replaced them at the first opportunity with those they developed themselves. After considerable expense, their conclusions were invariably that they failed to properly analyze their needs before buying the program. As a result, they may have wasted time and money and gained less benefit than they had hoped for.

Consulting Assistance - If a consulting organization has successfully developed and installed maintenance management information systems for clients in your industry, they are worth considering. However, the basic responsibilities of maintenance remain. Maintenance must be involved. If they are coached through the development process they must fully understand the implications. Make sure maintenance hasn't been 'led by the nose' with no idea of what the consultant has 'created' for them. In this regard, the biggest single factor against consulting assistance is their tendency to inject a whole new philosophy rather than simply get on with the task. Handled properly, with full maintenance participation, and effective 'in-house' support, professional consulting assistance can produce a satisfactory product in record time because they know where the obstacles are and how to get around them.

Monitor Development Progress

During the development process, it will be important to monitor progress and verify that the system will produce the information required. Therefore, a development plan and a time table are important. Monitoring must be done by maintenance unless it is an integrated system, in which case a team should monitor progress. There will be delays, particularly in the fully integrated system. Generally, the work order portion, once equipment files are loaded, will test out quickly. The most probable delays will be in its integration with accounting programs. These always prove troublesome because data must be transferred precisely. It is not unusual for maintenance to be able to get labor hour data quickly but, have to wait a year or more to get labor and material control costs. Since these delays are probable, the development plan should have options allowing use of certain portions of the system for work control. For example, if labor hours are reported correctly, they will accumulate fastest on the most troublesome equipment. Adding labor and material costs will not change the outcome. [1].

But, with the right data distributed correctly, maintenance can operate with partial system availability.

Summary - The time available for development has considerable impact on the method of development. Generally, those most desperate will do no 'homework' and try unsuccessfully to use a 'package'. Those doing their 'homework' but, pushed for time will use a 'package' as a transition and eventually develop their own program. They will learn what to avoid from the 'package' experience. Those with plenty of time will do their 'homework' but probably develop their own system over several planned, progressive phases. Those most likely to use professional consulting assistance are those who have made several abortive efforts, used several 'packages', admit they don't know how and are now convinced they must now do it properly. It will be important to the success of the development effort that the total plant be involved, All will be affected by the success of the maintenance department, the quality of its work and the information that impacts both.

1. Images, copyright New Vision Technologies, Inc.

34

Implementing the Information System

Preparing Personnel for System Implementation

Quality Training Means System Acceptance - Acceptance of any new information system will depend on how well the users understand how it can help them to perform their jobs. Thus, the education of system users is a key factor in successful implementation. The maintenance manager should review the system training program to ensure it is complete. When training begins, he should take an active role and insist that each supervisor do likewise. This will demonstrate commitment to the system and encourage users to master its use quickly. At the outset of the training program, the duties of key personnel should be associated with the information they will need. Then, those aspects of the system that personnel must have to perform their duties should be emphasized in the training program. Maintenance supervisors, planners and craftsmen function as a team and, since they will be the principal system users, they should be trained together. Since the supervisor is responsible for getting the work done, on time and with a quality effort, work control is his primary responsibility. Therefore, the system training should explain how he will:

- Review outstanding jobs
- Determine routine PM services to be done
- Make decisions on jobs to be done
- Assign crew members to jobs
- Provide job instructions.
- Monitor job status
- Close out completed, unplanned jobs

Typically, the planners training should include:

- Use of work orders for planned jobs
- Determination of work order status
- Determination of work order cost
- Observation of the backlog
- Observation of the use of manpower

Watch out for a tendency for the supervisor to try to delegate his work control responsibility to the planner. To ensure that the work control task remains in the hands of the supervisor, the planner should focus on major planned and scheduled jobs. After a planned job has been placed on an approved schedule, the supervisor would make the work assignments for a specific day and shift. If the time frame of a job has been agreed to, the supervisor should schedule the job and then assign it to his crew members. Typically, a PM service that requires no equipment shutdown and can be accomplished anytime within the week due. Get the supervisor to develop and assign simple jobs like routine dynamic equipment inspections (while equipment is running).

Individual crew members training should include:

- Entry of equipment specifications data
- Use of equipment specifications data
- Updating of equipment specifications data
- Review of repair history data.
- Entering new work

Start their training with useful systems and information. Craftsmen need to know part numbers. Therefore, train them first on using equipment specifications. Soon they will be using the computer regularly and confidently.

The Problem of 'Computer Literacy' - The degree of 'computer literacy' will differ among personnel and must be considered in the training. Often, the planner will have had previous computer experience and possess reasonable administrative skills. Usually, electricians and instrumentation personnel will have been exposed to the use of the computer as a diagnostic tool and they can easily adapt. Mechanical craftsmen seldom have use for the computer in their work but, as younger craftsmen enter the work force, it is more common to see 'computer literacy' based on their use of a personal computer outside of work.

Implementing the Information System

Some maintenance supervisors present a problem for system implementers because they lack this 'literacy' and are apprehensive about the use of the computer. Typical, when paper work control systems existed exclusively, the supervisor was unquestionably in charge. However, when the computer displaced these systems, many supervisors shared or yielded work control in favor of the planner. The planner could use the computer and the supervisor could not.

These circumstances must be reversed. The supervisor's primary responsibility is work control. The information system is the maintenance communication network with the supervisor at its hub. Training must give this strong emphasis.

Maintenance supervisors prefer an active field role over tending to administrative matters. Yet, making job assignments for crew members and following up on job progress are essential supervisory tasks even though administrative in nature. Therefore, the supervisor cannot ignore the value of the computer. The best course of action is to install the system for the supervisor and his crew as a team. Involve the craftsmen in entering new work in the computer especially equipment deficiencies they have found during the equipment inspections. This way, expertise among crew members contributes to the supervisors progress in learning to use the computer effectively. As the supervisor and his crew start to use the computer, they will demonstrate its value to themselves and will soon overcome his reluctance to use it.

Reporting Field Data - Training should include the reporting of field data by crew members and its verification by supervisors. If the field data isn't correct or is incomplete or late so is the information it produces. No matter how well conceived an information system may be, the critical aspect of its success is the accuracy, completeness and timeliness of the field data provided to that system. Field data includes all data reported by maintenance personnel as they carry out work or perform related functions:

- Maintenance craftsmen report man-hours, record equipment numbers, describe repairs, indicate failure codes etc.
- Maintenance planners open and close work orders
- Warehousemen post equipment and work order numbers against issues
- Purchasing agents transpose work order or equipment numbers onto purchase orders

Supervisors verify most data whether it is reported on a time card on the computer screen.

There is considerable room for error and there are many people who can make errors which seriously affect the quality of information. The most effective way to ensure that field data is reported accurately and completely is to train the personnel who report it. The key person is the maintenance craftsman. He is the actual person who does the work and is in the best position to accurately describe what he has done. In most instances, the craftsman is willing to report data. However, he must understand the information that his data will produce, what it will be used for and why it is important.

Training for craftsmen must not be limited to showing them how to fill in the blanks on the format displayed on the screen. They must be shown what their data produces. Showing the craftsman the actual reports and letting him use the display screens will contribute substantially to his appreciation of the importance of the information.

Obviously, in a very large work force, each craftsman cannot be given individual training. A training procedure must be worked out. Often the best way to do this is to show the craftsmen a facsimile of the reports that will be produced and encourage some discussion on how the information can be used. As actual reporting begins, it is necessary to familiarize personnel with codes that will be used in reporting. Helpful techniques include:

- Listing codes on the reverse side of a field reporting document
- Providing for selection from an on screen table file

These techniques help avoid unnecessary mistakes. It is also helpful in successfully initiating field reporting to accomplish it in phases. For example:

- If reporting will include components, it is best to omit component reporting until equipment reporting is mastered
- If system development will be delayed, start practice reporting early to gain proficiency

Don't overlook training for warehouseman and other administrative personnel. The main point to remember is that those who do not understand the purpose of what they are reporting will, most likely, not do it well.

The Supervisor as a Trainer - One of the best methods of ensuring that craftsmen are well trained is to require supervisors to carry out the training. The supervisor is motivated to do a good job because he knows his crew will be a critical audience.

Implementing the Information System

The best way to find out if supervisors understand the system is to listen to them explain it to their crews. As they do so, the quality of their understanding as well as their commitment can be observed.

Using the Planner - Maintenance planners will often have been involved in system development and be familiar with details. They can be used to help explain data entry requirements to supervisors and material control personnel.

Documentation - Training materials must be well documented and easily available to personnel. Table files from which the user can make selections are very helpful as are help screens which offer clarifying notes on system use.

Loading Files and System Testing - Make sure that all of the files are loaded before training begins. There is nothing more confusing than the attempted use of 'bogus' data, equipment numbers or descriptions for the person learning to use a new system. Use the actual equipment numbers and work descriptions. Training is more realistic when actual data is used. Prior to loading files in the programs, test data should be tried to ensure the data is properly transported between systems and that it produces the desired information. System files should be loaded carefully, especially equipment numbers and components. Smaller existing files such as failures codes or trade groups usually present no problems. Special assistance may be needed from MIS when previous files such as repair history or equipment specifications (part numbers, drawing references etc.,) must be converted and transferred.

Vendor Training - Many vendors of software packages may not have had specific experience in industry. The poorer ones will explain the system to the clerk and leave. The best ones will load the files and train everyone because they know they have a quality product and they want a solid endorsement from a satisfied customer.

Production Personnel - Production supervisors will have a requirement to enter new jobs into the system and make inquiries on the status of jobs. They must be included in the training effort so they can do these things competently. Production managers will require information on performance indices and costs. They should be shown the reports and have the use of the information explained to them. By starting as early as possible on the training of production personnel, they will be in a better position to offer comments on how the information can help them. Often, these comments will influence alterations in the basic maintenance program as it is modified to utilize a higher quality of information.

Develop a Plan for Information Use - The vital questions of who gets what information, when or how often they receive it and what actions they are expected to take must be addressed so specific training of real value is provided. Typically:

- The maintenance manager requires summaries that describe overall performance and cost control allowing him to judge accomplishments and direct corrective actions
- General supervisors need details of weekly accomplishments like PM schedule compliance to direct improvement actions
- Supervisors need work details to develop jobs for their crews along with feed back on job progress
- Planners need to know the status of major jobs and explain variances from their recommended job plans
- Craft personnel need to be able to look up part numbers and query repair history to prepare for upcoming jobs
- Maintenance engineers must track costs to find troublesome equipment, narrow problems with repair history then field check to develop corrective actions
- Operations managers want to know how much maintenance services have cost and obtain an explanation of current and future costs
- Operations supervisors will want to enter new jobs into the system and determine the status of outstanding jobs [1].

Successful Implementation Steps

Once maintenance has developed its information system, the more important task of successfully implementing the system awaits. Often, a competent maintenance information system fails to deliver because its users were not properly prepared to utilize its capabilities.

Implementing the Information System

As a result, information is incomplete, of poor quality and decision makers fail to control work properly. Here are ten steps that can help ensure this does not happen. See Table this page.

Step 1 - Confirm the soundness of the maintenance management program. Most unsuccessful maintenance programs can be traced to inadequate program definition and a lack of education for the personnel who must carry them out. Even if the most effective information systems are superimposed on a confusing or ill defined program, they will fail. The information system is merely the communications network of the maintenance program. Therefore, verify that the maintenance program is properly defined before commencing system implementation.

- Confirm Maintenance Program
- Confirm Roles of Key Personnel
- Phase out Previous Programs
- Load Files and Modules
- Verify Network and Hardware
- Verify Field Data Sources
- Establish 'Core' Training Group
- Conduct the Training
- Establish Schedule and Objectives
- Monitor and Verify Accomplishments

Step 2 - Ensure the roles of key personnel are correct and understood. Information systems have three objectives: Identify work that is required; control the work and measure the success of the work performed. As the maintenance program is executed, key personnel perform one or more of these functions. Thus, the information system provides the tools by which they carry out their roles. All users of the work order system define work through the work order system. Some work is identified by the superintendent as he observes high costs to identify equipment requiring attention. The maintenance engineer identifies other work based on failure trends summarized in repair history. Work is controlled primarily by the maintenance supervisor as he directs the efforts of his crew. The success of work performed is measured by a planner who notes the comparative cost and timely completion of major planned jobs. Similarly, the superintendent notes cost variances while the supervisor observes the cost of units and components in his area of responsibility. A plant manager checks performance using indices like cost per unit of product. Specific information is directed to designated personnel who, in turn, make decisions and take corrective actions. It also helps to develop a list of information needed by key personnel to help identify the actions it will require of them.

Responsibilities for action must be confirmed before information requiring action is directed to these individuals. Thus, early confirmation of the roles played in managing maintenance will help in the effective utilization of the information.

Step 3 - Phase out all previous, conflicting procedures to avoid unnecessary difficulties as you install any new information system. The information system your organization has just adopted is now the 'official' communications network of your maintenance program. If it has been carefully developed it need not be supplemented by word processed listings of jobs or PM services scheduled on a spread sheet. Make a clean break and give the new system a chance to meet your performance criteria unencumbered by outdated, ineffective habits and procedures. Be especially aware that people problems will require attention. Typical:

One maintenance superintendent was surprised that a new, highly acclaimed information system failed to be able to schedule lubrication services. A closer investigation revealed that six competing procedures, archaic remnants from several previous programs, were still in use. Their continuing use precluded reliance on the new system and, as a result, services were being scheduled haphazardly with no follow up on schedule compliance.

Any new, competent system must be given a fair test. To allow personnel to continue to use and depend on previous procedures denies that test. As a result, the system may fail to deliver quality information.

Step 4 - Load all files and modules and confirm the equipment numbering scheme. Training will be confusing if personnel are asked to pretend that 'ABC' is a conveyor. Therefore, load all files and modules before training begins. Unless the training yields immediate, useful results, the new program will be off to a shaky start. Make certain that the equipment numbering scheme is sensible. To illustrate:

Two processing plants built 6 years apart used the construction codes of two different contractors to number their equipment. Not only were the schemes different but, each failed to divide equipment by type, unit and component. A subsequent attempt to install a 'package' system was delayed considerably while steps were taken to realign equipment numbering procedures. The focal point of all maintenance work is equipment. If the work on equipment is to be effectively managed, equipment must be given a logical numbering scheme. This will make the implementation of any information system much easier.

Implementing the Information System

Step 5 - Verify that the hardware and networking arrangements are functional. The program works, the training has been carried out and everyone is ready to request, plan, schedule, assign, control and measure work.

But, suppose the system won't convey information adequately or at all. Not only does this situation create doubt in the minds of its future users but, it will set the implementation time table back considerably. It is avoidable. Test and then test again. Make sure that everything works.

Step 6 - Verify that the field data that will fuel the system is complete, accurate and timely. The work order system is the means by which field data are directed into the information system and converted into useful decision making information. However, if these data are incomplete, incorrect or late, little worthwhile information will be created. In addition, still fewer correct decisions will be made or proper actions taken. Quality field data is the catalyst of good information. To illustrate:

In a large smelter complex, maintenance supervisors 'created' and entered the labor data for their crews based primarily on assignments made at the start of the shift. This procedure neglected all of the verbal orders crew members received as well as the work they identified and completed on their own. Review of the resulting costs reports often showed massive amounts of material installed 'without the aid of human beings', material was charged by no labor was reported. In effect, the supervisors labor data was conjecture thus, inaccurate.

In this instance, crew members, as the best source of information on work they performed, should report the field labor data. The supervisor should verify it with questions and logic. Don't waste time, money and energy on the installation of a new information system unless you are able to develop the best field data you can get.

Step 7 - Establish a 'core' group of maintenance personnel that will train the work force. The 'core' group must receive an in depth education on the entire information system since they must be able to answer any question on the system. Also, prepare them for the unusual, since they may encounter resistance and criticism which, while not aimed at them, may affect their objectivity of getting on with the training. If you have purchased a 'package', don't rely on the vendor for training. Few will provide training and it may be superficial, generic and expensive. Therefore, if you want the system to be used properly and effectively, you will have to develop and utilize your own instructors.

Step 8 - Conduct adequate training. The amount of training varies according to the backgrounds of individuals. If you are training an instrumentation crew that uses the computer daily in its work, expect to spend a week with them. If neither the supervisor nor the crew have ever used a computer, expect to spend several weeks with them.

These times assume that personnel have a well defined maintenance program and they understand it. They also presume that the information system is easy to use. Your 'core' (in house) team should move from area to area. Training should include eight hours of classroom familiarization training per individual followed by at least 21 consecutive days, around the clock, working with the personnel in the area.

If you do not supply this minimum training, then you must be prepared to accept mediocre system performance as a result of 'trial and error' substituted for quality training.

Step 9 - Develop an implementation schedule with specific performance related objectives. Set reasonable targets with realistic objectives. Typical:

- At the end of the first week, every crew member will be able to enter new work in the system, print work orders assigned to them and report actual hours against each work order on the scheduling format
- At the end of the first week, each supervisor will know how to prepare a crew schedule assigning each member a full shift of clearly identified jobs
- On the completion of the second week of training, each supervisor will be able to query repair history and review costs against jobs, equipment and cost centers

Don't omit anyone from the training. Since the information system is the 'official' communications system, all must be able to use it competently. Don't let this happen:

Nine weeks after system implementation training was completed, a maintenance general supervisor was asked to pull together the repair history for a troublesome unit. When his information was questioned, it was discovered that he obtained most of it from hand written log books rather than from the more complete, accurate and up to date history of the new program. Further questioning revealed that he knew little about the program. He had excused himself from the training expecting that others would produce information for him.

Implementing the Information System

Unfortunately, some of his supervisors also assumed a casual attitude based on their observation of his actions, causing still further problems. Train everyone, no exceptions.

Step 10 - Monitor system use and accomplishments. System performance will be important. It is not enough to create information. The information must be put in the hands of the proper people who, in turn, must make decisions and take corrective actions. Monitor what happens. Typical:

- If compliance with the PM program is improving, then confirm that fewer emergencies are occurring and more work is being planned and scheduled
- If more work is being planned, see if worker productivity is improving
- If productivity is supposed to be better, find out whether fewer overtime hours are being used
- If compliance with the weekly schedule is not improving, then be able to pinpoint the cause:

    Maintenance didn't get materials
    Production didn't release units
    Maintenance failed to allocate labor for jobs

- If selected units show greater maintenance costs, expect to find meaningful narratives of what happened in repair history $_2$.

Summary - The appropriate information system will yield performance improvements only if effectively implemented. Effective maintenance management is linked to a good system, effectively used by well trained personnel operating within a clearly defined maintenance program. The education of system users is a key factor in successful implementation. Acceptance of the information system will depend on how well the users understand the way it can help them to perform their jobs. If these points are emphasized in the system implementation, it will be more effective.

1. - 2. - Images, copyright, New Vision Technologies, Inc.

35

Work Order Systems

Definition and Purpose

The work order system links accounting and production statistics together to produce information allowing maintenance to manage and administer its functions. The work order system is also a communications network by which work is requested, classified, planned, scheduled, assigned and controlled. It links with the accounting system to obtain labor and material data and with the production system to obtain statistics like tons or operating hours. The processing of these collective data produce decision making information. See Figure 35-1.

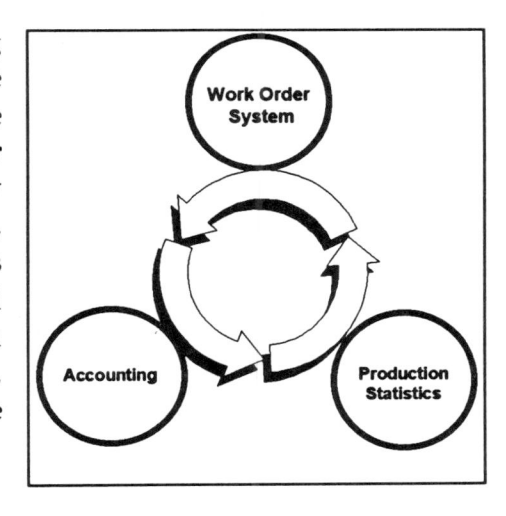

An effective work order system will:

- Cover all types of work performed
- Document job cost and performance
- Accumulate costs to jobs, units or functions
- Provide for control of routine, repetitive functions
- Allow the use of verbal orders
- Provide simple controls for simple actions
- Provide controls suitable to the work done
- Link with production statistics to establish useful performance indices
- Link with accounting documents to produce resource use information

As a communications network, the work system:

- Orders service
- Identifies work required
- Relates work to a unit of equipment
- Identifies a function to be performed
- Establishes the type of work, its timing and importance
- Identifies labor, material and equipment requirements for jobs
- Makes provision for job approval
- Ties field data to accounting
- Facilitates the use of standards
- Provides data for the accounting system
- Serves as basis for scheduling work

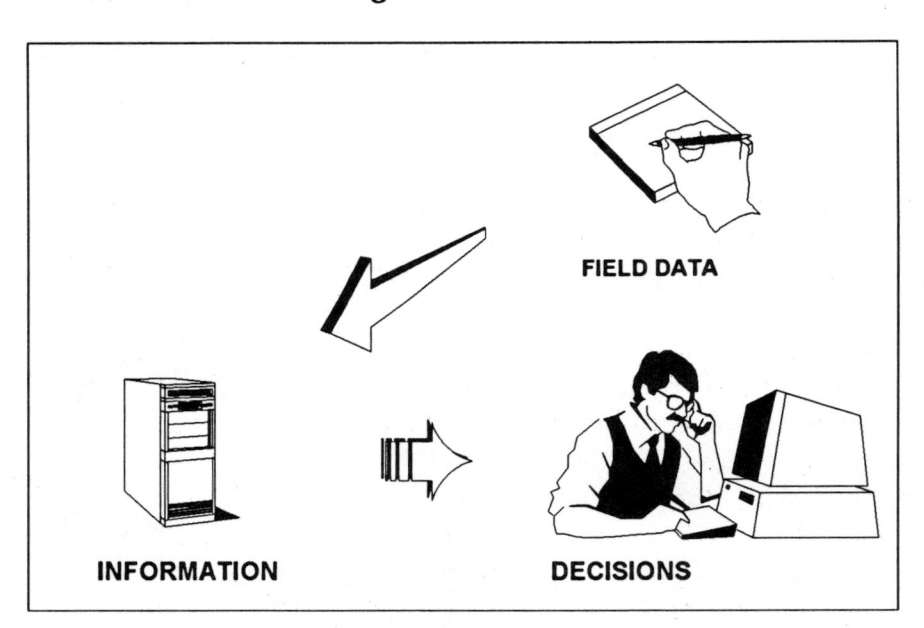

Figure 35-1. The work order system guides field data through system programming where it is converted into information for use by decision makers. [1].

Work Order System Elements - A work order system includes elements to cover all types of maintenance as well as non maintenance work (like construction). Each has a specific work control objective:

- Maintenance Work Order - Controls planned and scheduled work
- Maintenance Work Request - Requests and controls unscheduled repairs
- Verbal Orders - Often used for emergency work when its urgency does not permit time to submit a work request or prepare a work order

Work Order Systems

Work order system elements continued...

- Standing Work Order - A reference number identifying routine, repetitive actions such as PM services or shop cleanup
- Engineering Work Order - Controls engineering project (non maintenance) work like construction

The maintenance work order (MWO) is used to plan and schedule major jobs such as overhauls or replacement of components like engines. Because these jobs are important and costly, job cost and performance are measured to ensure effective control. The MWO establishes the cost and performance goals of each job and, as work is performed, labor and material data are collected to yield job cost and performance. When the job is completed, the work order is closed. Then, cost and performance can be compared with estimates. Next, the job cost is closed against the equipment on which work was performed. See Figure 35-2.

MAINTENANCE WORK ORDER

MWO #	Sec	Unit	Comp	Ct	Unit Name	Comp Name
514706	024	CM01	100	S	Continuous Miner	Cutting

Description of Work: **RPL CUTTER HEAD MOTOR**

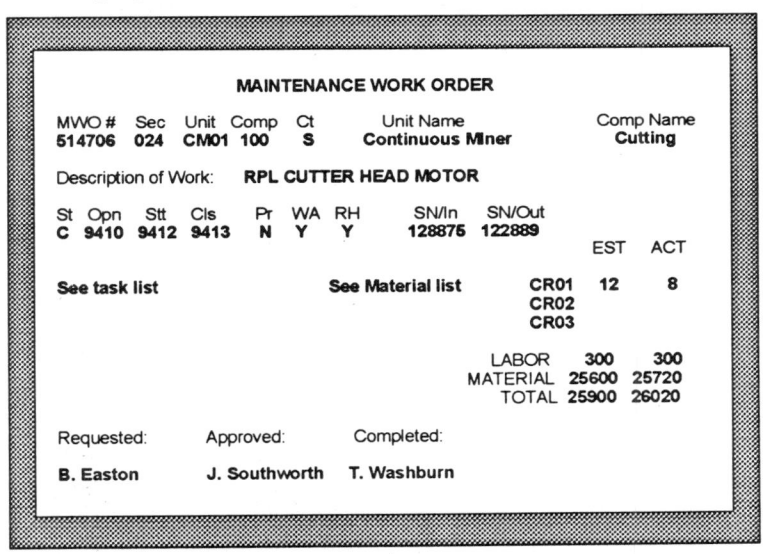

St	Opn	Stt	Cls	Pr	WA	RH	SN/In	SN/Out		EST	ACT
C	9410	9412	9413	N	Y	Y	128875	122889			

See task list   See Material list   CR01 12 8
                                     CR02
                                     CR03

LABOR   300   300
MATERIAL 25600 25720
TOTAL   25900 26020

Requested: Approved: Completed:
B. Easton   J. Southworth   T. Washburn

Figure 35-2. The MWO establishes estimated resource needs, job cost and duration. As the work is carried out, actual resource use and costs are accumulated and summarized when the job is completed.

The maintenance work request (MWR) is used to control unscheduled repairs. These are simpler, less costly jobs on which it is not necessary to isolate cost and performance. However, the jobs must be accounted for and many will contain information that should be placed in repair history (the narrative of what happened to the equipment). Simplicity and ease of use are important and the MWR is often considered as a simple, structured message format. The requester merely identifies the equipment and the repair needed.

Unlike the MWO, which isolates the cost of each job, an MWR only accumulates cost against the equipment. Unscheduled repairs are done at the discretion of the supervisor who fits them in at first opportunity. See Figure 35-3.

```
                    MAINTENANCE WORK REQUEST
Sec   Unit   Comp  Ct              Cr       Pr    MH    YrWk
024   CM01   100   U               ME       72    10    9421
---------------- Alt Z Shows all values for above files ----------------
Description of Work:   RPL CONVEYOR CHAIN           +


Assigned to:     J.P Brown
Requested by:    C. Johnson


----------------------------------------------------------------
         F1 Accepts    F2 Delete - Re enter    F9 Help    F10 Exit
----------------------------------------------------------------
```

Figure 35-3. The MWR format should be as simple as possible while requiring the requester to identify the equipment and the action required.

The standing work order (SWO) is used to control routine, repetitive functions like training or shop clean up, or to identify groupings of equipment that exhibit a low maintenance cost such as replacing light bulbs in offices. It is a reference number and it charges the activity to the department in which it took place. It should not be used to describe a major unit of production equipment, such as a haulage truck, as this would then require a standing work order for each component like engines or tires. This procedure soon becomes impractical since the use of a logical equipment numbering procedure has already provided unique numbers for these units and their components. See accompanying Table of Standing Work Orders.

SWO	Description of activity
901985	Safety meetings
901986	Training
901987	Shop clean up
901988	Snow removal
901989	Grass cutting
901990	Building repair
901991	Roof repair

Work Order Systems

Verbal orders are used primarily when the urgency of emergency work does not permit time to submit a work request or prepare a work order. Verbal orders are a fact of life for most plant maintenance departments. 'Outlawing' verbal orders seldom works. Therefore, some provision must be made to ensure the work is effectively controlled and information about the job is not lost. Verbal orders often correspond with emergency repairs. When they are used, there must be a means for the maintenance worker to report what he has worked on. If repair history is desired, there must also be a way for him to describe what he has done. The most common solution is to report the unit number, the action taken (for repair history), man-hours used etc., on the time card or other means of reporting field work. If it is subsequently determined that the cost of the job is required, a work order can be opened after the work is done but, before the labor data is processed.

The Engineering Work Order (EWO) covers capital funded project work such as construction, equipment installation, modification and equipment relocation. It is used by either maintenance or a contractor. See Figure 35-4.

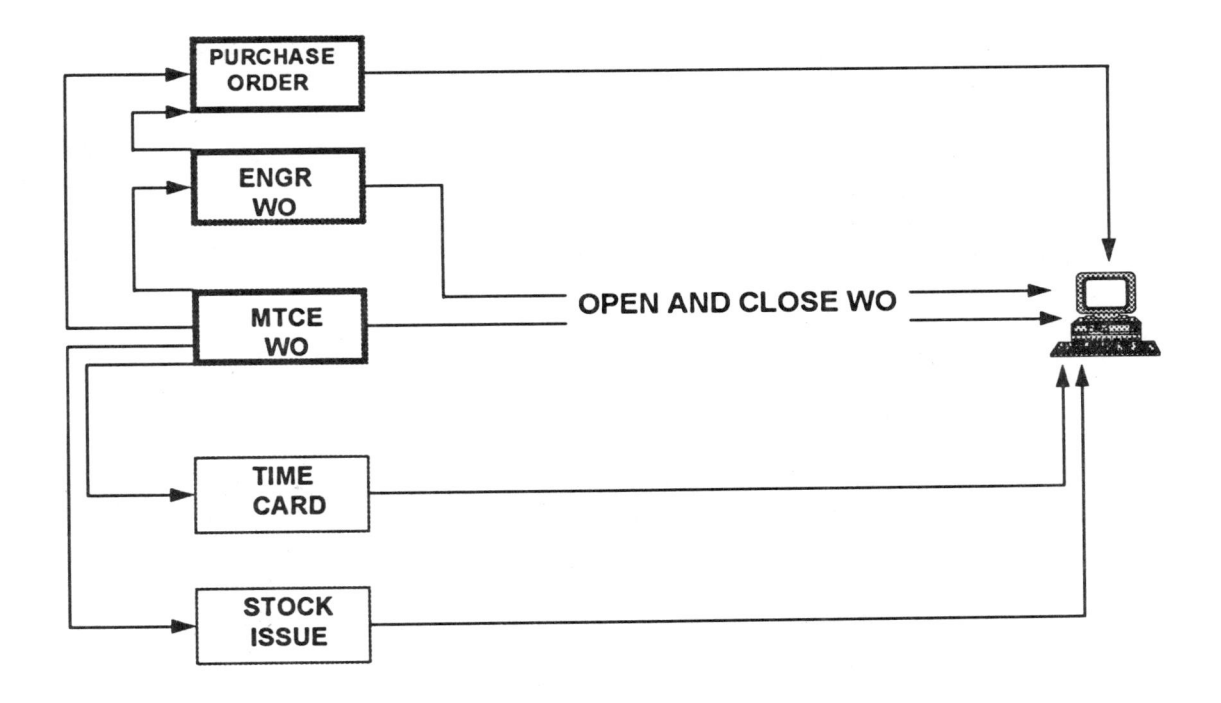

Figure 35-4. The EWO links with the maintenance work order when maintenance performs the work. When the work is performed by a contractor, the link is the accounting system.

One engineering work order (EWO) is used to control the elements of a project, some of which would be performed by maintenance using an MWO and others performed by a contractor using a purchase order (PO). See Figure 35-5.

Work Description	Week 24	Week 25	Week 26	Week 27
MWO Site preparation	██████████████	██████		
MWO Installation	██████████████	██████████████	██████████	
PO Contractor		████████████	██████████████	██
MWO Power		██████████	██████████████	
PO Instrumentation		██████████████	██████████████	██
MWO Testing			████████	██████████

Figure 35-5. The project manager observes the progress of each step of the project by noting the status of work orders and purchase orders linked to the project engineering work order.

Work Order Element versus Type of Work - The type of work dictates the degree of control required. Jobs of short duration need little manpower, are inexpensive and require minimum control. Thus, a simple work order element is dictated. See Figure 35-6.

☐ = PLANNED AND SCHEDULED MAINTENANCE
▒ = UNSCHEDULED REPAIRS
▨ = EMERGENCY REPAIRS

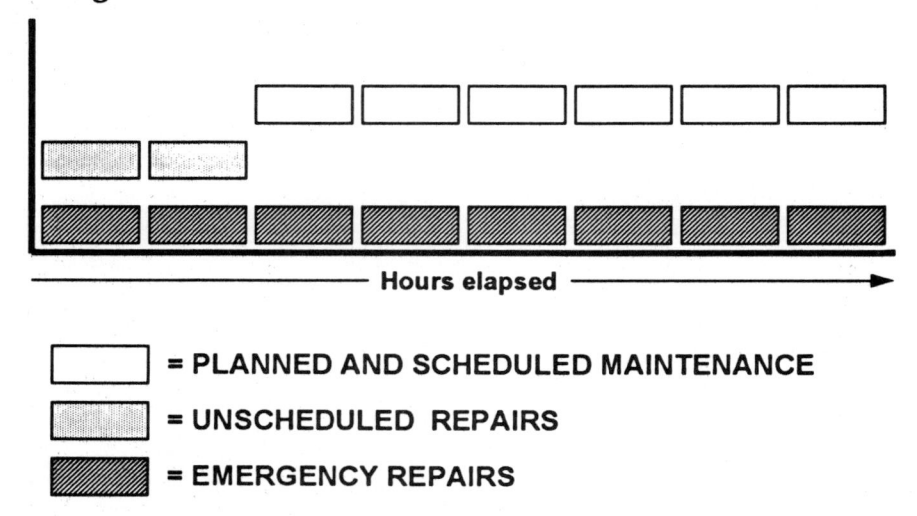

Figure 35-6. Jobs of 2 hours are primarily unscheduled repairs. Jobs from 3 hours upwards are usually planned and scheduled maintenance. Emergency jobs can be of any duration.

Work Order Systems

In the development of a work order system, each element must have a practical relationship with the type of work being controlled. For example, if the element for the control of unscheduled work is too complex or time consuming, it will be abandoned, leaving a void in the control of work and the information collected. Frequently, systems requiring a work order for every job, no matter how trivial, soon create an administrative burden for supervisors. Verbal orders then predominate and much valuable information is lost.

Volume of Jobs versus Cost - The greatest volume of jobs are done in less than 2 hours. These account for the least amount of maintenance cost. These jobs should be controlled in the simplest way using, as appropriate, message forms or verbal orders. But, big planned and scheduled jobs must be managed since they represent the major cost of performing maintenance work. See Figure 35-7.

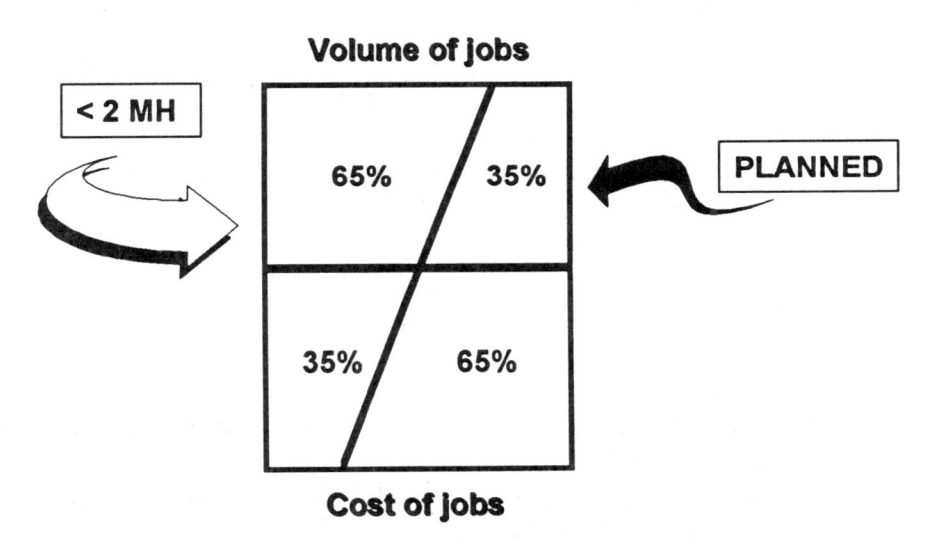

Figure 35-7. Sixty five percent of the volume of jobs are accomplished in less than 2 hours and account for only 35 % of costs. About 35 % of the total jobs done are planned however, they represent 65 % of maintenance cost.

Interface of Work Order - Accounting - Production Control Systems

It is useful to observe the relationship between the elements of the work order system (MWO, MWR, Verbal orders, SWO and EWO), the accounting system (time card, stock issue card and purchase order) and the production control system. Data from these three systems are brought together, processed and converted into information. In turn, the information is used to make decisions affecting the control of work, the use of resources, cost and performance within maintenance and across the plant. See Figure 35-8.

Industrial Maintenance Management

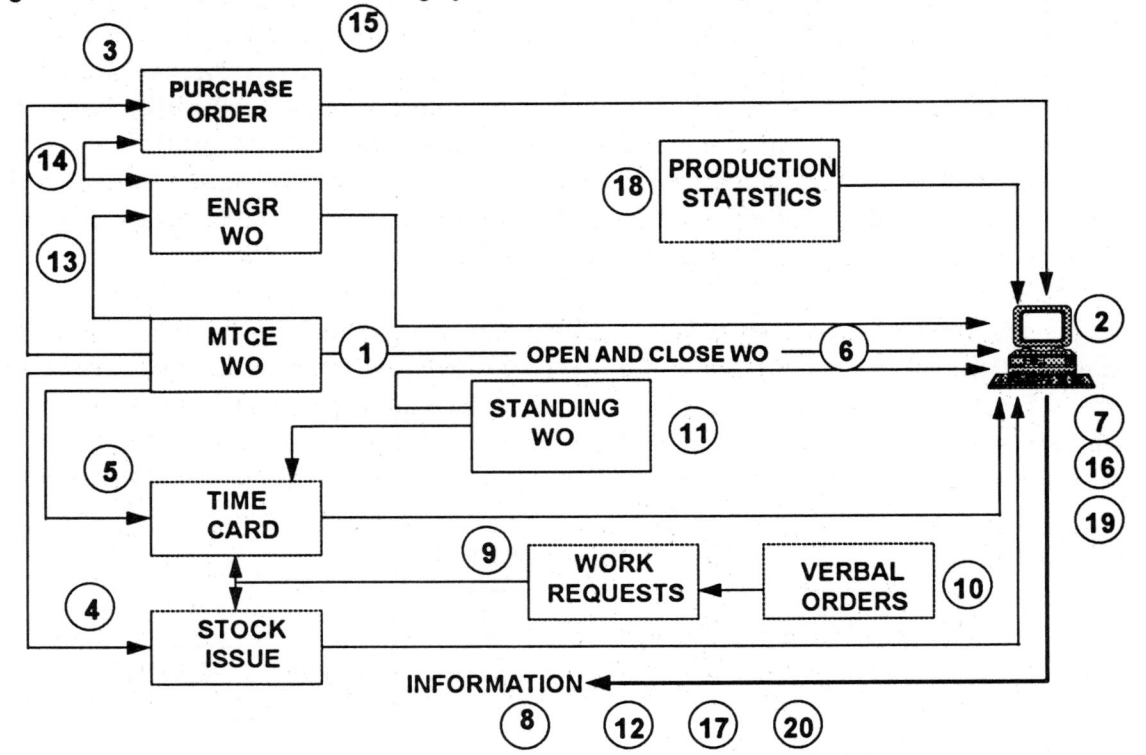

Figure 35-8. Work order - accounting system overview shows systems interface.

Explanation of overview:

1. The MWO is opened.
2. The system ties the MWO to the equipment and the job and, authorizes the expenditure of money to do the work.
3. The MWO number is placed on all related PO's so that the system will charge purchased materials to the job.
4. The MWO number is placed on related stock issue cards (SIC) so that the cost of stock material will be charged to the job. Concurrently, the SIC data debits the inventory control system, reducing the on hand balance by the number of items issued.
5. When the job is scheduled and work assigned, all personnel working on the job report man-hours used on their time cards. The cost of this labor is then charged to the job. In addition, the number of man-hours, by craft are summarized and compared with the estimates made at the time the MWO was planned. Concurrently, the labor data on the time card is forwarded to accounting to provide payroll and accounting information.
6. On completion of the job, the work order is closed.
7. The system now summarizes job cost and performance.

Work Order Systems

8. Information on the cost and performance (man-hours actual versus estimated and comparative job duration) of each MWO is now provided. Then, these costs are closed against the equipment on which the job was performed to appear on the detailed cost report. Similarly, if descriptions, failure codes etc., are required for repair history, they are compiled on the closing of each MWO.

9. In the interim, the MWR (work request) is used to request unscheduled repairs. As work is done, labor is reported against the unit on the time card while stock issues are recorded on the stock issue card. These costs accumulate against the unit not against the individual job. Keep in mind that the purpose of the work order (MWO) is to isolate the cost and performance of a major job. Since the MWR is a simple, low cost job, cost and performance on the job is not required.

10. Similarly, as emergency repairs are initiated with verbal orders, work is done and labor is reported against the unit on the time card while stock issues are recorded on the stock issue card. As with the MWR, these costs accumulate against the unit not against the individual job. However, if the emergency repair turns out to be a major one in which cost must be determined, a MWO may be opened

11. The SWO is opened for an extended period, usually one year. It is linked with the time card and cost accumulated so that, for example, the cost of training or grass cutting month and year to date will be known.

12. Costs for unscheduled (MWR) and emergency repairs (verbal order or MWO's) are now shown on the cost detail report accumulated against units of equipment. Similarly, the cost of SWO functions are summarized against cost centers.

13. In the event that maintenance will perform work as part of an engineering project, the necessary MWO's are cross referenced with the EWO for the project.

14. If a contractor is to perform all or part of the project, the EWO is linked with a PO (purchase order).

15. The PO is the means by which the contractor may invoice the plant thus relating his charges to the PO and the EWO (project).

16. The system now summarizes the cost of all MWO's and PO's to the EWO.

17. This permits the project manager to observe project costs from all sources.

18. Costs are now tied with production statistics.

19. The maintenance cost per operating hour for each haulage truck or for the fleet, for example, can be observed and compared.

20. Finally, an overall information picture is produced as the result of these systems being linked together.

Work Orders versus Cost Control

When a work order number is assigned to a job, it establishes a temporary relationship with the equipment on which the work will be performed. It also establishes a relationship with details of the job such as: estimated hours by craft or estimated material cost. A work order is necessary only when it is necessary to isolate the cost and performance of a job. See Figure 35-9.

Figure 35-9. The assignment of a work order number to a job establishes a temporary relationship with the equipment and the job details such as estimated hours or cost.

As materials are charged to the job, material cost accumulates. See Figure 35-10.

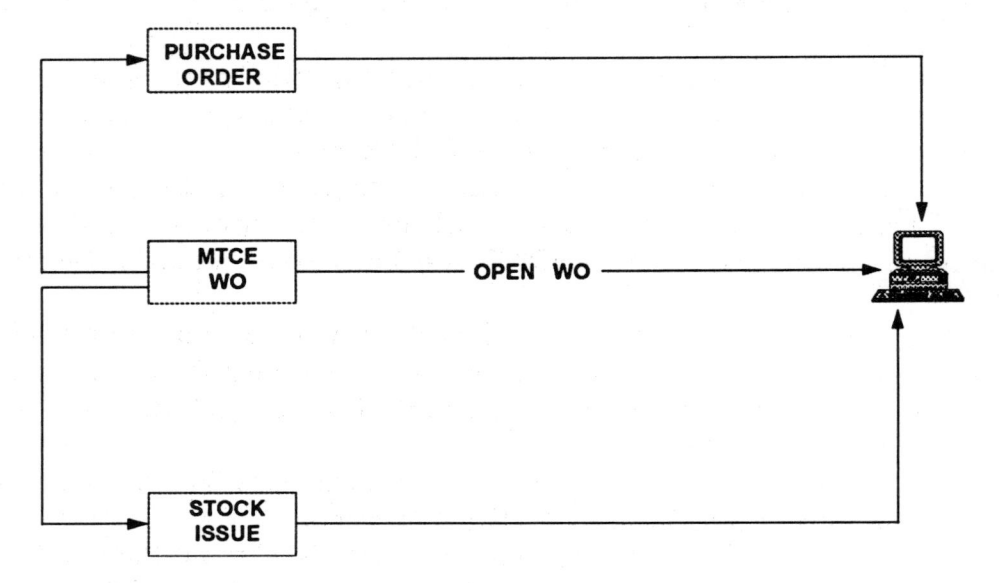

Figure 35-10. By placing the work order number on the purchase order or the stock issue card, material costs will accumulate against the job.

Once materials are assembled, the work is scheduled, assignments made and work completed, man-hours are charged to the job as men perform work. Thus, labor costs and man-hours accumulate against the job. It is then possible to compare labor cost and actual man-hours with estimates when the job is completed. After the job cost is compared with the estimate, the costs are closed against the equipment on which the work was performed. See Figure 35-11.

Work Order Systems

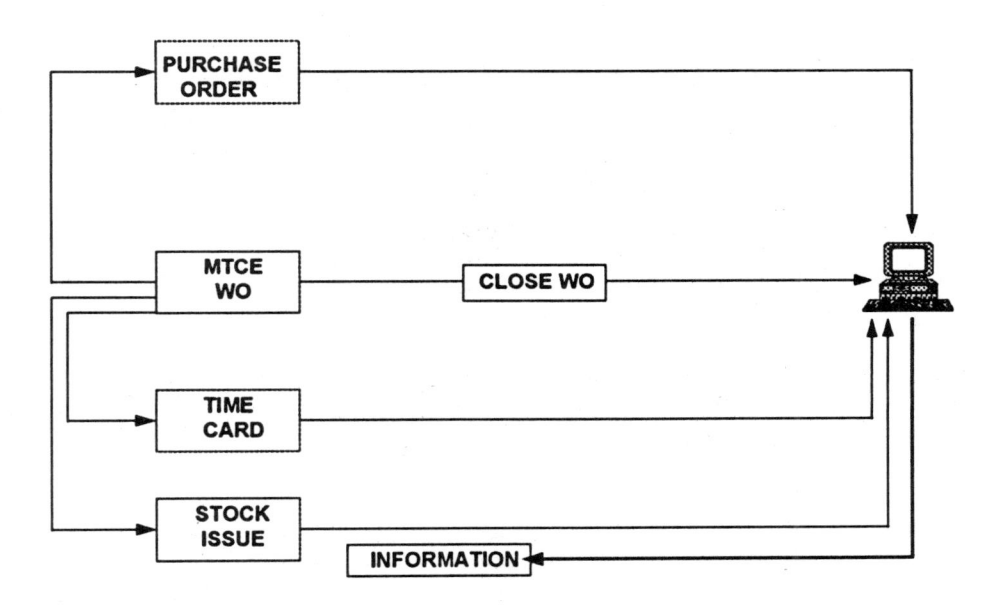

Figure 35-11. By placing the work order number on the time card, actual man-hours and their cost accumulate against the job. On closing, job costs are transferred to the equipment.

When labor and material data are reported against a unit of equipment, cost accumulates directly against the equipment and individual job costs cannot be determined. If job cost is required, a work order is necessary. See Figure 35-12.

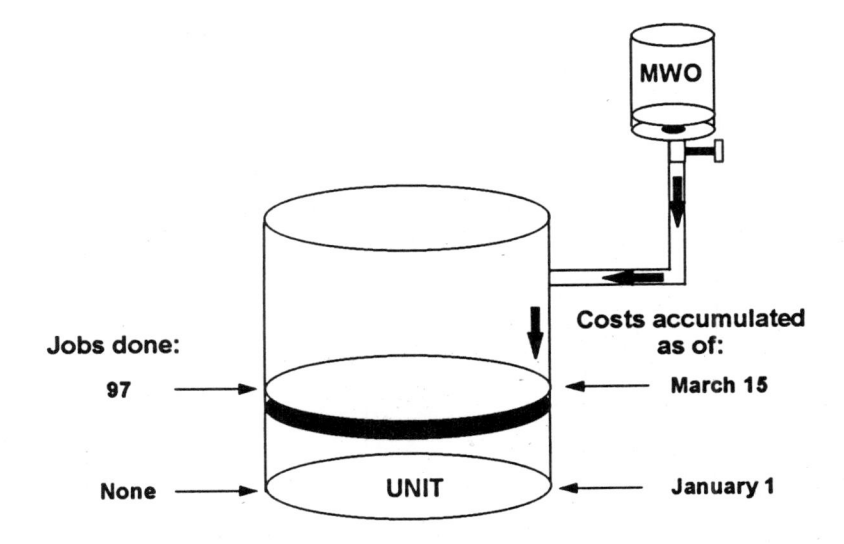

Figure 35-12. The MWO collects jobs costs much like a smaller tank (MWO) connected with a bigger tank (UNIT). When the MWO closes, the job cost closes to the unit where it accumulates with other unplanned job costs that were reported directly to the unit (no MWO).

Numbering Equipment and Maintenance Functions

A logical and consistent numbering scheme should be applied to fixed and mobile equipment as well as to functions. The numbering scheme should include:

- A department or cost center
- A type of equipment or . . .
- Function within the cost center
- Unit numbers for each piece of equipment
- Components as sub elements of units
- Sub divisions of functions

All of the equipment and function numbers are then brought together in a master equipment - function file sub divided by cost center. Equipment numbering examples follow. See Table of Illustrative Equipment Numbering below.

CC	TP	UNIT	CP	
06				Cost center 06, Grinding
	05			Conveyor
		011		Conveyor 011
			05	Belt
06	05	011	05	Belt, conveyor 011, CC 06

CC	TP	UNIT	CP	
06				Cost center 06, Grinding
	90			PM services, CC 06
		910		Lubrication
			03	Route 03
06	90	910	03	Route 03, Lub., CC06

CC	TP	UNIT	CP	
10				Haulage
	50			Loader
		020		Loader 020
			01	Engine
10	50	020	01	Engine, Ldr. 020, Haulage

Work Order Systems

Failure Coding

The failure code is used to identify the cause of failure for individual jobs. A typical failure code listing is illustrated below. See Failure Code Table below.

Failure Code

01	Fair, wear and tear
02	Lubrication
03	Blockage
04	Alignment or adjustment
05	Electrical
06	Hydraulic
07	Accident or fire
08	Mal operation
09	Parts failure

Over an extended period, the failure trend is analyzed for numerous jobs to determine whether a pattern of failures exists. The pattern, in turn, dictates a specific corrective action. For example, if improper equipment operation is a major portion of equipment failures, more or better operator training is indicated. See Figure 35-13.

```
01 - Fair wear and tear
02 - Lubrication
03 - Alignment
04 - Hydraulic
05 - Electrical
06 - Parts failure
07 - Accident, fire
08 - Operator error
09 - Misuse of equipment
```

PERCENT OF FAILURES

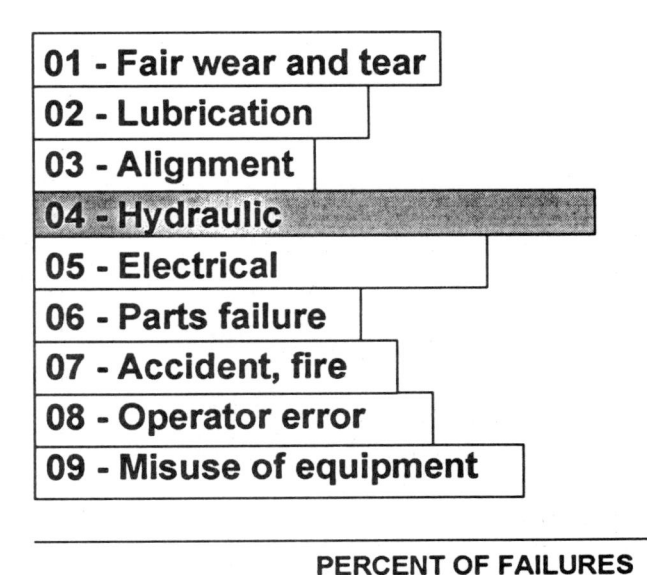

Figure 35-13. Hydraulic failures indicated the highest percent of failures when the fleet repair history failure trend was analyzed.

Work Description

Each description of work should be given a short title consisting of an action verb and an object. This provides an additional file search allowing the development of detail for standardizing jobs. For example, of numerous possible jobs able to be done on a wheel motor, a standard job description for RPL ARMATURE (Replace Armature) would allow a planner to isolate all of these jobs for a fleet. Next, he would isolate all work orders with this title for the wheel motor component and then have the computer produce a standard bill of materials for this job or a list of all man-hours by craft for every repetition of this job. As a result, he could construct standards for jobs very quickly. See Table below.

ADJ	Adjust	RBL	Rebuild
ALN	Align	RLC	Relocate
CLN	Clean	RPR	Repair
FAB	Fabricate	RPL	Replace
ISP	Inspect	RPN	Reposition
IST	Install	RMV	Remove
LUB	Lubricate	SPL	Splice
MAC	Machine	SVC	Service
MOD	Modify	TST	Test
OVH	Overhaul	WLD	Weld
PMI	PM Inspection	WNT	Winterize

The use of standard titles also speeds the visual search of repair history. About 22 verbs describe most maintenance actions.

Categories of Work Help Define Priority Schemes

Generally, there are five categories of maintenance work, including:

• Scheduled Maintenance - Extensive major repairs such as: rebuilds, overhauls or major component replacements requiring advanced planning, lead-time to assemble materials, scheduling of equipment shutdown to ensure availability of maintenance resources including: labor, materials, tools and shop facilities.
• Unscheduled Repairs - Unscheduled, non emergency work of short duration often called 'running repairs'. Work that can be accomplished within approximately one week with little danger of equipment deterioration in the interim. Repairs are usually performed by one person, often in two hours or less, and in about 40% of instances parts are not required.

- Preventive Maintenance - Any action which can avoid premature failure and extend the life of the equipment. It includes equipment inspection, testing and monitoring to avoid premature failures and lubrication, cleaning, adjusting and minor component replacement to extend equipment life.
- Emergency Repairs - Immediate repairs needed as a result of failure or stoppage of critical equipment during a scheduled operating period. Imminent danger to personnel and extensive further equipment damage as well as substantial production loss will result if equipment is not repaired immediately. Scheduled work must be interrupted and overtime, if needed, would be authorized in order to perform emergency repairs.
- Routine Maintenance (repetitive work) - Janitorial work, building and grounds work, training, safety meetings or shop clean up and highly repetitive work such as tool sharpening.

In addition, there are four types of non maintenance work including:

- Construction - The creation of a new facility or the changing of the configuration or capacity of a building, facility or utility.
- Installation - The installation of new equipment.
- Modification - The major changing of an existing unit of equipment or a facility from original design specifications.
- Relocation - Repositioning major equipment to perform the same function in a new location.

These categories of work are pertinent to effective priority setting. First, consider the objectives of a priority scheme. They are to:

- Identify the importance of a job compared with all others.
- Establish the time within which the job should be completed.

It makes little sense to assign a priority to an emergency job when such work should be done immediately (based on most definitions of emergency repairs). Priority setting is made more difficult by trying to include all types of jobs in the scheme. Rather, each type of work defines its own time frame.

- Complete emergency repairs within one day.
- Complete unscheduled repairs within one week.
- Complete PM services the same week as scheduled.

Thus, the definition of a priority (above), and the division of maintenance work into well defined categories can help solve the problem.

By using these time periods, all categories of work (except scheduled maintenance) are covered in the simplest, most appropriate way. Each planned and scheduled job would be given a priority rating prescribing its importance and the number of weeks within which it should be done. These weeks are designated as:

- Open week - The week the job is opened
- Start Week - The week the job should start
- End Week - The week by which the job should be completed

The time between the start week and the end week is called a 'window of opportunity'. See Figure 35-14.

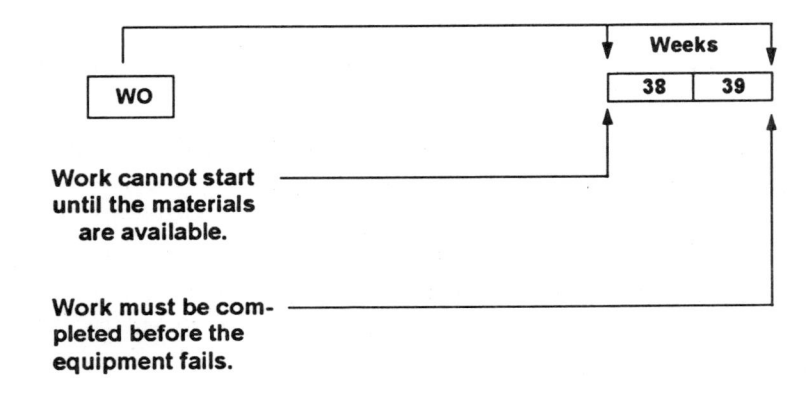

Figure 35-14. The job cannot begin until the materials are available but, must be competed before the equipment could fail if the work were not done.

When planning is completed, jobs are placed on a tentative schedule by priority, according to the recommended maintenance target dates (start week and end week) and presented at the weekly scheduling meeting with operations. Operations would review the recommended jobs and consider them according to the availability of equipment. Some jobs may have to be negotiated by weighing the risk of failure and resulting unscheduled downtime versus the contribution of the equipment to meeting production targets.

Of the jobs accepted by operations, labor would be allocated according to the priority assigned each job. Some jobs may have to be deferred because the amount of available labor for the scheduling period has already been allocated to jobs with higher priorities. By carefully following the priority setting scheme in awarding labor, operations soon understands that a correct, thoughtful application of realistic job priorities will be the best way to ensure that labor is awarded and their work is done on time.

Work Order Systems

Collectively, the jobs to which labor is allocated constitute the approved weekly schedule. For planned and scheduled maintenance, a typical joint production and maintenance priority rating scheme is shown below. The most effective priority setting schemes are those in which operations and maintenance jointly assign a priority rating. To ensure that there is common agreement, both parties should define the meaning of the ratings they would use in a specific circumstance. See Figure 35-15.

Production Rating	Maintenance Rating
10 One of a kind equipment	9 To ensure safe operation
8 Reduce production capacity	7 Overhaul
6 Reduce output	5 Replace major component
4 No effect on production	3 Rebuild component
2 Improve working conditions	1 Paint, clean up

Priority Number = Product of ratings

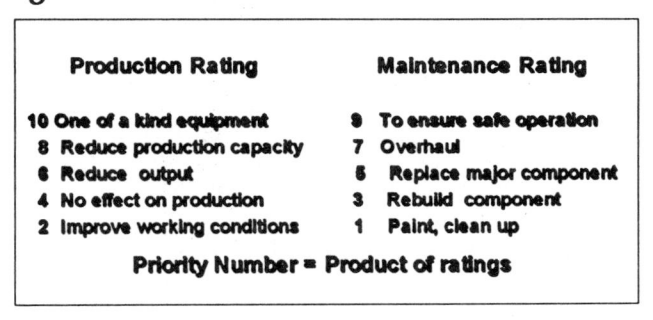

Figure 35-15. The product of the production rating and the maintenance rating is the priority number. Priority numbers range from a high of 90 to a low of 02.

Handing Work That is Not Planned and Scheduled - All work that does not require planning and scheduling is normally sent directly to the maintenance supervisor for assignment to crew members or, if a team organization is being used, for accomplishment by team members. Typically, this work includes:

- Preventive maintenance
- Emergency repairs
- Unscheduled repairs
- Routine maintenance (done at the supervisors discretion)

Applying the designated time frames, the supervisor would:

- Complete the PM services the week they are assigned.
- Complete the emergency repairs within one day.
- Completed the unscheduled repairs within one week.
- Perform the routine maintenance at his discretion.

The supervisor's crew strength and composition should be arranged so that he can accommodate PM services, do some planned work and keep up with unscheduled repairs. If emergency repairs occur, he must interrupt other work to take care of them.

It is also a good plan to tell operations that their declaration of an emergency condition automatically authorizes maintenance to use overtime if required. Thus, operations, who pays the bill, will counsel its personnel to avoid the frivolous assignment of an emergency work classification when it is not warranted.

Priorities Applied to Non Maintenance Work - Since non maintenance project work will compete for the same crafts used on major maintenance jobs, there must be a comparable priority setting arrangement for projects. See Figure 35-16.

Rating	Description
9	Projects to meet safety or regulatory needs
8	Primary utility upgrade
7	Plant equipment installation based on ROI*
6	Department equipment installation based on ROI*
5	Pollution control revisions or additions
4	Needs to improve morale
3	Plant image, employee or community related needs
2	Discretionary work

Figure 35-16. Square rating numbers to obtain the priority number. Note that the range of project priority numbers (high 81, low 04) fits within the range of maintenance priority numbers (high 90, low 02) so that, at the time of labor allocation, maintenance jobs and projects can compete fairly for available labor. * ROI - Return on investment.

Time to Completion

Since the priority number also represents the time within which the work should be accomplished, a suitable target period should be established for each priority range regardless of whether it is major maintenance or project work. See Figure 35-17.

Priority Range:	Open WO: within:	Start:	Complete:
90-70	1 week	1 week later	The following week
69-40	2 weeks	2 weeks later	Within 4 weeks
39-20	3 weeks	4 weeks later	Within 8 weeks
19-02	4 weeks	8 weeks later	Within 16 weeks

Figure 35-17. The highest priority jobs would be opened within one week and started the next week. Their target for completion is the week after.

Work Order Systems

One of the major problems created with poor priority setting scheme is the build up of work orders in the backlog going back for an extended period. In some extreme instances, jobs have been in the backlog log for many months. Generally, if a job is in the backlog more than 16 weeks or 4 months, there is about a 90 % chance that:

- The job has been accomplished piecemeal
- It is no longer required
- The priority has changed

Therefore, these jobs should be identified and reviewed with operations. Also, the backlog of all jobs should be purged after the first year of operation with any new system. This precludes future inaccuracies carried forward from poor estimates made while estimating techniques were still being learned.

Bringing the Work Order Elements Together

The focal point of all of the work order elements is the weekly schedule. After the weekly scheduling meeting is conducted with operations, all of the major jobs (MWO) and PM services requiring equipment shut down are approved. Then labor is allocated, material delivery verified and supporting equipment, transportation or rigging confirmed. The schedule is given to the maintenance supervisor who will perform the work. He scans the outstanding jobs (MWR) that are ready plus dynamic PM services. These jobs are added to the schedule to take advantage of the downtime created by the weekly schedule of major jobs. This rounds out the crews schedule for the week coming up. See Figure 35-18.

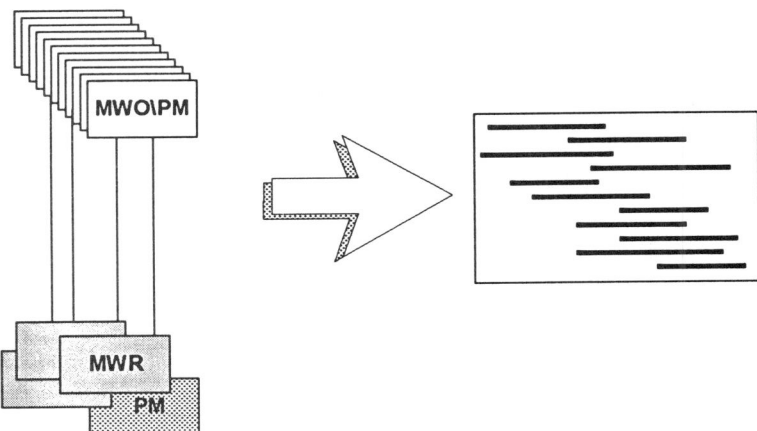

Figure 35-18. The weekly schedule brings together scheduled jobs and PM services to which the supervisor may add other jobs to take advantage of the downtime created by the schedule.

As the week unfolds, the supervisor prepares daily or shift schedules for his crew, gradually working through the jobs.

If emergencies occur, he pulls personnel from the unscheduled jobs (MWR) first, then the lowest priority scheduled jobs (MWO) to perform the emergency repairs. Each 24 hours there is a daily coordination meeting with operations to adjust the schedule in the event of operational delays.

Each day, using the information system, the supervisor, planner and other key personnel observe the status of jobs. They note the degree of completion, on time completion, costs etc.

Toward the end of the week, when the scheduling meeting for the next week is held, the planner or other key personnel would report on schedule compliance, performance gains or losses, backlog reduction etc.

Summary - As the communications network for the maintenance program, the work order system must provide for requesting, planning, scheduling, assigning, controlling and measuring work performed by maintenance. It must be able to handle all types of work with control measures that match the work being done. As the elements of the work order system link with the accounting system and production control statistics, they must be easy to use and able to focus field data and convert it into decision making information to successfully manage maintenance.

36

Utilizing The Information System Effectively

Decision Makers Information Requirements

Various maintenance organization levels have a requirement for information to permit them to make decisions related to their jobs. Similarly, the time frame within which they must apply their decision making dictates the nature of the information they require. See Figure 36-1.

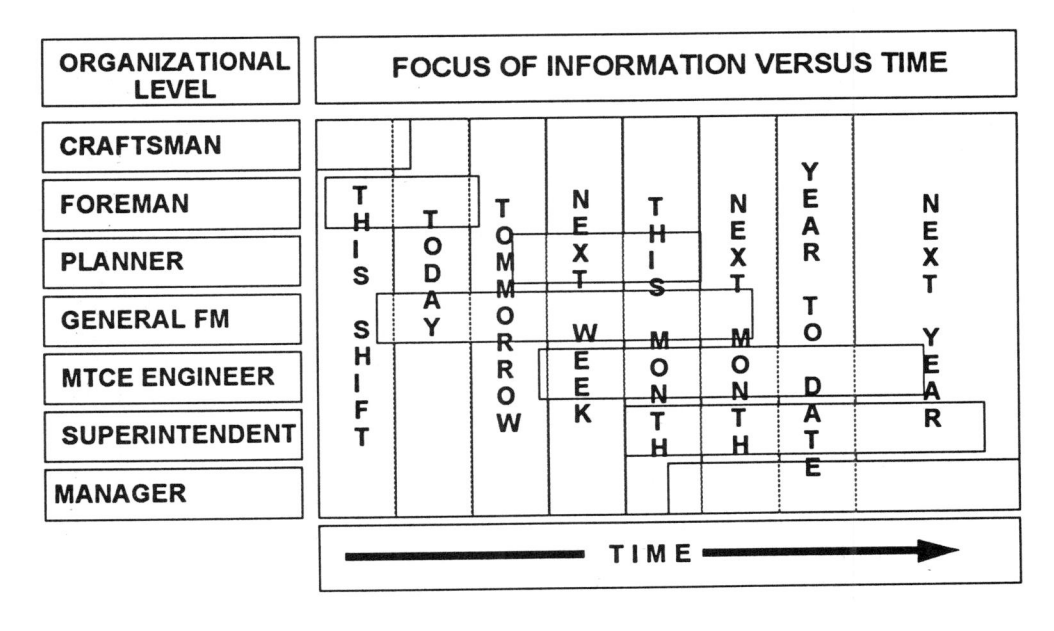

ORGANIZATIONAL LEVEL	FOCUS OF INFORMATION VERSUS TIME
CRAFTSMAN	
FOREMAN	THIS SHIFT / TODAY / TOMORROW / NEXT WEEK / THIS MONTH / NEXT MONTH / YEAR TO DATE / NEXT YEAR
PLANNER	
GENERAL FM	
MTCE ENGINEER	
SUPERINTENDENT	
MANAGER	
	TIME →

Figure 36-1. Each organization level makes decisions within a particular time period according to their responsibilities and the information they need.

Maintenance Manager - The maintenance manager is interested in the time period extending from the current month into next year. Thus, their most useful information will include summaries and projections. Typical:

- A monthly absentee summary for the maintenance department
- Compliance with the vacation policy last month
- Total maintenance costs year to date
- A forecast of major component replacements for six months

Maintenance Engineer - The maintenance engineer is interested in a time span from the current period into the next year as well as the past historical picture. Typically, this would include:

- The life span of engines of all Type 1 units in Fleet A
- Trend of lube failures on conveyor rollers in the past 6 months
- The specific repairs made on unit 16 in the past 6 months

General Supervisor - The general supervisor must be aware of the current situation but, because his supervisors are overseeing the current shift and tending to the week at hand, his information requirements are less intense during this period. Similarly, the planner is working on the week following with only a need to verify the timing of proposed major events. Thus, the general supervisor's focus is to ensure his personnel are prepared for major events in the near future. Typically, his information focus would include:

- The % completion of an overhaul as of the current week
- Compliance with the PM program last week
- PM services not completed in area B this week
- The % improvement in worker productivity in the past 6 months
- The % of MH able to be used on non maintenance work
- The current backlog of all crafts
- A report detailing the cost for each unit in a cost center

Planner - The planner focuses on the preparation of major jobs and obtaining the resources to carry them out. Typically, this information includes:

- The cost of replacing a specific major component
- The availability of a specific major assembly that is in stock
- The MH available, by craft, for planned work next week
- The delivery status of a major assembly required for an overhaul

Utilizing the Information System Effectively

Supervisor - The supervisor is concerned with the current shift primarily with continuing interest in getting ready for tomorrow. Therefore, his information interests span today and tomorrow. Generally, they include:

- The current cost and performance status of a major job
- A labor use summary showing the use of labor by type of work
- Mechanics awarded overtime in area B this week
- The dramatic rise in the electricians backlog since last quarter

Craftsman - Finally, individual craftsmen have information needs, as well. Their needs are more of a technical nature, including:

- A standard bill of materials for a specific, repetitive job
- A list of interchangeable bearings
- The utilization of specific spare parts

Information for Control of Preventive Maintenance

Regular, consistent equipment inspections, testing and monitoring assure that equipment deficiencies are uncovered before failure. Therefore, the first benefit of PM is reduced emergency repairs. Concurrently, as deficiencies are discovered earlier, there is more lead time before the repair must be made. Thus, the second benefit of PM is an ability to plan more work.

Using a Standing Work Order for PM - Since preventive maintenance services are routine and repetitive, the computer is ideal for scheduling them. A standing work order (SWO) is used to denote a PM service. Typically, the standing work order number would equate to the cost center in which the service will be carried out and the service to be performed. Each SWO would show the frequency, the craft performing the service and the man-hours required:

- Frequency - 2 weeks
- Craft - Mechanic
- Man-hours - 3 man-hours

Scheduling PM - Each PM service might be set up, for example, in a scheduling program by identifying the equipment to be serviced, the starting year week (9401 is the first week of 1994) and the frequency in weeks. If the starting week is 9401 and the frequency is one week, the computer would schedule the PM service each week starting with year week 9401.

Once all of the PM services have been loaded for fixed equipment, the user would identify the year week for which services are to be scheduled. The computer would respond with the services due. See Figure 36-2.

```
Sec     Unit    Comp    YrWk    Description

024     CM01    001     9424    PM Weekly inspection
024     CM02    001     9424    PM Weekly inspection
024     CM02    004     9424    PM Monthly service
025     CM04    004     9424    PM Monthly service
025     SC08    002     9424    PM Two week inspection
```

Figure 36-2. Five PM services are due during year week 9424 on three different units of equipment.

Corresponding checklists may be stored electronically and printed by crew members after the PM service assignment has been given to them.

Scheduling Mobile Equipment PM - PM services for mobile equipment are carried out at variable intervals: 250 hours, 500 hours etc. The service is specified for the unit of equipment and the current operating hours entered along with the desired operating hour interval. By updating the operating hours, the PM status of all units can be shown. For each unit, the current meter hours, the hours of the next service as well as the hours until the service and the total hours are displayed. A negative figure indicates the service is overdue. From this data, the units of equipment that are to be scheduled are selected. The user would schedule the units with the greatest number of negative hours first. After the PM service scheduling has been coordinated with operations, the supervisor would be notified of the services due so that assignments can be made to crew members. See Figure 36-3.

```
                    MOBILE PM - WEEKLY STATUS

Unit        Unit Name   Current   Next  Svc   Until   Total   Requester
                        Meter                 Svc     Hours

024CM01000  Miner 1     3725      3665  A     -60     3725    Johnson
024CM02000  Miner 2     4244      4305  A      61     4244    Johnson
024FD01000  Feeder 1    7621      7625  B       4     7621    Johnson
024CM04000  Miner 4     1243      1270  A      27     1243    Johnson
```

Figure 36-3. For each unit, the current meter hours, the hours of the next service, the service required as well as the hours until the service and the total hours are displayed. A negative figure indicates the service is overdue.

Utilizing the Information System Effectively

PM Compliance - At the end of each week the number of services completed should be measured and a compliance report sent to management. Services not completed should be rescheduled. Preventive maintenance services hold the promise of reducing emergency repairs and permitting maintenance time to plan work. The computer aids in achieving these objectives through effective scheduling and follow up on schedule compliance

Controlling Planned and Scheduled Work

Control of planned and scheduled work is helped by using the computer to support planning and scheduling. It is an effective way to handle the volume of jobs as well as the details of their status, performance and cost. Planning, scheduling and job execution are carried out in a cycle. See Figure 36-4.

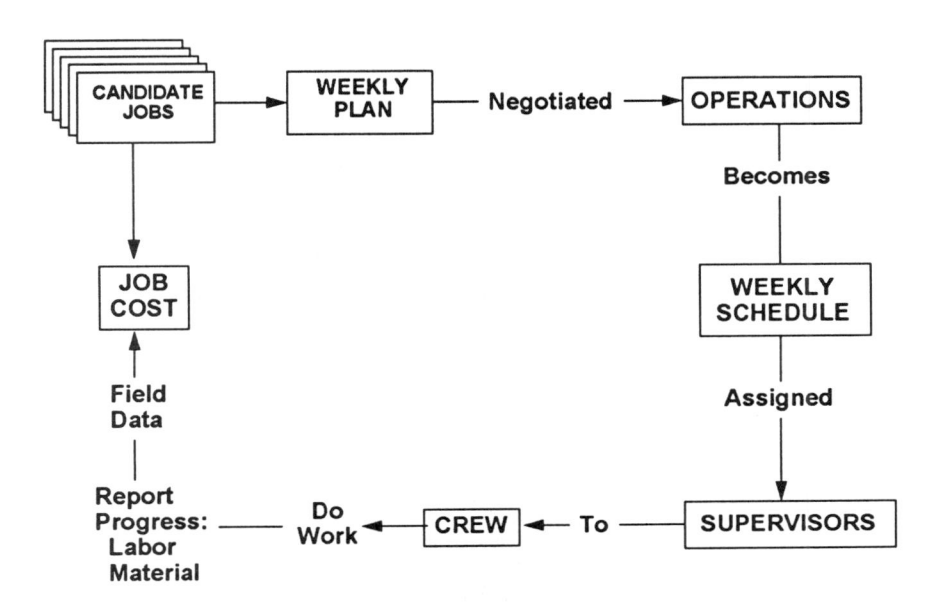

Figure 36-4. A group of jobs forms a weekly plan for negotiation with operations. Upon its approval, the plan becomes a schedule and is assigned to supervisors to carry out. As crew members do the work, they report progress on labor and material use. This is the basis of information on job cost and performance.

The Preliminary Schedule - The maintenance planner starts by organizing a group of candidate planned jobs into a plan for a time period such as a week or a shutdown. Maintenance supervision then reviews the plan prior to negotiating it with operations to obtain their concurrence in shutting down equipment so the work can be accomplished. As necessary, the importance of the jobs to maintenance is negotiated with operations against the possible loss of the equipment productive capacity during the execution of the proposed plan.

The Approved Schedule - Once agreement is reached, the plan is approved and converted into a schedule. In turn, work is assigned by supervisors to their crews who do the work and report progress. The information system converts reported field data, like man-hours used or materials installed, into information such as job cost and performance. At subsequent scheduling meetings, schedule compliance is reported. While much of the work to be planned will be derived from the PM equipment inspections program for the first year, subsequently, these inspections will help confirm the timing of actions like the periodic replacement of major components. Cost reports and repair history will also identify troublesome equipment some of which require planned work.

Control of Periodic Maintenance - Over a period of time, major equipment components exhibit deterioration characteristics dictating a predictable life-span and creating a general pattern of possible replacement. Through equipment condition monitoring (inspection and testing), the status of deterioration is checked to help determine whether the component should be replaced and how soon. Data on component life span is recorded in repair history and analyzed for patterns by component by equipment type. As life span patterns emerge, projections are made of the approximate timing of the replacement of components unit by unit. Actual timing of component replacement is checked against the latest PM inspection results. This overall technique is called forecasting.

The user identifies the equipment and the component to be replaced then specifies the year week when the activity is to start and the frequency of the action. Similarly, intervals can be variable such as operating hours or tons rather than fixed weekly intervals.

As weeks pass or operating hours accumulate, forecasted actions are identified as being due for a particular future week. These actions then become candidate jobs for a future weekly schedule after the actual status of the component is verified via the latest PM inspection.

Accompanying each forecasted action are standards specifying how the work is to be done (task list), a standard bill of materials (to expedite ordering) and a tool list to reserve special tools required.

This documentation is similar to the check list that accompanies PM services. Both the check list and the standards are stored electronically within the computer program and displayed or printed as needed.

Job Status

The most important information for the control of planned and scheduled maintenance is the Work Order Status Report. This report allows the user to determine the status of a work order, from inception to completion and to summarize cost and performance on its completion. Once planning has been initiated, the planner records data to include, estimated:

- Man-hours by craft
- Labor cost
- Material cost

Opening the Work Order - The planner then opens the work order establishing a temporary link between the work order number, the job, the equipment on which the work will be carried out and job details such as estimated manhours or material costs. See Figure 36-5.

```
           MAINTENANCE WORK ORDER STATUS REPORT

MWO#      Sec   Unit   Comp   Ct   Unit Name          Comp Name
426820    024   CM01   100    S    Continuous Miner   Cutting

Description of Work: RPL CUTTER HEAD MOTOR                +
St    Opn    Stt    Cls    Pr   Wa   FC   RH   SN/In    SN/Out
S     9410   9412          90

Work Order Notes                                    EST      ACT
                                        TR1         12
See Material List                       TR2
                                        TR3
                                        LABOR       300
                                        MATERIAL    25600
                                        TOTAL       25900

Requested:         Approved:         Completed:
B. Easton          J. Southworth
04/03/94           04/16/94
```

Figure 36-5. The work order shows the detail at the time the job is opened in the maintenance information system and a link is made with the accounting system.

Work Order Cost Accrual - As planning continues, the planner uses the purchase order to obtain direct charge materials and the stock issue card to obtain stock parts for the job. By placing the work order number on each of these documents, the cost of the materials is debited against the job.

Similarly, the time card compiles labor costs from man-hours reported as craftsmen perform the job. Once the availability of materials is confirmed, the planner places the MWO on the schedule, obtains approval of the schedule at the weekly scheduling meeting and sends the approved schedule to the supervisors. Each MWO can be seen on the computer by any user. Work has now begun and we see the status of the MWO in the next stage. See Figure 36-6.

```
                MAINTENANCE WORK ORDER STATUS REPORT

    MWO#     Sec    Unit   Comp    Ct   Unit Name          Comp Name
    426820   024    CM01   100     S    Continuous Miner   Cutting

    Description of Work: RPL CUTTER HEAD MOTOR                    +
    St    Opn    Stt    Cls    Pr   Wa   FC   RH   SN/In     SN/Out
    S     9410   9412          90   No             128875    122889

    Work Order Notes                                     EST      ACT
                                              TR1        12       8
    See Material List                         TR2
                                              TR3
                                              LABOR      300      200
                                              MATERIAL   25600    21000
                                              TOTAL      25900    21200

    Requested:          Approved:         Completed:
    B. Easton           J. Southworth
    04/03/94            04/16/94
```

Figure 36-6. The work order in progress has now accumulated both material and labor costs as well as actual man-hours. All of these are compared with the planner's estimates.

Completing the Work Order - As the work is scheduled and performed, each craftsman places the work order number on his time card opposite the number of hours that he worked on the job. This step accumulates the actual man-hours by craft as well as the cost of labor used. The employees number indicates his craft and hourly rate of pay. In turn, the information system summarizes these costs and related data, such as actual man-hours by craft, and debits them against the job. As the job is completed, the supervisor adds details such as:

- The serial number of the component installed (SN/In)
- The serial number of the component removed (SN/Out)
- The failure code (FC)
- Whether the job satisfies a warranty (WA)
- If the job is to be placed in repair history (RH)

When he signals the planner that the work is completed, the work order may be closed. Then, cost and performance are summarized. See Figure 36-7.

```
              MAINTENANCE WORK ORDER STATUS REPORT

    MWO#     Sec   Unit  Comp    Ct  Unit Name          Comp Name
    426820   024   CM01  100     S   Continuous Miner   Cutting

    Description of Work: RPL CUTTER HEAD MOTOR              +
    St   Opn    Stt   Cls   Pr   Wa   FC  RH   SN/In       SN/Out
    S    9410   9412        90   No                128875  122889

    Work Order Notes                                  EST      ACT
                                         TR1         12
    See Material List                    TR2
                                         TR3
                                         LABOR       300      300
                                         MATERIAL    25600    25720
                                         TOTAL       25900    26020

    Requested:        Approved:        Completed:
    B. Easton         J. Southworth    T. Washburn
    04/03/94          04/16/94         04/18/94
```

Figure 36-7. The work order has now been closed. The cost and performance details are summarized and compared with the planners original estimates.

Closing out the Work Order - Since the work order (MWO) has now isolated the cost and performance of a major planned and scheduled job, the closing performs two functions:

- The costs of labor and material are closed against the unit of equipment.
- The job may be recorded in repair history to establish a pattern of repairs to help confirm problems and determine corrective actions.

Work Order Administration

The computer also permits a planner to administer between 75 - 100 work orders in various stages of planning. Some are ready to schedule, while others are awaiting materials. Others may be overdue for a variety of reasons. A status is assigned to each work order, and it allows the planner to focus on those work orders in a specific stage in the planning sequence. Some, for example, are awaiting materials and clearly not ready for scheduling, while others are totally planned awaiting only the allocation of labor.

Assigning Work Status - The various status's include:

- Waiting for Materials
- Waiting for Labor
- Waiting for Shutdown
- Waiting for Completion of Shop Work
- Ready
- Scheduled
- Deferred

Planning Performance

Generally, it is not advisable to spent too much time comparing the estimate and actual cost, man-hours or timing on single completed work orders. Don't try to hold the planner responsible, the poorer ones will change estimates to match the actual and claim the scope of the job changed.

Measuring Scheduling Compliance - A periodic spot check is much better. Specifically, examine each weekly schedule and try to set goals for better overall performance. For example:

- For the first 6 months set a goal of estimated versus actual hours or cost for all work orders on the schedule of $\pm$ 12%.
- Establish a target schedule compliance of 60% (6 of 10 jobs scheduled are completed).
- Next, set a higher goal of $\pm$ 10% and 80 % compliance.

These goals are realistic and they challenge the efforts of everyone involved, not just the planner. The computer makes it possible for a planner to plan and schedule a large number of jobs in various stages. It also provides a means of scheduling forecasted major jobs and determining the status of each job.

Handling Emergency and Unscheduled Work

In an effective maintenance program, emergency repairs should not exceed 8 - 10% of manpower used, while unscheduled repairs should be no more than 15 - 20%. The information system helps to limit these jobs indirectly by scheduling PM services, which:

- Limit emergencies by finding problems before the equipment fails and . . .
- Uncover unscheduled repairs before they develop into more serious jobs

Utilizing the Information System Effectively

Controlling Emergency Repairs

Verbal Orders - The use of verbal orders to initiate emergency repairs is a common practice that few maintenance departments have been able to change. In fact, the very nature of the work suggests that maintenance must act first and document later. The best initial solution lies in the development of guidelines:

- If the emergency repair results in a job that warrants cost tracking, a work order should be opened 'after the fact'. Typically, the work order is selected from a block of numbers set aside for this purpose. Thus, personnel doing the work merely record the number and get on with the task. Subsequently, the work order can be opened.
- If the job is of inconsequential cost but, it meets repair history criteria, direct entry into repair history is appropriate.
- If substantial materials are required but very little time used (a $ 500 circuit board requiring 10 minutes of labor), materials should be charged to the unit and a repair history entry made. Labor use might not be reported.

Work Order versus Equipment Number - Generally, work order systems that rigidly require a work order for every job have difficulty dealing with emergency repairs and the verbal orders that often trigger them. In some extreme (and unrealistic) circumstances there are even efforts to outlaw verbal orders.

The most practical solution is to design the information system from the outset so that it will accept field data either against a work order number (to isolate job cost and performance) or against the equipment number (which accumulates data against the unit). In the latter case, there must be a system feature that allows repair history to be collected without opening and closing a work order. Generally, the most effective way is to use the time card or work report to convey repair history data into the data base. Alternately, the system might provide a means for direct entry of field data into repair history.

Handling Unscheduled Repairs

If an information system can deal successfully with emergency repairs, it can deal with unscheduled repairs as well. However, too many unscheduled repairs are simply done with no record whatever. Often, they are a source of valuable information for repair history which, if not recorded, is lost. The main problem centers around the method by which these jobs are recorded.

Managing Unscheduled Jobs - While many unscheduled repairs are the result of verbal orders, unlike emergency repairs, they are seldom done immediately. Over 90% of requests for running repairs (which often do not require action for a week or more) are made directly to supervisors or crew members.

To avoid forgetting the request, many supervisors assign the work immediately and, as a result, detract manpower from other work that is more urgent and important. Too often this happens because there is no easy way to record the jobs and assign them later. Thus, the information system must contain a means whereby either the supervisor or his crew members can record jobs that do not require immediate action.

A simple entry screen which allows a supervisor or any crew member to enter jobs directly into the system solves the problem of how to handle the volume of running repairs that are often lost or mishandled. Using the same procedure, operations personnel may enter requests directly as well. Features such as table files save time and ensure correct equipment numbers are recorded.

Subsequently, these jobs must be able to be identified by equipment, craft, status and when they should get done. Unscheduled jobs are filed by equipment and further sorted by craft, status (waiting for shutdown or materials), proper timing etc. This arrangement allows the supervisor to screen these jobs and combine them with other work as he makes assignments to crew members. Similarly, a crew member might select several jobs that match equipment he will be working on, providing the supervisor has assigned a status signaling these jobs are ready to be done. In addition, operations personnel can also research the status of jobs requested previously.

Flexible Work Order Systems - The handling of emergency and unscheduled repairs can be difficult for inflexibly designed work order systems. However, because they represent the greatest volume of work done, it is prudent to determine that the program under development, or the package under consideration will handle this work competently.

Controlling Labor

Maintenance controls cost primarily through improving their control of labor. In turn, this improvement impacts on the plant's profitability and its ability to compete and survive.

Factors Influencing Cost - Other factors, such as improper equipment operation, affect the cost of maintenance but, maintenance has little control over them. However, maintenance has considerable influence over the control of their own labor and information contributes significantly to the quality of that control. The control of labor is a key factor in the ability of maintenance to manage costs.

Craftsman Best Source of Job Information - Good information on the control of labor is derived from correct field data. The best source of data regarding work performed is the person performing the work, the craftsman. Questionable, less successful methods of obtaining field labor data should be avoided. Typical, a supervisor of a 10 man crew, with each man performing 6 - 8 jobs during the shift, produces guesswork when he tries to report labor. This yields erroneous data which, in turn, produces incorrect information.

Ideally, individual maintenance employees should report what they do. Maintenance supervisors should verify the data. A well designed maintenance labor reporting document should include the:

- Employee number
- Date and shift
- Equipment worked on
- Work order performed
- Description of work done
- Man-hours of regular and overtime used
- Coding of man-hours (temporary pay grade)
- Coding of absentee data
- Coding of the reason for overtime
- Employees signature
- Supervisors verification

Methods of Labor Reporting - Currently, most systems provide for on screen labor reporting and subsequent verification by supervisors before labor data is processed. This is the employees bill for his work. It is also the primary data source for labor control information for both maintenance and accounting.

Ease of Reporting - The design of the maintenance program has impact on the ease of labor reporting. Typical is the avoidance of a work order for every job. If, for example, labor can be reported against either a work order or a unit of equipment, several advantages accrue:

- Verbal orders (essential in some industries) can be used successfully.
- Repair history is available without a work order.
- Time is not wasted opening and closing work orders.
- Craftsmen account for time more accurately.

Thus, the concurrent development of field labor reporting procedures along with system development is advantageous.

Labor Information Objectives

To successfully control the use of labor, information must satisfy five objectives:

- Determine work force size and composition
- Measure the utilization of Labor
- Assess overtime use
- Control absenteeism
- Improve productivity

Manpower Needs - Each type of work done by maintenance has a different manpower requirement. Therefore, the work load is best measured by first defining the different types of work to be performed. See Appendix B - The Maintenance Work load.

In turn, the definitions will help to prescribe the type of work and the amount of manpower needed to carry it out. Typically, the work load includes:

- Preventive maintenance
- Emergency repairs
- Unscheduled repairs
- Planned and scheduled maintenance
- Routine activities (like training)

PM Workload - Preliminary estimates are then made of craft man-hours required to carry out each element of the work load. Some portions of the workload can be computed. For example, a PM inspection done every 2 weeks requiring 3 mechanic man-hours per repetition would require 78 mechanic man-hours annually. Similarly, estimated craft man-hours for all PM services would yield a total craft manpower estimate for the entire PM work load.

Utilizing the Information System Effectively

Scheduled Maintenance Workload - Past planned and scheduled maintenance work provides a picture of historical manpower use by craft. These can be assembled to provide estimates of current needs. To illustrate:

- An overhaul due in September requires 210 mechanic man-hours.
- A drive motor replacement due this month requires 22 electrical man-hours.

Routine activities yield other data. For example:

- Each electrical craftsman will undergo 65 hours of training annually.
- One full time utility man is required for tool repair.

Manpower Needs - When this data is assembled, it provides a starting point for estimating the number of personnel that will be required. See Figure 38-8,

```
Cost Center 06                    Crafts
Type of Work      MW     ME     EL     WL     LB    Total

PM                       1260   1080                2340
Scheduled         3060   2160    900   2880   2700  11700
Unscheduled       1620   1620    360    540    540   4680
Routine                  1260   1080                2340
Emergency          720    900    180    180    360   2340

Total MH          5400   7200   3600   3600   3600  23400
```

Figure 36-8. By assembling estimated craft man-hours by type of work, an overview of needs can be assembled. (MW=millwright, ME=mechanic, EL=electrician, WL=Welder, LB=Laborer)

These man-hour requirements are then converted into an equivalent number of craft personnel by dividing each by 1800 MH/man/yr. See Figure 36-9.

```
Cost Center 06                    Crafts
Type of Work      MW     ME     EL     WL     LB    Total

PM                       .7     .6                   1.3
PLANNED           1.7    1.2    .5     1.6    1.5    6.5
UNSCHEDULED        .9     .9    .2      .3     .3    2.6
ROUTINE                  .7     .6                   1.3
EMERGENCY          .4     .5    .1      .1     .2    1.3

Total Men         3.0    4.0   2.0     2.0    2.0   13.0
```

Figure 36-9. An estimated total of 13 men would be required to perform maintenance in cost center 06 with a craft distribution as shown.

The result is a preliminary crew of 13 made up of:

```
3 Millwrights
4 Mechanics
2 Electricians
2 Welders
2 Laborers ... and a supervisor.
```

Having established the preliminary crew of 13 made up of 5 different crafts, the labor utilization data is used to confirm craft man-hours worked against types of work.

Distribution of Manpower by Work Category - A typical distribution is:

Preventive Maintenance	10%
Scheduled maintenance	50%
Unscheduled repairs	20%
Routine activities	10%
Emergency repairs	10%

Using labor utilization information, the actual distribution of craft man-hours is compared with the target levels (above). For example:

Type of Work	Target	Actual	±
PM	10	3	- 7
Scheduled	50	27	- 23
Unscheduled	20	40	+ 20
Routine	10		
Emergency	10	20	+ 10

As more data are gathered and performance comes closer to the target distribution, the work load can be confirmed. Typical:

Type of Work	Target	Actual	±
PM	10	7	- 3
Scheduled	50	45	- 5
Unscheduled	20		
Routine	10		
Emergency	10	18	+ 8

If a traditional maintenance organization with a supervisor and crew is used, the number of supervisors required must be determined so that control of the crew members activities can be assured.

Utilizing the Information System Effectively

Number of Supervisors Required - To establish the number of supervisors required, determine the shifts requiring supervisory coverage.

```
Areas   Annual Shift Coverage    Shifts per year

  3     52 weeks X 5 shifts per week    780
  4     52 weeks X 7 shifts per week   1456

Total Shifts Coverage Required         2236
```

Then, determine the shift coverage capability of per supervisor.

```
5 shifts per week X (52 weeks - 7 weeks of vacation, holi-
days and compensatory time) = 225 shifts coverage per super-
visor per year.
```

Next, divide the total shift coverage required by the shift coverage capability of a single supervisor.

```
2236 shifts/225 shifts per year per supervisor = 9.9 or 10
supervisors.  Since there are only 7 areas, there should be
7 full time supervisors and 3 relief supervisors.
```

Establishing Supervisory Coverage - Using this data as a guide, the maintenance organization must then decide whether some shifts actually require a maintenance supervisor. For example, a plant with a small crew on each afternoon and night shift might place the crews under the operational control of the operations shift supervisor. However, if this is done, the operations supervisor must be aware of the work the crew has been assigned and under what circumstances he will interrupt their assigned tasks (emergencies for example). Similarly, crew members must understand this working relationship.

Labor Utilization

The effective use of labor by category of work is an excellent barometer of the quality of labor control. See Figure 36-10.

When adequate man-hours are spent on PM, the result is a:

- Reduction in manpower used on emergency and unscheduled repairs
- An increase in manpower used on planned and scheduled maintenance

Planned work is better organized and work is performed more productively.

MAINTENANCE LABOR UTILIZATION REPORT
WEEK 40 ENDING SEPT 30

DEPARTMENT 207

CRAFT	PREV MTCE	SCHD MTCE	UNSC RPRS	EMER RPRS	NON-MTCE	TOTL WKFC
MILLWRIGHT	55	276	102	88	67	588
MECHANIC	46	244	96	66	42	494
ELECTRICIAN	23	122	54	32	46	277
WELDER		72	58	21	66	217
INSTRUMENT	18	32	30	12	45	137
PIPEFITTER		48	12	9	12	81
LABORER		121	6	5	41	173
TOTAL MH	142	915	358	233	319	1967
% DISTRIBUTION	7	47	18	12	16	100

Figure 36-10. Man-hours reported against categories of work yield a picture of the effectiveness of labor utilization. In the illustration, 7% of manpower spent on preventive maintenance has resulted in 47% of manpower being used on scheduled maintenance while holding manpower spent on unscheduled and emergency repairs to 18 and 12% respectively.

When used in with labor utilization data, the backlog provides a means to adjust the size and composition of the work force. See Figures 36-13, 14.

Controlling Overtime

Overtime use should be classified by reason for use to determine how effectively it is being used and controlled. Concurrently, information on overtime confirms whether established overtime approval policies are complied with and overtime is used effectively. See Figure 36-11.

MAINTENANCE OVERTIME SUMMARY
WEEK: 40 ENDING: 30 SEPT

OCC CODE	DESCRIPTION	1 CALL-IN	2 EMERGENCY	3 SCHEDULED	4 UNAUTHORIZED	5 CONTRACT	TOTAL
01	RIGGER					5.0	5.0
03	DIESEL RPRMAN	10.0					10.0
04	WELDER/MECHANIC			20.0			20.0
05	MECHANIC			35.5			35.5
10	FITTER						3.0
17	MACHINIST		6.0				6.0
20	ELECTRICIAN		2.0				2.0
25	WELDER				12.0		12.0
37	BOILER MAKER		1.0				1.0
	TOTAL OVERTIME MH	98.0	46.0	80.0	14.5	18.0	265.5
	% DISTRIBUTION	38.2%	17.9%	31.1%	5.7%	7.1%	100.0

Figure 36-11. A weekly report of overtime use by craft and reason helps to determine how effectively overtime is being used and controlled.

Utilizing the Information System Effectively

Controlling Absenteeism

Poorly controlled absenteeism can be a major detriment to the effective control of labor. Therefore, the same level of control applied to the use of labor must be applied to absenteeism. Typically, an effective absentee report shows the use of overtime by reason and craft for a weekly period. See Figure 36-12.

```
                    MAINTENANCE ABSENTEE SUMMARY
                    WEEK: 40  ENDING:  SEPT 30

                PER  UNN  VAC  JRY  LVV  INJ  NAT  BRV  LTE  DSP  OTH  TOT
DESCRIPTION     BUS  BUS  ION  DTY  ABS  URY  GRD  MNT  RPT  ACT  ER   AL

MILLWRIGHT       12   6   80         8    2                             108
MECHANIC          4   2   40    2    8              8                    64
WELDER            1   3   40              8                   2       1  55
MACHINIST         4                                           4           8
ELECTRICIAN       5   8   80         4    1                   2    2  1 103
INSTRUMENT            1                        40             1    1  1  44
PIPEFITTER        1   4   40         3    4                   2       1  55

TOTAL MH         27  24  280    2   23   15   40    8        11    3  4 437
% DISTRIBUTION    6   5   64    1    5    3    9    2         3    1  1 100
```

Figure 36-12. The absentee report allows the reader to assess the effectiveness with which absenteeism is being controlled.

Improving Productivity

Productivity measures the quality of the control of labor. Although actual measurements are most commonly made using random work sampling techniques, the information system should contain indices that allow maintenance to observe macro trends in productivity. Typically macro productivity trends include:

- Labor cost to install each dollar of material
- Man-hours to achieve each utilized operating hour
- Man-hours of PM for each man-hour of planned work

Labor control is vital to effective maintenance because it is the primary way that maintenance can control the cost of the work it performs. It starts with finding out how many personnel are needed and then ensuring they are used effectively. Effective labor use is the result of getting PM done, planning and quality supervision. The ingredient that binds all of these actions together is quality information.

Using the Backlog Effectively

The backlog is an effective management tool for the control of labor providing it is understood and correctly applied. The backlog is the number of estimated man-hours of work required to perform all identified, but incomplete, planned and scheduled work. It is an index of the degree to which maintenance is keeping up with the generation of work as well as a means of determining the proper size and composition of the workforce.

Interpreting the Backlog - Generally, a rising backlog means that more work is being generated than the maintenance workforce can do. If the rise is long term, more personnel may have to be added. If the rise is shallow and uneven, the use of overtime may be adequate to control it. When the backlog is expressed in estimated man-hours by craft a rising backlog indicates too few personnel; an unchanged backlog indicates there are enough personnel and a falling backlog indicates an excess of personnel. See Figure 36-13.

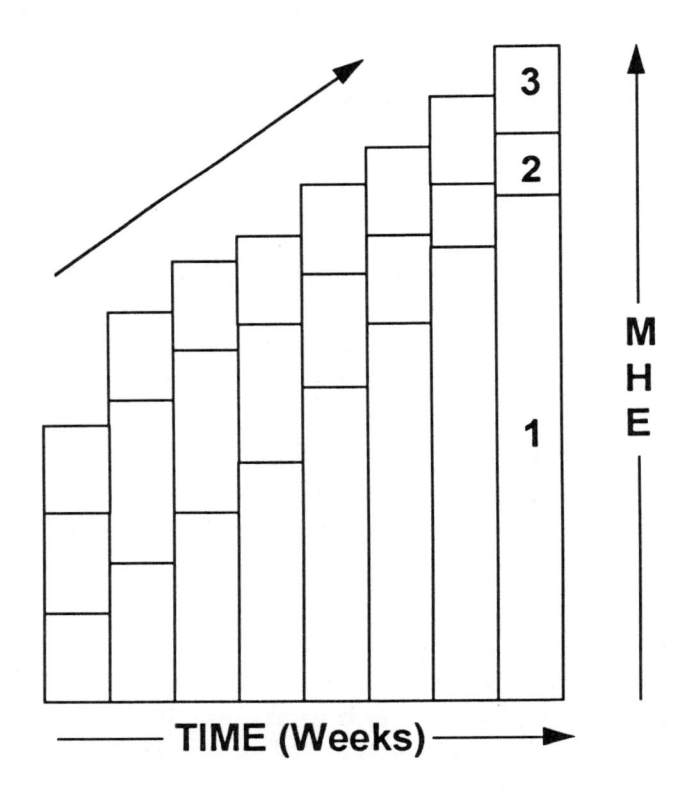

Figure 36-13. Backlog data graphed over an 8 - week period reveals an overall increase. However, backlog data suggests a need to add craft 1, reduce craft 2 and no changes in craft 3.

Utilizing the Information System Effectively

Before work force changes are made as a result of backlog data, other possibilities should be considered to reverse trends. Consider:

- Using temporary contractor help
- Authorizing overtime
- Deferring work to reduce the backlog

If none of these steps help then, personnel adjustments should be made.

Backlog Content - The backlog includes only planned and scheduled work and estimated man-hours by craft. It excludes:

- Emergency work because it must be done immediately
- PM services whose man-hour requirements are the same week after week
- Unscheduled repairs not requiring estimated man-hours

Backlog Computation - The backlog exists for individual planned work orders. When a work order is opened, its estimated man-hours are added to the backlog. As actual man-hours are reported against the work order, the backlog is reduced. When the work order is closed any remaining estimated man-hours are canceled. The backlog of work is like a tank of water which fills as the work order is opened, empties partially as the work is done and empties when the work order is closed. Typically, backlog information is presented in a format much like the Backlog Summary Report for a one week period. This permits the user to see the changes from the start of the week (STWK) through the end of the week (NDWK) for all crafts. Priorities are shown to indicate the backlog distribution by importance of work. See Figure 36-14.

```
                    BACKLOG SUMMARY REPORT
                    Week: 40  Ending: 30 Sept
                      MAINTENANCE BACKLOG

      PRIORITY (90-70)  PRIORITY (69-50)  PRIORITY (49-25)  PRIORITY (24-02)   ALL PRIORITY
      ST   +    -   ND   ST   +    -   ND   ST   +    -   ND   ST   +    -   ND   ST    ND    ±
CRAFT WK   WK   WK  WK   WK   WK   WK  WK   WK   WK   WK  WK   WK   WK   WK  WK   BL    BL

MECH  148  13   40 121  271   47   28 290  110   27   12 125   27   12    6  33  556   571  +15
ELEC   96  14   28  82  181   29   66 145  107   31   16 122   42   17   12  47  427   396  -31
WELD   57  10   13  54  201   47   28 220   62   46   21  87   16   21    2  35  336   396  +60
RIGR  128  46   51 123  142   96   57 181   91   28   14 105   81   46    1 126  442   535  +93
FITR  287  49  121 215  108   40   21 127   47   17   12  52   42   12    2  52  484   446  -38
HLPR  147  28   62 113  122   47   21 148   28   30   10  48   17    5    3  19  314   328  +14

TOTL 1247 128  218 1157 1582 317  417 1482  562 107  172 497  210   98  105 203 3601 3339 -262
```

Figure 36-14. During week 40, the overall backlog dropped by 262 hours while the backlog for riggers climbed by 93 hours and the electrical backlog dropped by 31 hours.

Backlog versus Priority - Effective priority setting is necessary for meaningful backlog data and, the priority scheme should meet specific criteria:

- Priorities should be set jointly by maintenance and operations.
- Each rating should have a definite meaning.
- Work of a certain priority should be completed within a specific period.
- The resulting priority numbers should be used to establish the order in which work will be done and specify how labor will be allocated.

Proper Backlog Level - To help determine the proper backlog level, the number of weeks to complete work of a particular priority range would be established using random sampling. Assume that data were:

Priority range	Work completed in:
90-70	2 weeks
69-50	4 weeks
49-30	8 weeks
29-02	16 weeks

This means that a work order of a specific priority range should be able to be planned, opened, scheduled and completed within this these periods.

With the backlog, there can be no generalizations such as, 'six weeks of backlog is acceptable.' Such a statement raises the question, 'for which craft?' The proper level of backlog is the amount of planned and scheduled work a single maintenance craft can keep under control while effectively carrying out the remainder of its work load. It must be controlled with the amount of labor available after PM, unscheduled and emergency repairs have been performed.
1.

The correct backlog level can be established by isolating backlog data for each craft against the time period for completing each priority level of work.

Utilizing the Information System Effectively 301

Observe Figure 36-15 as you follow the steps listed below.

Construct a rectangular grid using the priority range and the established target weeks of completion as parameters:

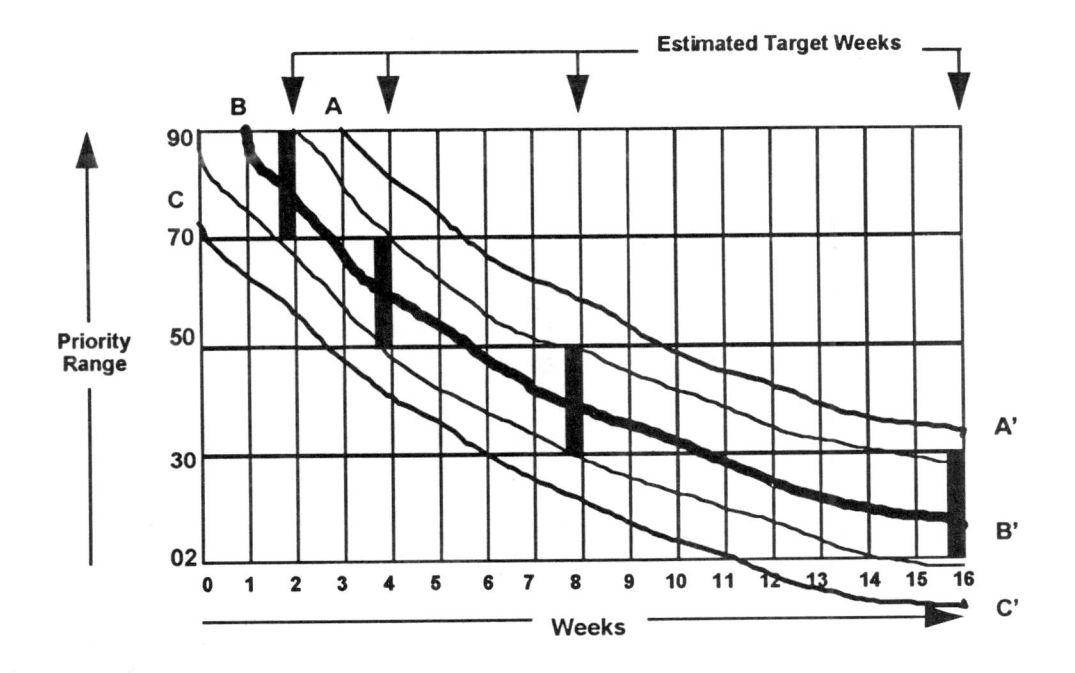

Figure 36-15. Verify the sustainable backlog level with statistical sampling.

- Step 1 - Construct a band (light lines) showing the desirable range of time within which work should be completed for the craft group in question.
- Step 2 - Use a random sampling of about 100 completed work orders to identify work orders either containing this craft or done exclusively by this craft.
- Step 3 - Plot, by priority, the opening week and the closing week of each work order. Plot the opening week at zero (0) and the closing week on the scale of 2 to 16 weeks. The result will be a medium line as with A-A' or C-C' or a heavy line as B-B'.
- Step 4 - If the line is as A-A', the backlog exceeds the capability of the craft to carry it out.
- Step 5 - If the line is as B-B', the backlog is within the capability of the craft to carry it out.
- Step 6 - If the line is as C-C', the craft capability exceeds the backlog requirements.

After this determination has been made, review the backlog report data and determine the average 'end of week' backlog for the craft in question over a period of from 4 to 6 weeks. To illustrate, assume the graph drawn for a specific craft group fell in the B-B' range. This would mean that the average backlog for the craft, as shown on the backlog reports, is approximately correct for the group.

Net changes (±) in the estimated hours can be converted into the approximate number of personnel (by craft) required to control the backlog. Assume that the data on 6 weeks of backlog revealed estimated man-hours as shown below.

Week	Backlog MH (estimated)
1	284
2	310
3	201
4	205
5	220
6	190
Average	235

To determine the weeks of backlog divide the average backlog (235) for this craft by 40 (MH/man/week). This yields 5.9 or 6 weeks of backlog.

Overdue Work Orders - Many organizations struggle with work orders that seem to remain in the backlog forever. At fault usually, is a poor or ignored priority setting scheme. When a work order has not been completed within 16 weeks:

- The work may have been done already, piecemeal.
- The work was given too high a priority.
- The equipment may no longer exist.

If you discover that the work is no longer necessary:

- Stock materials set aside for the job should be returned to stock.
- Determine whether purchased materials or fabricated and machined parts can be used on another job.

Many of the work orders not completed within a reasonable period may not have had a high enough priority to warrant labor allocation. When this happens, consideration must be given to the disposition of materials for these jobs.

Utilizing the Information System Effectively

Disposition of Unused Materials - Consider:

- If there were no materials, advise the requester to resubmit with a higher priority if the work is still required.
- If there are materials, the requester should be advised that the materials are on hand and the work could be done if the priority were higher. The work order should not be closed since the materials charged to it constitute a cost accrual.

There should be an annual review of all open work orders with originators to cancel work or change priorities. Also, with any new planning group, the backlog should be reduced to zero at the end of the first year since their labor estimating skills are probably poor, and the erroneous backlog data is compounded by carrying it into a second year.

Mathematics of the Backlog

Three relationships exist between the estimated and the actual man-hours as the work order is completed:

- Actual man-hours > estimated man-hours
- Estimated man-hours > actual man-hours
- Actual = estimated man-hours

The backlog computation rules, on closing the work order are:

- Actual man-hours > estimated, the backlog = 0
- Estimated man-hours > actual hours, the remaining estimated hours are canceled
- There is no negative backlog

Computing the Backlog - Consider Figure 36-16 (next page). It illustrates a 9 week period (weeks 10 - 18) and shows 9 work orders (A - I) that are opened, worked on and closed. Three crafts are identified: mechanics, millwrights and electricians, each with an estimated number of man-hours. The single heavy line indicates the planning duration beginning with the week that the estimated man-hours for the work order are entered into the backlog. The double heavy line indicates the weeks the work is being performed and shows the actual man-hours used. An asterisk (*) indicates the week that the work order was closed and shows the backlog man-hours remaining. You are to compute the backlog, week by week, starting with the backlog data shown in the lower chart

304   Industrial Maintenance Management

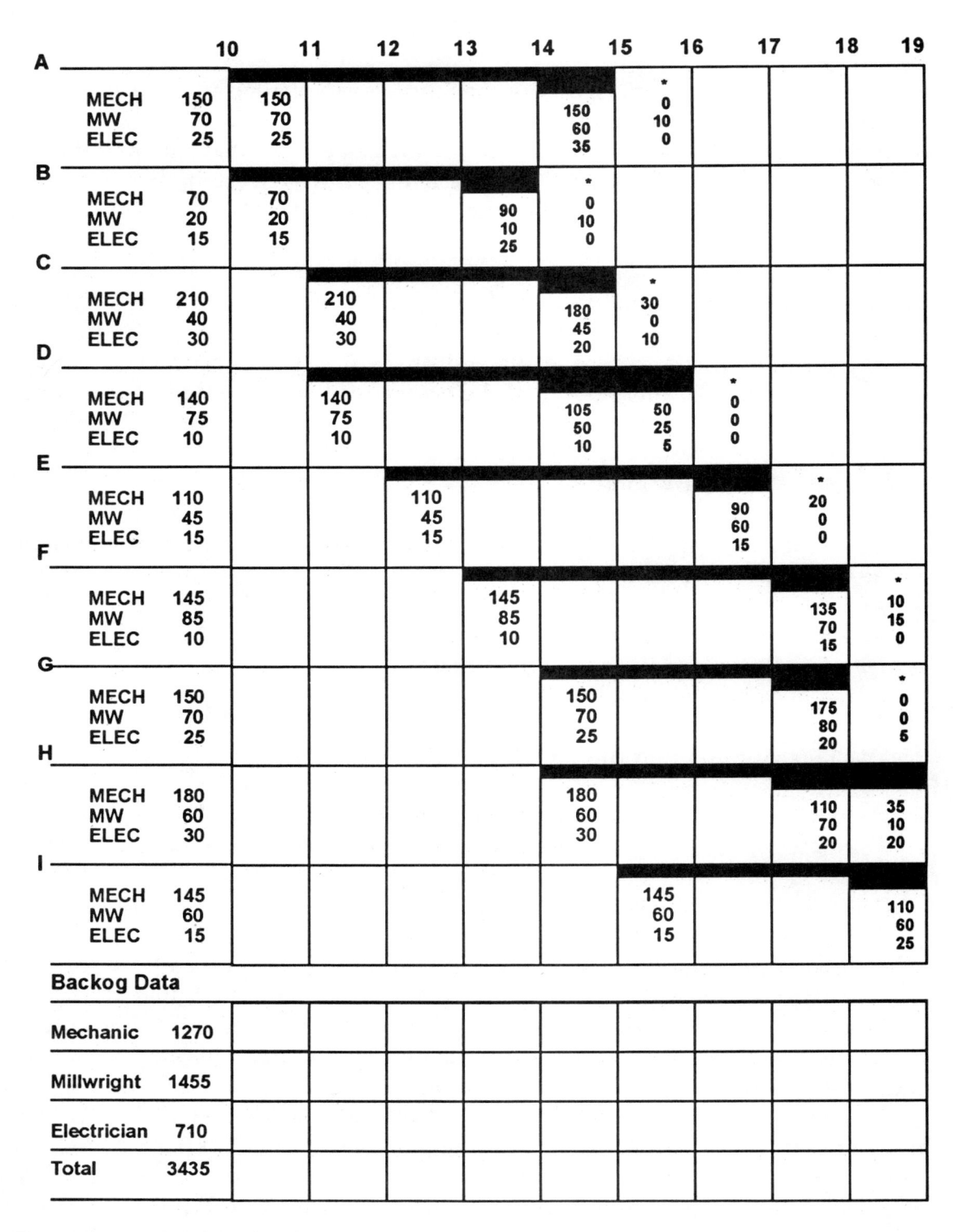

			10	11	12	13	14	15	16	17	18	19
A	MECH MW ELEC	150 70 25	150 70 25				150 60 35	* 0 10 0				
B	MECH MW ELEC	70 20 15	70 20 15			90 10 25		* 0 10 0				
C	MECH MW ELEC	210 40 30		210 40 30			180 45 20	* 30 0 10				
D	MECH MW ELEC	140 75 10		140 75 10			105 50 10	50 25 5	* 0 0 0			
E	MECH MW ELEC	110 45 15			110 45 15				90 60 15	* 20 0 0		
F	MECH MW ELEC	145 85 10				145 85 10				135 70 15	* 10 15 0	
G	MECH MW ELEC	150 70 25					150 70 25			175 80 20	* 0 0 5	
H	MECH MW ELEC	180 60 30					180 60 30			110 70 20	35 10 20	
I	MECH MW ELEC	145 60 15						145 60 15			110 60 25	

Backog Data

Mechanic	1270										
Millwright	1455										
Electrician	710										
Total	3435										

Figure 36-16. This computation considers most of the possible backlog situations.

Utilizing the Information System Effectively

Backlog information helps to determine whether maintenance is keeping up with the generation of new work, and it provides a means of adjusting the size and composition of the work force as workloads change.

Correct answers for computations made are shown on the page 322, the last page of this chapter.

Controlling Materials

Maintenance work may be delayed by a temporary lack of labor but, without materials the work may not be able to take place at all. Therefore, the control of materials is as important to maintenance as is the control of labor. In many mines, material control is not a maintenance function. Rather, maintenance, as the biggest user of materials, is more often a customer. Thus, just as maintenance is a service to production, material control is a service to maintenance. Maintenance must acknowledge this by ensuring that effective material control service is high on their list of essentials. Therefore, the maintenance and material control systems must function in close harmony. The material control system performs three functions:

- Inventory control
- Purchasing
- Spare parts manufacture and component rebuilding

Inventory Control Information - These functions are mutually supporting so that if parts are not in stock, they may be purchased or manufactured. The inventory control system provides information on:

- Stocked parts and substitutes
- Stocked parts related to types of units and components
- Quantities on hand and Quantities set aside for identified work
- Number on order and Reorder points and order quantities
- Unit costs
- Re supply sources
- Costs of issued parts
- Turnover rate
- Number of issue days on hand for critical parts
- Reconciliation of records and physical counts
- Stockage cost
- Value of stock by code and user area
- The status of actions for re stockage

Stock Information - Maintenance helps determine what will be stocked and the initial quantities. Thereafter, the inventory control system adjusts these quantities based on their use. Maintenance supervisors, planners and craftsmen require three levels of stock material information:

- Craftsmen must be able to identify parts required for routine repairs.
- Supervisors must be able to determine materials required for unscheduled or emergency jobs They would advise planners on materials required for planned and scheduled jobs
- Planners must be able to determine materials needed for major planned jobs

Parts Identification- The identification of parts is a problem for maintenance. The average supervisor is proficient in identification but, many craftsmen are not. As a result, supervisors often obtain parts while crew members wait, unproductively. Current systems can help craftsmen solve this problem. See Figures 36-17 and 18.

```
   (INQUIRY)            Equipment ID - 702-21-750           94/06/01
                           List of Components
   ----------------------------------------------------------------
   001      Electric motor              017     Idler shafts
   005      Reduction units             018     Drive belts
   008      Sprockets (roller chains)   019     Transfer belts
```

Figure 36-17. The user inquires by equipment ID, selects the component and enters the data in the equipment specification program.

```
   (RESPONSE)   702-21-750   # 21 Coal Conveyor             94/06/02
                       011   Electric motors
   ----------------------------------------------------------------
   MOTOR ITEM # (          )
   MAKE:  CBE       MODEL:  9F3266N           HORSEPOWER         50
   RPM:   1760      FRAME:  254U        VOLTAGE:  440   AMPS:  13.5
   OTPT SHAFT DIA:  1.375"  DE BEARING #:   40BC0J
   ODE BEARING #:   358C0J
   ----------------------------------------------------------------
```

Figure 36-18. The equipment specification program responds, for the component selected, with details of the component. Similarly, it could respond with part numbers.

Utilizing the Information System Effectively

Equipment Specification Program - An equipment specification program allows a craftsman (or other user) to specify equipment type and component, and obtain a listing of corresponding stocked parts or component details. Some equipment specification programs permit a visual presentation of components as well. The craftsman views an 'exploded view' of a component assembly allowing him to select the correct part and review its assembly as well. He can print the diagram and take it to the work site if needed. See Figure 36-19.

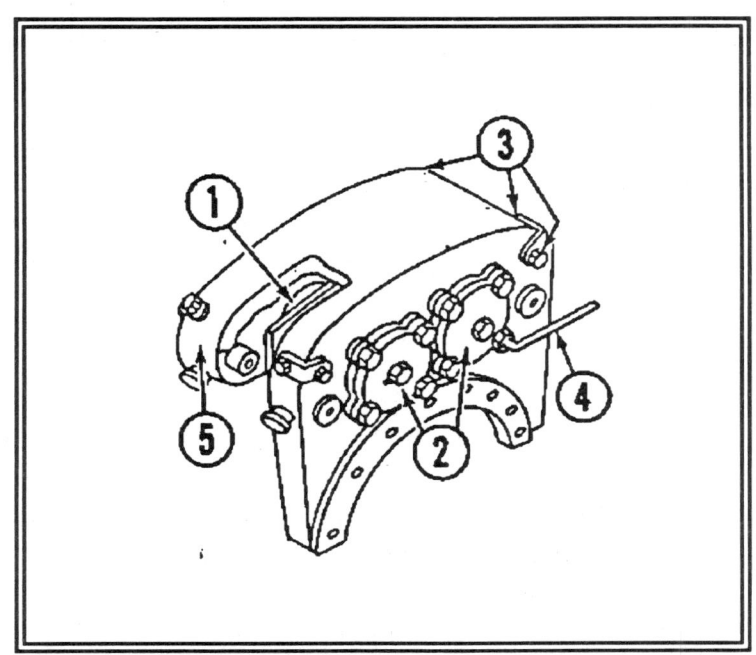

Figure 36-19. Graphics speed the identification of materials and simplify the process. Thus, crafts personnel and team members can identify materials more readily.

Purchasing Information

Few mines want to incur the cost of stocking materials if they can purchase them as needed. Therefore, the cost of stock materials use versus purchased materials has fallen significantly. Generally, the advent of better information has made a difference. For example, the need for major assemblies can be better anticipated. Thus, orders can be placed to suppliers with assurance that on time delivery will result. Inventory control systems of several mines can communicate. Thereby, stocked materials not available at one site can be 'borrowed' from another. While maintenance should avoid the role of the 'professional' purchasing agent, they often have a requirement to identify materials that must be purchased. If there is not time to plan the anticipated work, maintenance supervisors may have to make inquiries directly of suppliers.

However, only the purchasing agent can commit the funds. Thereafter, maintenance should have access to purchasing information allowing them to obtain the status of any purchase order. This is especially true of the planner, who handles most purchased material needs.

Parts Manufacturing

When parts or fabrication are required for planned jobs, maintenance field planners coordinate with shop planners to get the work done. However, when stocked parts are to be manufactured, it is the warehouse supervisor or the purchasing agent who works with the shop planner. Therefore, a close working relationship and compatible systems are desirable. Generally, parts being manufactured for stockage require a determination of minimum quantities because the 'set up' costs can be significant. Similarly, maintenance should not misuse the capabilities of their shops to manufacture a part for $ 600 that could have been purchased for $ 12.

Commercial Shops

Commercially manufactured stocked parts, are the domain of the purchasing agent acting on needs expressed by the warehouse. However, when major components are rebuilt by a commercial shop, there is a need for maintenance involvement. Generally, components to be rebuilt are delivered to a classification point by maintenance. Once classified, components are sent to the authorized vendors shop. These transactions are controlled with a purchase order. Vendors deliver the rebuilt components to the warehouse where they are placed back in stock. Once issued, maintenance should be able to track component performance by serial number in repair history.

On site Parts Delivery

The ability of the warehouse to deliver stock or purchased materials to the work site contributes to maintenance productivity and performance and it should be encouraged.

Standard Bills of Materials

A link exists between the maintenance work order, stock issued and materials purchased for specific planned jobs. If any of these jobs are repeated, a standard bill of materials can be developed. Then, upon determining that the job will be due again, the same materials may be ordered in advance.

Merging Systems

There are a number of possible combinations of maintenance and material control systems. Among the combinations:

- The development or purchase of a fully integrated system
- The linkage of existing maintenance and material control systems
- The development or purchase of a maintenance system and its linkage with an existing material control system and vice versa

It is important to weigh the economics of these choices. For example:

- One organization found that the cost of writing communications software to link their existing maintenance and material control systems was only 15% of the cost to develop an integrated system, and 20% of the cost to purchase a system.
- Another organization decided to defer electronic handling of material transactions between systems because both departments needed more time to overcome 'cultural' problems. In the interim, less expensive linkages became available.
- A third organization purchased only the maintenance portion of an integrated 'package' program because their existing material control program was functional and effective. The cost of the communications software reduced the overall cost of integration over the package by 45%.

After the maintenance and material control departments have established a solid working relationship and solved any 'cultural' problems, the stage is set for the mutually beneficial use of their systems. Thereafter, a careful assessment of needs must be made including a determination of how the systems must function together. Concurrently, the economics of the options as well as the capabilities of personnel to use them effectively must be weighed.

Controlling Cost and Performance

The ultimate objective of good maintenance performance is to ensure plant profitability. In turn, achievement of this objective depends on quality information shared between management, operations and maintenance. Generally:

- Management needs trends and summaries to assess the overall effectiveness of maintenance.

- Operations needs assurance that their operating funds are well spent and quality work is being done.
- Maintenance needs to identify work, then control the work and measure how well they have accomplished it.

Performance Indices - Performance indices and trends are the principal means by which management can assess performance. Among them:

- The units of product per maintenance dollar reveal the overall contribution of maintenance the plant productive effort.
- The maintenance cost per operating hour assesses how well maintenance contributes to the reduction of downtime.
- The labor cost to install each dollar of material reveals how effectively maintenance uses its manpower to perform work.

While none of these indices are the result of a formal, structured report, the data from which they are developed are. This linkage allows managers to observe the trends the indices point out and, with confidence, ask maintenance to identify and deliver corrective actions when the indices reveal unfavorable trends.

Specific Maintenance Performance - In addition to trends, other management indices should identify specific maintenance performance. Typical:

- Compliance with the preventive maintenance schedule
- Compliance with the weekly schedule of major planned and scheduled work
- More PM = reduced emergency repairs ?

These indices reveal whether the maintenance program is being carrying out effectively and provide management focus for directing corrective actions. For example, a plant manager was concerned that compliance with the PM schedule was declining. On attending the weekly schedule meeting, he learned that poor compliance was attributable to operations not making equipment available on time.

Maintenance Cost - Management is also interested in actual maintenance cost versus budgeted costs on a current month and year to date basis. This assessment reveals the overall quality of maintenance management. While trends will reveal a stable, improving or declining performance, they do not do not pin point problems.

Performance Trends - An unsatisfactory performance revealed by a declining trend must be followed by a detailed maintenance investigation to identify the problems and make corrections. Thus, management uses indices to require explanation by maintenance and corrective actions. As corrective actions are taken, management continues to observe the indices and notes the impact of the corrective actions on future trends.

Performance Information for Operations

A primary interest of production will be the cost of the maintenance service it receives. However, they will not want to be overcome with details. A cost summary showing actual versus budgeted costs for each cost center is an excellent starting point. This information will alert them to troublesome areas and open the way for suggesting solutions. Once the troublesome areas are identified and discussion has begun, confirmation of problems is in order.

Sharing Information

Maintenance can confirm which units and components are at fault for high costs. In addition, the nature of what happened can be revealed by repair history. However, little will happen unless joint action results from use of the information. For instance:

- If operator error results in excessive damage to equipment, high maintenance costs will not be reduced until training produces more competent operators. Thus, operations should scan the failure summary contained in repair history.
- Similarly, if operations fails to make equipment available for scheduled servicing or repairs, they must be held accountable for excessive maintenance costs. Therefore, it is in the interest of operations to note PM compliance and, if poor compliance is their fault, act to improve it.

Downtime

Downtime is non productive time for equipment. It is expensive and unrecoverable. Its causes must be uncovered and reduced. While inadequate maintenance is a contributing factor, operations contributes as well. Too often, causes of downtime are incorrectly attributed to maintenance because operations is usually the 'score keeper'. Therefore, downtime performance data should be identified and implemented jointly by maintenance and operations.

Generally, downtime is attributable to maintenance when:

- Maintenance has been notified and is still responding. This measures the effectiveness of maintenance communications.
- Maintenance is performing the work to correct the problem. This measures the efficiency with which maintenance gets its work done.
- Equipment is down while materials are being procured. This measures the quality of material control.

Downtime is attributable to operations when:

- Equipment operators are absent or late.
- There are production delays
- Equipment is not put back into service promptly on completion of repairs.

Questionable areas must be resolved face to face, such as, operations ignored a PM service and the equipment failed.

Refinements

Once the maintenance program is in place, refinements relating to the way in which the maintenance services are carried out will be important. For example:

- Compliance with the PM service schedule indicates concern in assuring reduction of emergency repairs and downtime. Management, operations and maintenance should all observe it.
- Compliance with the weekly schedule indicates interest in the effectiveness with which maintenance has accomplished major jobs. Again, all three parties should observe it.

Interest in Quality - A sure sign of operations interest in the maintenance program is their demand for quality work. There are a number of ways it can be achieved. For instance, at a remote mining operation, the 'pit' operation did minimal equipment servicing and were dependent on a distant garage for all major servicing and repairs. Since equipment operators were on an incentive program, they became increasingly critical (often vocally) of the quality of work being performed by the garage. Supervisors worked long and hard to solve the problem with little progress. Finally, operators volunteered to travel to the garage and road test each unit before it was transported back to the pit. The first few tests were confrontational.

Utilizing the Information System Effectively

But, the operators and the mechanics started to communicate and the complaints soon disappeared. Management marveled at the result of a little bit of direct quality control.

Similarly, a face to face quality effort usually brings good results. Typical:

- In a concentrator, operations recommended that maintenance craftsmen have operators inspect completed repairs. The results, measured in quality work, were dramatic. For the first time, craftsmen got the satisfaction of being told directly that they had done a good job.
- In a large smelter, maintenance supervisors asked production supervisors to test run equipment before the maintenance crew left the work site. Maintenance crews took this opportunity to point out that equipment clean up before work began could help them. The result was better overall cooperation and better clean up, illustrating that quality control can work both ways.

Performance Information Helpful to Maintenance

Maintenance is ultimately judged on its ability to minimize downtime, reduce cost and produce quality work. To do this they must identify work, control the work and measure how well they accomplished it. These needs span both the work order and information systems. Thus, to identify work quickly and accurately, maintenance must:

- Have an easy to access work requesting system
- Be able to respond to verbal requests
- Have a strong detection oriented PM program
- Convince operations to report problems promptly
- Have Continuous information on the status of jobs
- Obtain timely cost information
- Get up to date information on the use of labor

And, to measure the effectiveness of work accomplishment, maintenance must have:

- Knowledge of job cost and performance
- Information on worker productivity
- The ability to compare actual and estimated job costs
- Access to actual versus budgeted costs for maintenance

Thus, the success of any maintenance information system is the result of effective conceptualization, development, implementation and information utilization by those who manage the plant, those who receive maintenance services (operations) and by maintenance, as they perform the service. The successful improvement of maintenance cost and performance relates directly to the availability of information and its effective use by management, operations and maintenance. Therefore, the design of the maintenance information system must acknowledge these information needs and system implementation must deliver the needed information. It follows that education on information use must be built into the utilization process.

Maintenance Engineering

A primary objective of maintenance engineering is to ensure the maintainability and reliability of equipment, buildings and facilities. Maintenance engineering is necessitated by the increasing complexity of production equipment and the resulting demand for the utilization of modern technology to monitor equipment performance and conduct repairs.

Improving Equipment Performance - Recommendations for improving equipment performance are based on quality information. Some is direct information like repair history and some indirect like field investigation. Repair history data, for example, indicates a possible failure pattern while field investigation reveals the exact cause.

Maintenance engineering also performs a variety of advisory services including:

- Development of the PM program
- Recommending non destructive testing techniques
- Development of standards for major jobs
- Helping planners establish the scope of planned jobs
- Recommending standard repair techniques

In addition, maintenance engineering monitors new equipment installations to ensure that they are operable before being placed in service. Thereafter, they ensure the equipment is maintainable through testing and they verify that drawings, instructions, wiring diagrams, spare parts lists etc., are available. Typical:

As one contractor started to withdraw his crew from an equipment installation job, the maintenance engineer pointed out that the location of a retaining wall made it impossible for maintenance to inspect and lubricate the equipment.

Utilizing the Information System Effectively

Developing Standards - Maintenance engineering also develops standards based on field investigation and engineering measurements. Once in use, the standards are verified by information on labor, material and tools used on the corresponding work orders. For major repetitive jobs, similar information can be used to help to develop historical standards. Labor standards can be developed by observing the number of actual man-hours by craft required to complete a specific repetitive job. Similarly, a standard bill of materials can be developed from historical material use data. With this information, maintenance engineering can confirm their engineered standards and develop new historical standards for other jobs.

Problem Solving - Information helps the maintenance engineer to narrow problems and identify possible solutions. Field investigation then reveals the exact cause and yields recommendations to correct the problems. Performance and equipment reliability are major areas of concern for the maintenance engineer. Of special interest is the performance of components. Typical, manufacturing plant was using 6 vendors to rebuild 44 different components. When they experienced a surge of component failures they could not associate the failed components with any one vendor. Maintenance engineering developed serial number tracking within repair history resulting in the identification of a vendor who had only replaced seals but not worn internal parts. This caused the plant to establish standards for component rebuilding which, in turn, reduced failures. Knowledge on the life span of components helps to forecast when components should be scheduled for future replacement. See Figure 36-20.

REPAIR HISTORY

FC	MWO#	Sec	Unit	Comp	Tr	MH	YrWk	Description of Work	SN (In)	SN (Out)	OpHrs
1	64236	024	CM02	400	EL	8	9243	RPL TRAM MTR OFF SIDE (JOY)	112123	312413	1102
1	65135	024	CM02	400	EL	6	9301	RPL TRAM MTR OFF SIDE (JOY)	223335	112123	1234
1	65223	024	CM02	400	EL	8	9305	RPL TRAM MTR OFF SIDE (JOY)	124498	223335	1286
1	65465	024	CM02	400	EL	5	9329	RPL TRAM MTR OFF SIDE (TRAM)	344456	124456	1589
1	65553	024	CM02	400	EL	5	9408	RPL TRAM MTR OFF SIDE (TRAM)	129032	344456	1703
1	65664	024	CM02	400	EL	6	9403	RPL TRAM MTR OPP SIDE (JOY)	590234	129032	3307
1	65701	024	CM02	400	EL	8	9408	RPL TRAM MTR OPP SIDE (NAT)	452903	590234	3490

Figure 36-20. By using repair history to measure the elapsed operating hours (or weeks) from the installation to the replacement of components, it is possible to forecast future replacements.

Specialized Information - In another instance, the maintenance engineer suggested a periodic special edition of the regular cost report in which component cost by type of equipment was summarized independently of units of equipment.

As a result, troublesome, costly components were identified and isolated for further observation. See Figure 36-21.

```
                              EQUIPMENT COST BY TYPE/COMPONENT
                                       FEBRUARY 1994
                    ------- CURRENT MONTH ----------    -------------- YEAR TO DATE
                    MANHOURS      LBR    MTL    TTL     MANHOURS      LBR    MTL     TTL    AVG CST
         TYPE CP    RT      OT    CST    CST    CST     RT      OT    CST    CST     CST    MONTH
         51 LHDS-PRODUCTION
         00 GENERAL     1021.5  29.0  10666  10895  21571   2595.0 150.5  28933  25756   54697  273424
         01 ENGINE      490.5   39.0   5930  30993  36936   1119.0  84.5  13546  52324   65884   32933
         02 CONVERTER    20.0    1.0    217     24    241     66.0   3.0    755    295    1050     522
         03 TRANS       358.5   24.5   4198  12732  16932    797.5  55.0   9312  12892   22217   11102
         04 DRIVE TRAIN 592.0   28.5   6668  17294  23968   1291.5  87.0  15131  29702   44841   22416
         05 TIRES       132.5   11.5   1543  10634  12187    289.5  26.5   3432  58180   61622   30805
         06 BUCKET      479.5   23.0   5504   2800   8310    770.5  30.5   8715   8389   17110    8549
         07 HYDRAULICS  625.5   44.5   7331   7166  14509   1664.0 102.0  19438  11477   30929   15457
         08 BOOM         43.0    0.0    485   1458   1946    200.0   3.5   2306   4002    6312    3151
         09 BRAKES      495.0   39.5   5939   4460  10411    913.5  71.5  11014   7330   18358    9172
         10 FRAME       902.0   60.0  10448  10292  20749   1814.5 105.5  21091  15099   36204   18096
         12 ELECTRICAL  249.5   15.0   2945   3172   6125    539.0  31.5   6384   8555   14949    7467
         13 SERVICE      10.5    0.0    112      0    112     38.5   0.0    405      0     405     198
         TOTAL         6411.0  315.5  61996 111920 173997  12096.5 751.0 140462 234001  374578  187210
```

Figure 36-21. The special cost report signaled a need to investigate possible fleet-level problems with engines, drive trains, tires and frames.

Verification of improved equipment performance can be confirmed from reports such as the Fleet Cost Performance Report. See Figure 36-22.

```
                              FLEET COST/PERFORMANCE REPORT
                                         APRIL 1994
         PRODUCTION      8 CU.YD. LHD
         TYPE/    CLNUP   MTC   SVC   IDL   DLAY  SETUP  WORK    %     %    BKTS   BKTS/  TTL  BKTS/
         UNIT     HRS     HRS   HRS   HRS         HRS    HRS    AVL   UTL           W-HR   HRS  T-HR

         510516   3.3    271.5  7.0   15.5  20.0   0.5   231.8  50.6  87.2  9302   40.1  549.8  16.9
         COST/BUCKET  OPNS: 0.27   F&L: 0.01  TIRES: 0.56  MAINT: 0.58  TOTAL: 1.42   YTD:  1.42
         F & O/PER BUCKET DIESEL:  271 GAL/0.03  10-WT:  0 GAL/0.0   30-WT:   0 QTS/0.0   ATF:  8 GAL/0.0
         TRAM INFORMATION MILES:  326.74   FT/BUCKET:  185.55

         510571   8.0     87.6  15.8  44.8  28.0   0.0   368.5  84.2  84.3  15596  42.3  552.7  28.2
         COST/BUCKET  OPNS: 0.26   F&L: 0.02  TIRES: 0.10  MAINT: 0.24  TOTAL: 0.61   YTD:  0.61
         F & O/PER BUCKET DIESEL:  445 GAL/0.03  10-WT: 41 GAL/0.0   30 WT:  20 QTS/0.0   ATF:  2 GAL/0.0
         TRAM INFORMATION MILES:  680.68   FT/BUCKET:  230.4
```

Figure 36-22. Note the cost-related information including: cost per bucket (maintenance) and fuel and oil per bucket (diesel fuel). Other performance elements include maintenance hours, availability and utilization.

Observation of Labor Use - Since the maintenance engineer advises on both PM and the standards for planned work, he is interested in labor utilization to verify how well personnel are carrying out PM services.

If, for example, PM should represent 11% of manpower use but only 3.2% is reported, some PM services are not being done. Verification of actual performance yields recommendations to improve compliance.

Similarly, a check of backlog trends may reveal a need to adjust the craft composition of crews to ensure they can carry out required PM services.

Maintenance engineering tasks, whether carried out by an engineer or delegated to personnel with the work force require solid information. Excessive costs lead to the repair history which leads to a field investigation. Suspected problems require special reports to confirm the problem as maintenance engineering goes about its task of assuring the reliability and maintainability of equipment.

Controlling Non maintenance Work

While construction, equipment modification, installation or relocation are not maintenance, many maintenance departments perform this work. The conduct of non maintenance can create problems:

- Excessive non maintenance work may cause maintenance to fall behind in its regular work. Thus, limits are required.
- Maintenance is an operating expense while non maintenance is usually capitalized. Unless the difference is acknowledged, maintenance budgets will be exceeded.
- The work order system must include a way to control non maintenance work as the maintenance work order is inappropriate, especially when work is contracted.

These problems can be solved with good information, to illustrate:

- Information on labor utilization allows management to determine whether non maintenance work is excessive.
- Maintenance can segregate non maintenance work and avoid expensing work that should be capitalized.
- A separate work order element for non maintenance work precludes misuse of the maintenance work order.

Using the Engineering Work Order - The engineering work order (EWO) is the means by which non maintenance project work is controlled after it is approved and funded.

It is not unusual for maintenance to utilize contractors for this work and to even work with them on the job. Therefore, a sufficiently flexible procedure is necessary to manage non maintenance work.

Non maintenance work like construction requires a feasibility determination before it can be funded and considered along with regular maintenance work.

Once funded, the Engineering Work Order (EWO) can be used to link non maintenance work with the maintenance work order (MWO) and the information system. See Figure 36-23.

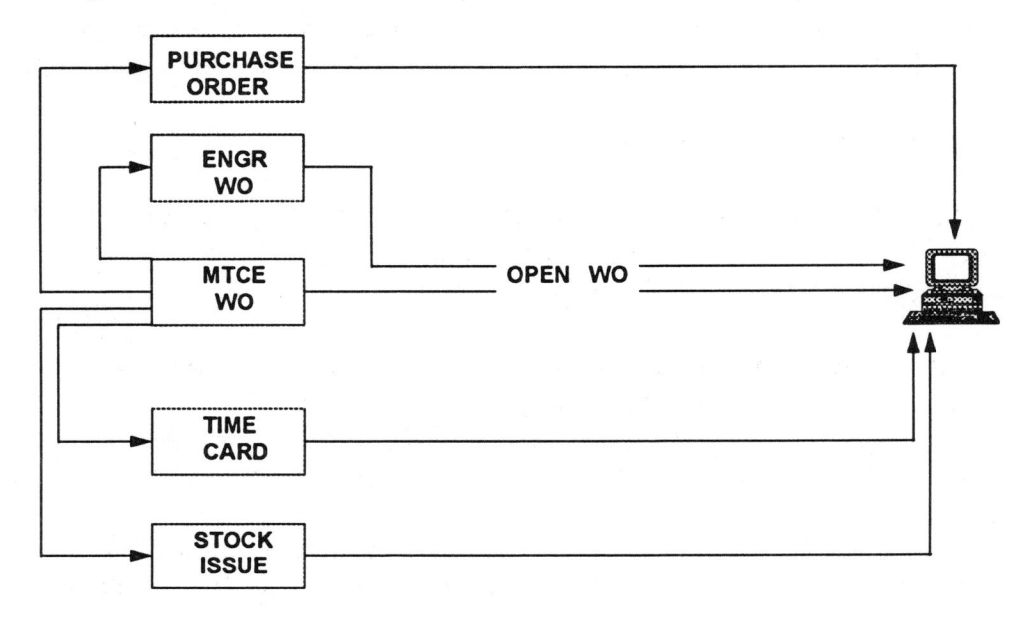

Figure 36-23. The Engineering Work Order (EWO) must fit within the overall work order system in order to develop in-formation necessary to manage non-maintenance projects.

By opening the EWO separately it establishes as a cost entity allowing work to be performed by either maintenance or a contractor, or both.

If a contractor performs the work, a purchase order is used to isolate contractor costs submitted via his invoices.

The project engineer would monitor the work on site and approve all invoices.

If the work is done by maintenance, a series of maintenance work orders would be prepared covering each project phase. The EWO becomes the focal point for the project.

Utilizing the Information System Effectively

The EWO covers a project to 'Install a 250 ' conveyor'. See Figure 36-24.

```
             ENGINEERING WORK ORDER

EWO#                 Sec      Unit    Comp   Ct
716820                07     05011     03     S
Description of Work:      IST 250' CONVEYOR
St     Opn    Stt         PR
9217   9222   9224        81
                                     EST      ACT
Work Order Notes             COSTS:
Test operation 06/12/94.     LABOR            37600
Electrical inspection 06/01/94  Stock         14210
Next cost review 05/11/94.   Purchased        26280
                             EQUIPMENT         5690
                             TOTAL            83780

Requested:          Approved:            Completed:
RL Jones            PD Williamson
04/03/92            04/16/92
```

Figure 36-24. The EWO is the focal point of project control along with the project plan..

The engineer uses a project plan to control project steps. See Figure 36-25.

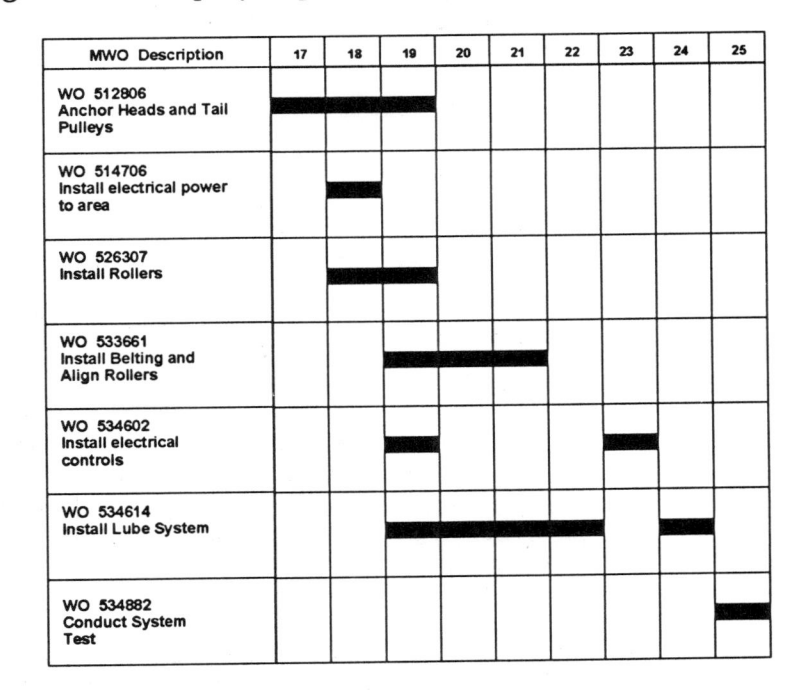

MWO Description	17	18	19	20	21	22	23	24	25
WO 512806 Anchor Heads and Tail Pulleys	■	■	■						
WO 514706 Install electrical power to area		■							
WO 526307 Install Rollers		■	■						
WO 533661 Install Belting and Align Rollers				■	■	■			
WO 534602 Install electrical controls				■			■		
WO 534614 Install Lube System				■	■	■		■	
WO 534882 Conduct System Test									■

Figure 36-25. The project plan includes 7 maintenance work orders.

320 Industrial Maintenance Management

The project engineer manages the project by observing progress on each the Maintenance Work Order status as they would look when he evaluates progress during week 18. See Figure 36-26, 27 and 28.

```
                    MAINTENANCE WORK ORDER
MWO#     Sec   Unit    Comp    Ct
514706   07    05011   03      S
EWO: 716820
Description of Work:  IST ELECTRIC POWER
St    Opn   Stt         Pr   Wa   FC   SN/In    SN/Out
9218  9219  9222        81
                                          EST      ACT
Work Order Notes                 Tr 1     125      141
See print 07-A14
                                 LABOR    3700     2820
                                 Stock    4950     3710
                                 Purchased 2100    1475
                                 TOTAL    10750    8005
Requested:        Approved:        Completed:
RL Jones          PD Williamson
04/03/92          04/16/92
```

Figure 36-26. The progress on MWO 514706 (Install Electric Power) shows about a 76% completion based on man-hours used versus estimated and 74% completed based on cost expenditure to date versus estimated cost.

```
                    MAINTENANCE WORK ORDER
MWO#     Sec   Unit    Comp    Ct
526307   07    05011   03      S
EWO: 716820
Description of Work:  IST ELECTRIC POWER
St    Opn   Stt         Pr   Wa   FC   SN/In    SN/Out
9218  9219  9222        81
                                          EST      ACT
Work Order Notes                 Tr 1     410       55
PO# 7708125
                                 LABOR    8200     1100
                                 Stock    1750      975
                                 Purchased 9255    9255
                                 TOTAL    20205   11330
Requested:        Approved:        Completed:
RL Jones          PD Williamson
04/03/92          04/16/92
```

Figure 36-27. MWO 526307 (Install Rollers) has just started as the progress is checked and it shows a 26% completion based on labor use and a 56% cost expenditure since purchased material costs have been committed.

Utilizing the Information System Effectively

```
                MAINTENANCE WORK ORDER
MWO#    Sec   Unit   Comp   Ct
533661  07    05011  03     S
EWO: 716820
Description of Work:  IST BELTING

St    Opn   Stt         Pr    Wa    FC   SN/In      SN/Out
9218  9219  9222        81               456744
                                    EST       ACT
Work Order Notes                    Tr 1      245
ALN ROLLERS; PO 7708126             LABOR     4900
                                    Stock     1510
                                    Purchased 13950
                                    TOTAL     20360
Requested:         Approved:        Completed:
RL Jones           PD Williamson
04/03/92           04/16/92
```

Figure 36-28. MWO 533661 (Install Belting) reveals no work performed but the commitment of funds for purchased materials is shown.

The project engineer would continue to monitor progress with frequent job site visits during which time the detailed job plan would be consulted.

With the availability of good information on non maintenance projects, the problems of misuse of manpower, incorrect funding and project control can be avoided and replaced with effective project management.

Summary - The availability of quality information can make a considerable impact on the way maintenance can be managed. Not only are we able to control all of the maintenance program elements like PM, planning, scheduling etc., but, we can control resources better as well, labor, materials and shop work.

Those outside of maintenance in operations, staff departments and management see differences as well. They are aware of how maintenance functions, they can better anticipate maintenance needs and they can obtain information on maintenance cost and performance.

But, the information revolution has had the most significant impact on the ease with which the information can be obtained. With powerful PC networks, the cumbersome mainframe equipment with its need to shutdown for nightly file maintenance may soon be a thing of the past. Similarly, MIS staffs may also fall victim to downsizing.

The same ease of obtaining information has also affected the ease with which information can be used:

- Craftsmen will identify and order their own material
- Teams will find work control easy and effective
- Planners will use historical data easily to develop new standards
- The maintenance engineer will easily carry out reliability engineering tasks
- Warehouses can be consolidated as a result of shared information

Future plant maintenance organizations will not only be smaller and more efficient but, they will be more knowledgeable, all due to better information.

Shown below are the correct answers to the backlog computation exercise shown on page 304. See Figure 36-15.

Backog Data

Mechanic	1270	1490	1840	1950	2025	1920	2000	1910	1495	1340
Millwright	1455	1545	1660	1705	1780	1750	1775	1730	1530	1455
Electrician	710	750	790	805	800	800	805	790	740	710
Total	3435	3785	4290	4460	4605	4470	4580	4430	3765	3505

1. Images, copyright New Vision Technologies

V

Supporting Maintenance

37

How Management Can Help

Advice to the Plant Manager About Successful Maintenance

The single, most important action that a plant manager can take to assure profitability is to identify and correct the problems that derive from poor maintenance. If maintenance is poorly executed, its excessive cost is not the only factor affecting profitability. The equipment you rely on will not enable you to deliver your product to the marketplace at a competitive price. 1.

Good maintenance, on the other hand, assures profitability.

But, the benefit of a quality maintenance program to a plant's profitability is lost if that program is not effective. Many potentially good maintenance programs are rendered ineffective if they must operate in an environment where the program can be ignored. If, for example, operations fails to make equipment available for PM services, the potential benefit of PM is lost. Yet, maintenance, as a service organization, cannot compel operations to comply with their program. However, this support and cooperation is the essence of a positive operating environment. Thus, a major concern for plant managers must be to ensure that maintenance can perform in a favorable environment. To assure the environment exists, your help is required.

Your assignment of a clear objective to maintenance and provision of positive guidelines about maintenance, for example, will help to ensure necessary cooperation and support is given to maintenance. While you can influence support and cooperation from operations and staff departments, maintenance must contribute as well. Consider, for example, whether maintenance has an adequate program ?

- Does preventive maintenance minimize emergency repairs ?
- How well do they plan major jobs ?
- Is there an effective major component replacement program ?
- What are they doing about reducing down time ?

The most frequent cause of poor maintenance is the failure to define the program effectively and educate personnel. Maintenance personnel, uninformed of their program, perform poorly. Similarly, operations is confused. But, other issues must be addressed as well:

- Are they organized properly ?
- Are their key personnel properly trained ?
- Is their information system adequate ?
- What are they doing about eliminating major equipment problems ?

These problems, are unlikely to improve unless the plant manager gets involved. Too often, plant managers point to maintenance and say, "We don't know what they don't know and, what we know about maintenance may not be so!" They may be too remote from maintenance problems to offer help. Yet, help is needed.

Remember, the single, most important action that a plant manager can take to assure profitability is to identify and correct the problems that derive from poor maintenance.

Evaluation and Corrective Action

An evaluation is the best way to establish the current maintenance performance level and identify the activities needing improvement. It is also the starting point for the improvement effort. However, the most important aspect of an evaluation is education of the very people who can create the improvements to assure profitability. See Chapter 42 - Conducting Effective Evaluations.

How Management Can Help

Some Check Points

Maintenance Objective - The objective you assign to maintenance must spell out what they do. They must focus on equipment maintenance but, will they support engineering projects ? If so, will the involvement interfere with basic maintenance ?

Organization - You must know what key personnel are doing to ensure they support the maintenance objective and focus on the right activities. If a team is to be formed, make sure the maintenance program is well enough defined to support it. Supervisory training can be a neglected area, especially training on the maintenance program. As a result, craft personnel, who could contribute, have no framework for helping. This void can be devastating if you are moving to a team organization.

Preventive Maintenance - PM is the principal way to avoid emergency repairs and unnecessary downtime. It results in more planned work, better productivity and lower cost. It extends equipment life and reduces capital expenditures. But, none of this will happen unless operations makes the equipment available so the services can be performed. You can help.

Labor Control - This is the primary way that maintenance can control the cost of the work they perform. You must know how effectively maintenance controls labor and whether they are improving productivity ? It will not suddenly get better when a team is installed or TPM is implemented. Improving productivity is a cultural change. If it is to be improved, you may have to see that it gets done.

Maintenance Information - You must make certain the essential information is available and provided to the right people. The work order system must work in concert with the accounting system. Field data that drives the system must be complete, accurate and timely. The biggest problem may be the supervisor who rejects the computer in favor of verbal orders. Each requires attention.

Maintenance Engineering - Reliability and maintainability of production equipment requires that steps be taken to find and eliminate problems. Guidelines are necessary to screen non maintenance work and ensure it is necessary, feasible and properly funded. These are your guidelines. Thereafter, the end result must be reliable equipment with its process continuity preserved. These necessary procedures must be verified.

Material Control - Purchasing and warehousing ensure that required materials are made available on time. Maintenance must specify what they need and when. The maintenance forecast, which prescribes future major component replacements, must allow purchasing to procure materials on time. Maintenance must have standard bills of materials for all major repetitive jobs to streamline material procurement. These are the practical ingredients that make this relationship work. They must be verified.

Performance - Review selected maintenance performance indices to determine the commitment behind the maintenance effort to control and improve performance. Your ability to guide corrective actions will be strengthened with your personal knowledge concerning maintenance. 2.

Summary - Improving maintenance often requires the direct involvement of the plant manager. An evaluation gives considerable impetus to the improvement effort. Strongly consider it.

1. - 2. Images copyright, New Vision Technologies, Inc.

38

The Operations Role in Successful Maintenance

The Operations - Maintenance Partnership

A Message to Operations - The principal objective of maintenance remains that of assuring the safe, effective operation of production equipment no matter the type of operation nor the way the plant has organized. [1]

Operations and maintenance have a common objective of plant profitability which, in turn, influences the mutual support required of their daily activities.

Teamwork has become a by word in the successful plant operation and nowhere is this more apparent than in the operation - maintenance working relationship. Operations personnel need reliable equipment to meet production targets on time, at least cost with quality product. Therefore, they must understand, support and actively participate in maintenance activities that can assure this happens.

Typically, in this relationship there are a number of basic requirements that operations must meet to support their end of the positive relationship. Among them, operations should:

- Have good knowledge of equipment maintenance requirements.
- Release equipment promptly for service or repair.
- Operate equipment carefully and correctly.
- Be familiar with the maintenance program.
- Help identify the timing and cost of future work like overhauls.
- Check the timely completion of PM services.
- Report problems promptly.
- Clean, adjust and service equipment.
- Be concerned with job cost and quality of work.
- Not overload maintenance with unexpected work or unreasonable demands.

Maintenance, on the other hand, can reinforce this positive relationship considerably by defining its program properly and advising operations of its key requirements. While we expect that operations may have to delay services and repairs for operational reasons, there are few circumstances in which they ignore the program, providing they are well informed of its requirements. Operations has the same aspirations as maintenance in sustaining profitability. Therefore, high on the list of requirements for maintenance in securing operations support is the proper definition of their program and the explanation of its requirements to them.

Within the maintenance program, there must also be a continuous flow of information to operations about maintenance requirements and operations feedback. For example:

- A weekly scheduling meeting appraises operations of specific maintenance requirements by unit, job and time and provides opportunity to negotiate the schedule.
- A daily coordination meeting adjusts the timing of events.
- PM schedule compliance tells operations how well both are following the program.
- The forecast advises operations of major upcoming events.
- A cost report provides both parties, and management with a picture of expenditures.
- The downtime report pinpoints troublesome situations that threaten production targets.

The Operations Role in Successful Maintenance

Equipment Management Program

The objective of the maintenance program is to keep production equipment in a safe, effective operating condition so that production targets can be met, on time, at least cost with quality product. Maintenance cannot accomplish this objective without the support of operations, the equipment users.

But, many plant organizations have sought a partnership beyond the maintenance operations partnership. They have involved every plant department in an effort to manage the total life cycle of equipment from design, selection, installation, operation, maintenance, overhaul, modification and ultimately, replacement. See Figure 38-1.

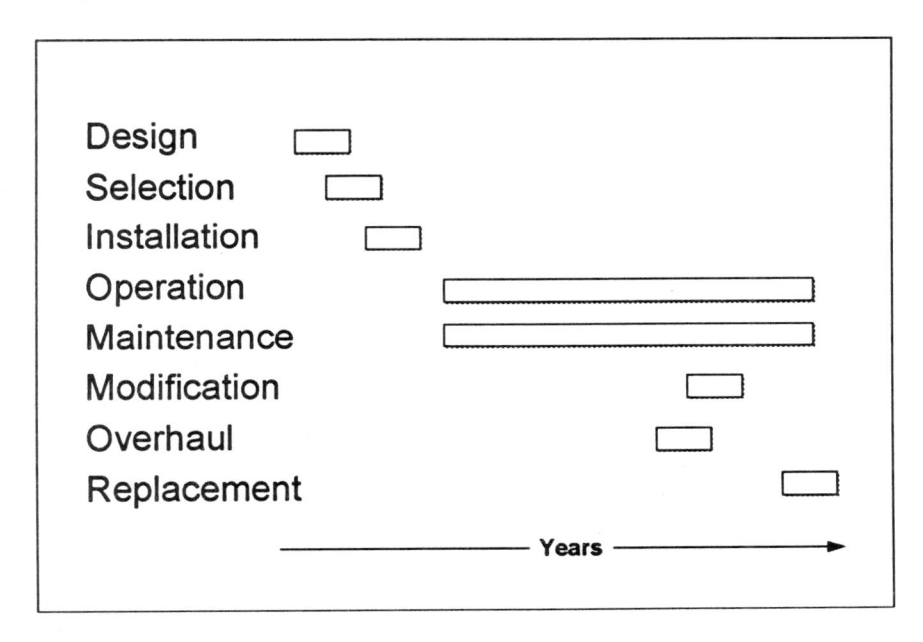

Figure 38-1. The equipment management program spans the total life cycle of equipment and involves every department in the process.

Total Productive Maintenance (TPM)

TPM pushes the focus of equipment into the maintenance - operations realm, and is considered a refinement to the equipment management program. Essentially, TPM charges operations and maintenance with a joint responsibility for the proper care of equipment in stating the operations applies 'first aid' while maintenance performs 'surgery'. See Figure 38-2. 2.

330 Industrial Maintenance Management

Target	Methods	Activities			Allocation	
		Restore deterioration	Inspect deterioration	Prevent deterioration	PROD	MTCE

- Maintenance (OEE>85%)
  - Normal production
    - Proper operation
    - Set up & adjustment
  - Routine maintenance
    - Cleaning
    - Lubrication
    - Tightening
    - Routine inspection
    - Minor servicing
  - Periodic maintenance
    - Inspection
    - Servicing
    - Overhaul
  - Predictive maintenance
    - Condition monitoring and machine diagnosis
    - Parts replacement and overhaul
  - Breakdown maintenance
    - Early detection and prompt reporting
    - Repairs
- Remedial
  - Reliability
    - Strengthening
    - Load reduction
  - Maintainability
    - Inspection procedures
    - Servicing procedures
    - Servicing quality

■ Department with primary responsibility

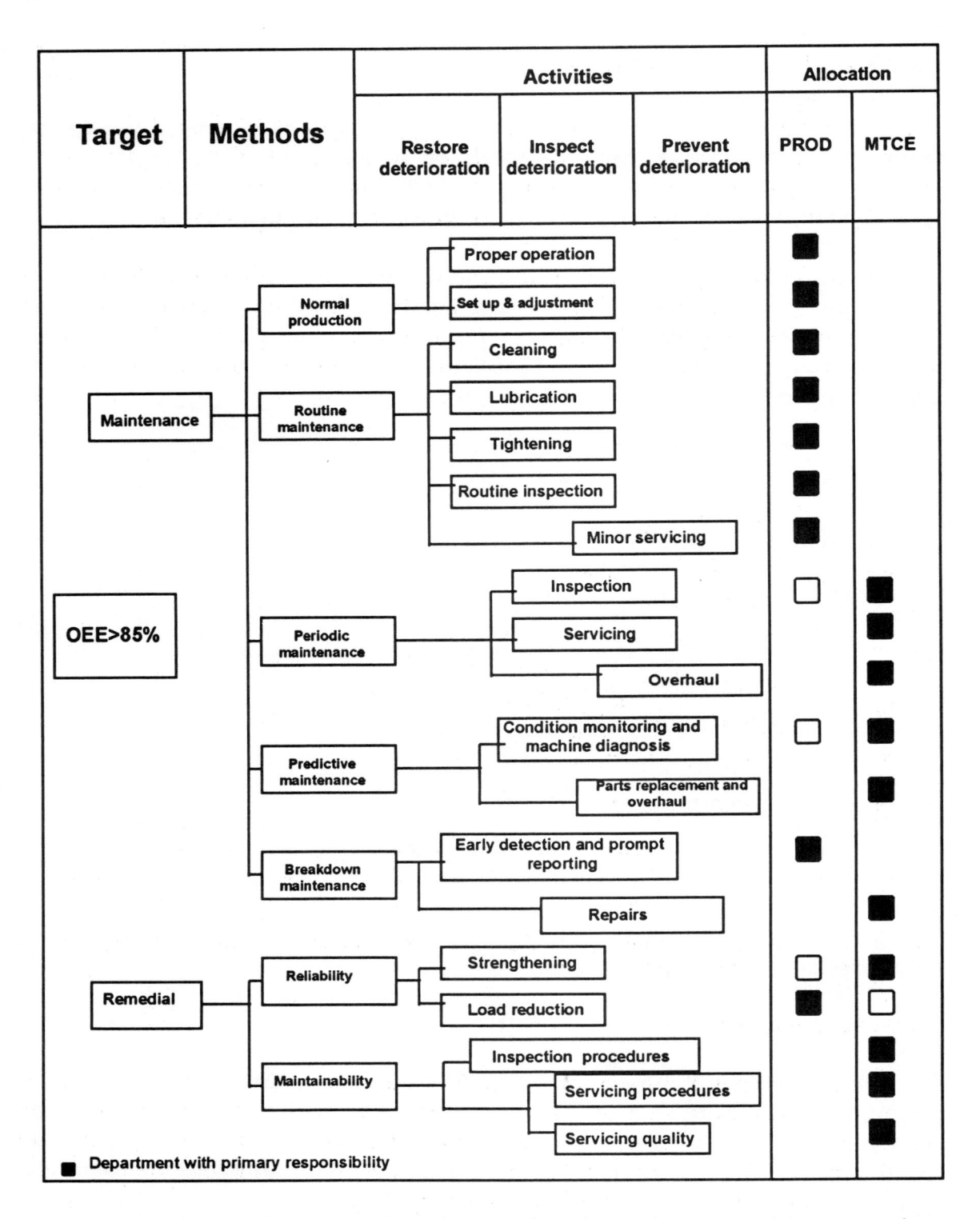

Figure 38-2. TPM specifies responsibility for reliable equipment between maintenance and production.

Operator Maintenance - One of the most difficult transitions in plant maintenance has been to open the door to operators so they can perform maintenance as stipulated in the Total Productive Maintenance concept of sharing maintenance responsibilities.

Traditionally, craft unions have jealously guarded their 'right' to perform maintenance. While, operator attention to cleaning and checking such as the haulage truck operators 'walk around inspection' (before, during and after the shift) was tolerated, the extent to which TPM suggests collaboration was not perceived, even a few years ago.

Downsizing Hastens Team Formation - While team organizations foster an operations-maintenance relationship and the logical division of maintenance responsibilities, some maintenance organizations often did not embrace it as quickly as they should have. In many instances, plants did not involve operations in programs like TPM until they realized that their survival would depend on downsizing maintenance. Only then did many plants take the TPM concept of shared maintenance responsibilities seriously.

Implementing Maintenance Performed by Operators - With the issue of operations responsibility for maintenance settled, the task of how to do it then became a major issue. Many plants have applied a very logical solution to the division of maintenance - operations responsibilities. The criteria was simple: the task must improve maintenance and the operator must be trained to perform it competently. Successful transitions involved craftsmen and their experiences were noteworthy. It was found that:

- Craftsmen will correctly identify activities able to be performed by operators.
- Cleaning and adjusting are logical choices but, minor component replacements require training.
- Craftsmen were not eager to turn over lubrication to operators because they felt it was too critical.
- Operators felt they could verify lube levels adequately but, did not want to perform the total lubrication function as it would involve them in the logistics of replenishing lubricants.
- Craftsmen were pleased to be asked to train operators and on finding a willing, competent operator often made judgments to trust him well beyond the expected division of tasks.
- Those instances of crafts involvement proved to be the most successful.

Summary - Little is gained by a plant as it witnesses operations and maintenance looking like 'jack asses pulling on opposite ends of the same rope' (equipment).

It is in the interest of the plant as well as of maintenance and operations to create an environment of cooperation that will allow the maintenance program to be carried out successfully. The previous chapter described the help that the plant manager can provide with clear direction, guidance and monitoring of the maintenance effort. In this chapter, the operations and maintenance departments make that cooperation and mutual support a reality. They understand each others requirements and share the responsibility for effective maintenance. When done well, success is measured in people satisfaction of a good job done plus, more tangibles results like lower downtime, reduced cost, better performance and assurance of plant profitability.

1. Image, copyright, New Vision Technologies, Inc.
2. M. Tajiri; F. Gotoh, TPM Implementation, McGraw Hill, 1992, p. 45.

39

Staff Support for Maintenance

Staff Departments Can Help Maintenance

A Message to Staff Departments - Just as maintenance is a service to operations, staff departments exist to provide a service to the operating departments, operations and maintenance.

Where maintenance is concerned, their primary support comes from material control departments, warehousing and purchasing. But, this material control relationship is also more complex in that it crosses department lines. For example, the shops that rebuild components and manufacture spare parts for inventory are in maintenance but, purchasing and warehousing are their customers.

The plant engineering department develops, engineers and manages the non maintenance projects often carried out by maintenance.

Accounting and management information services (MIS) provide the means by which maintenance controls its work and manages its resources.

Common to all of these departments is the realization that maintenance depends on them to provide reliable service. Yet, in order to do so, staff personnel must:

- Understand the maintenance program.
- Be service oriented.
- Act promptly on recommendations for improving services.
- Confer with maintenance to improve support.

Perhaps the most useful thing that maintenance can do to enhance the working relationship with staff departments is to involve them in the day to day execution of the maintenance program. For example:

- Ask material control to attend the weekly operations - maintenance scheduling meeting so they can be better informed about current and future needs for materials.
- Invite a member of the accounting department over to help explain the payroll system to craftsmen and, as a result, get fewer complaints (while at it, have the supervisor attend).
- Make sure the purchasing agent can access the forecast so he can see how many components he must have in the 'rebuild' pipeline and how quickly he must move them.
- Arrange to have planners, purchasing agents and warehouse managers sit together for a regular monthly meeting to become more familiar with each others procedures and problems (and how to avoid them). [1].

Summary - Staff personnel often feel left out of the maintenance picture until there is trouble. That trouble is often traceable to their inadequate involvement and knowledge of what maintenance is doing or what it needs, and when. When trouble occurs, staff departments spend enormous amounts of time and energy defending themselves against oversights and perceived omissions not always their fault. A sound policy for maintenance is to make absolutely certain they have told staff departments of their needs. Then, involve them directly in maintenance business like sharing information or attending scheduling meetings. The results will be worth the effort.

---

1. Images copyright, New Vision Technologies, Inc.

VI

Achieving Productivity and Performance

40

Measuring Maintenance Productivity

Measuring Productivity is Easier Than You Think

The maintenance department that checks productivity regularly is rare, perhaps even non existent. Despite this, productivity measurement remains the most effective way of verifying the quality of labor control. Numerous reasons are offered for not checking productivity, among them:

- Measurements will be misunderstood.
- The work force will suspect layoffs.
- Personnel will feel threatened.
- Measurements aren't accurate.
- "We don't trust industrial engineers".
- "We don't have time to check productivity".

Managers confronted with such resistance soon conclude it is counter productive to ask maintenance to verify its quality of labor control. In such an environment, even skilled industrial engineers are reluctant to conduct measurements despite strong management sanction. The result is that nothing happens. Productivity, if poor, remains that way. Worse yet, productivity that could easily be improved isn't:

- Current levels aren't known, and . . .
- The factors that inhibit productivity are not identified.

Why Not Let Craftsmen Check Productivity ?

Isn't that like 'putting the monkeys in charge of the bananas' ? Yet, it intrigues. How can it be done ? Briefly, if a measure is made of the factors that inhibit productivity, it is possible to deduce the level of productivity. Moreover, the factors that cause a loss of productivity can be clearly identified so that corrective actions can be initiated. If, as a result, work is made easier for craftsmen, aren't they among the winners ?

Craftsman Participation - It can be successfully argued that the best source of information on work performed by maintenance is the craftsman himself. Therefore, by extension, they are in a position to comment successfully why they have difficulty getting their work done. Generally, an industrial engineer can effectively measure about five productivity related factors: waiting, travel, clerical, idle and working. By contrast, the involved craftsman can, if properly trained and oriented, measure most elements causing him to lose time while performing work. Typical:

- Identifying parts
- Obtaining parts
- Waiting because there is no work available
- Incorrect instructions
- Tools are not available, broken or not enough
- Drawings are not up to date or can't be found
- Waiting for another craft
- Waiting for equipment to be shut down
- Waiting for lifting equipment
- Excessive clerical tasks
- Using archaic repair techniques
- More or better training is needed
- Union or labor related problems

In this age of involvement, the best weapon we have to reduce resistance is participation. Why not use it ? If used successfully, most of the reasons offered by maintenance for not measuring productivity soon disappear.

Technique

Random sampling is the key. Not every maintenance worker need be involved in the measurements. However, all workers should be aware of how the measurements are to be made, why and what they reveal.

Measuring Maintenance Productivity

To start, there must be total agreement that the improvement of worker productivity is a common objective of management (to utilize labor as effectively as possible) and of the work force (to be able to carry out work efficiently and with the satisfaction of knowing that quality work is being done).

Step one

- Meet with work force representatives (or union officials) and maintenance supervision.
- Openly discuss the reasons, on both sides, for improving productivity.
- State the policy of improving productivity to reduce downtime not reduce the work force.
- Assure personnel that the work force will not be reduced.
- Explain what will be done with the results of productivity measurements.
- Describe how data on measurements will be shared.
- Explain how joint decisions will be made in determining how to correct and reduce the factors inhibiting productivity.

Step two

- Identify the frequency of measurements (six months).
- Establish the periods when measurements will be made (weeks 21 and 22).
- Explain how supervisors will randomly select 10% of their crews to participate on alternate days during the measurement period.
- Demonstrate how each worker will select the longest job performed on the day he participates and record the time spent on each of the factors affecting productivity as well as the time spent working (See Figure 40-1.).
- Explain that no worker will be identified by name.
- Show how the data will be summarized to reveal the factors that must be improved, the degree and the priority of improvement.
- Reach agreement on how management and worker groups will analyze the data and establish a plan for correcting and improving the problems identified.
- Determine how results will be shared plant wide and comments welcomed
- Establish the period for the next measurement.

Step three

Provide a simple method of collecting the data and welcome constructive comments from all.

Step 1 Describe job worked on: *Replaced vent fan.*

Step 2 Indicate duration of job.

| 1 | 2◄ | 3 | 4 | 5 | ►6 | 7 | 8 |

Duration 4:50

Step 3 Shade in approximate duration of each delay.

Step 4 Indicate the reason for the delay.

| | 04 | 13 | 09 | 10 | | | |

Code	Description of Delay	Time
01	No work available.	
02	Equipment not shutdown.	
03	Area needed cleaning.	
04	Trouble identifying materials.	.15
05	Trouble identifying or locating diagrams.	
06	Waiting to obtain materials.	
07	Replace incorrect materials.	
08	Materials not available.	
09	Waiting for operator assistance.	.15
10	Waiting for another craft. (*electrician*).	.45
11	Transportation delay.	
12	Waiting for crane.	
13	Tools missing or broken. (*Ladder broken*)	.15
14	Poor lighting or ventilation.	
15	Administrative conflicts (meetings).	

Total delays 1:30

Comments: *Fill the steel rack every day.*

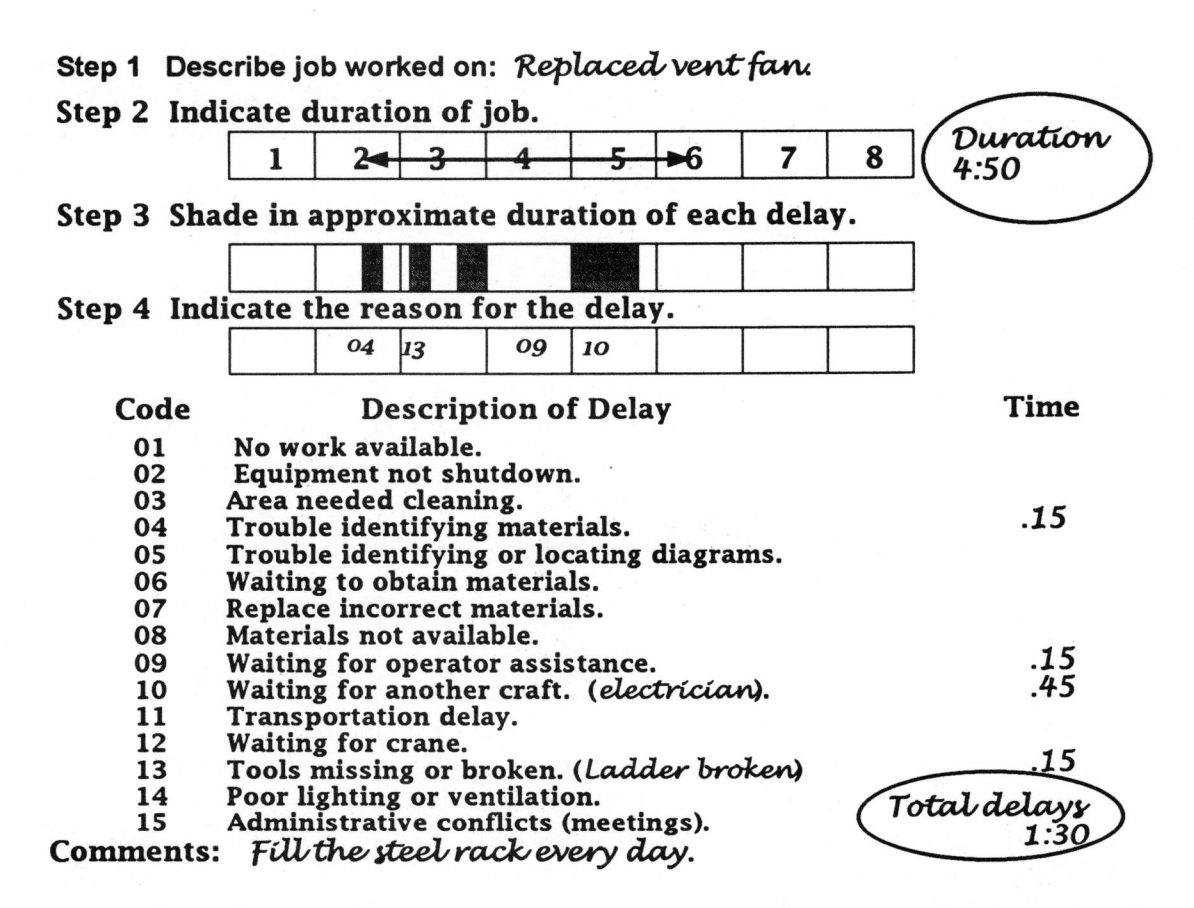

Figure 40-1. Total delays for this 4.5 hour job were 1.3 hours of which .45 hours were spent waiting for an electrician. Note the useful comment to "Fill the steel rack every day".

Summary - Productivity will not improve itself. Human beings must dedicate themselves to its improvement. Both management and the work force suffer from poor productivity. Sufficient leadership must be exercised to call attention to this need and then do something about it. The technique described is a possible starting point in improving productivity. It bridges the communications gap that has, for so long, blocked necessary action in measuring productivity and getting improvement started.

41

Establishing the Essential Performance Indices

Obtaining the Total Performance Picture

There are many indices of maintenance performance including labor use, downtime, cost, backlog reduction etc. Each measures a specific element. However, no single index of performance produces a total picture. Rather, a combination of indices is required to yield an overview of maintenance performance.

Purpose of Indices - Generally, performance indices are aimed at providing the manager with a comparison of current performance against trends or desired levels of performance. Thus, the indices must trace performance in an historical sense. In addition, the indices must be rooted in or derived from readily obtainable information or data. This permits maintenance to seeks answers to an improvement need, suggested by a poor current index, in its own information system. For example, if the cost per ton index reveals an unsatisfactory ratio, maintenance should be able to find the offending equipment in its cost reports and then isolate the problem in existing repair history. Thus, they can respond quickly with corrective actions or solutions.

Family of Indices - Each plant must develop a family of indices which best portray the overall performance picture. In this effort to identify the most effective indices, the plant must consider several approaches, among them:

- The information needs of different management levels
- The factors that most influence maintenance performance
- The ease with which data to compose indices can be obtained and applied

Use of Indices - The information needed by the plant manager versus the maintenance superintendent, or the maintenance supervisor differs in both its nature and intent. The plant manager, for example, needs summaries and trend information that tells him he is safe or headed in a safe direction. Yet, his time must divided among a multitude of interests and concerns. Thus, he has little time to study reports. But, he must be well informed. Consequently, he has the greatest need for indices that give him a comprehensive performance picture quickly. By contrast, the supervisor wants immediate information. He must know what happened on the previous shift and what new problems are created. He has little interest in trends and summaries. See Figure 41-1.

INFORMATION VERSUS MANAGEMENT LEVELS

MANAGER

Reports to management	Supervisory development
Costs versus budget	Community relations
Environmental compliance	Morale
Total quality control	Status of equipment

MAINTENANCE SUPERINTENDENT

Maintenance costs	Work priorities
Work force levels	Capital expenditures
Status of materials	Budget compliance
Supervisory capabilities	Quality of work

GENERAL SUPERVISOR

Maintenance costs	Schedule compliance
Productivity	Use of overtime
Availability of materials	Safety performance
Quality of planning	Level of craft skill
Control of labor	Adequate training
Use of materials	Adequate shops

SUPERVISOR

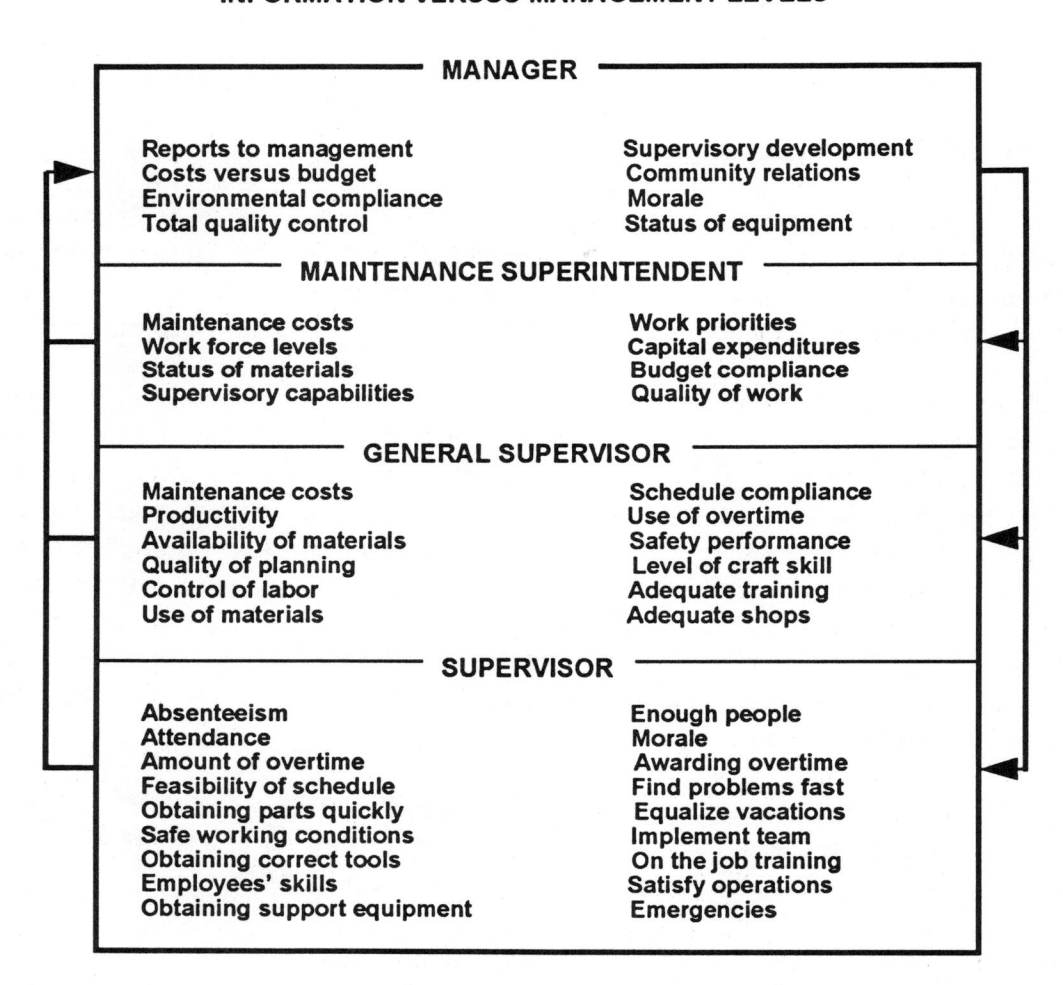

Absenteeism	Enough people
Attendance	Morale
Amount of overtime	Awarding overtime
Feasibility of schedule	Find problems fast
Obtaining parts quickly	Equalize vacations
Safe working conditions	Implement team
Obtaining correct tools	On the job training
Employees' skills	Satisfy operations
Obtaining support equipment	Emergencies

Figure 41-1. The plant manager has the greatest need for meaningful performance indices which give him a total performance picture in the least amount of time.

Establishing the Essential Performance Indices

Cost Versus Performance

There is little question that the cost of the plant operation, and the contribution of maintenance to that cost is of great concern to any plant manager. In this period of necessary cost cutting and downsizing, managers are striving to push costs down to levels that will permit profitability of the plant. Thus, any selection of performance indices must be rooted firmly to costs. No determination of the best indices can be made without appreciating the forces that drive maintenance costs. Consider:

- Total maintenance costs are controlled by the cost of material used rather than by the cost of labor. Most maintenance departments have a stable workforce thus, total labor cost variations are largely due to overtime.
- Material costs are determined by the volume of equipment that must be maintained and the speed at which equipment repair consumes spare parts and materials. Thus, equipment repairs characterized by repeated emergencies consume materials faster than equipment subjected to a deliberate, well ordered maintenance program. Therefore, a key maintenance cost reduction objective should be to reduce emergency repairs to slow the rate at which materials are consumed.
- Labor costs, of course, are incurred by the need to install materials. But, emergency repairs are done less productively, more labor is consumed and the work is often rushed resulting in lower work quality. Often emergency repairs are done simply to get the equipment running again and they must be repeated when equipment is needed less urgently. Thus, the expenditure of labor on emergencies is not only inefficient but, often it may be wasteful because the work must be redone.
- Not surprisingly, the only significant control maintenance has over the cost of the work it performs is the efficiency with which they install materials. Thus, better PM, leading to fewer emergencies and more planning, results in better labor utilization and less labor cost to install the materials.
- As work is carried out maintenance exercises better control over individual jobs that are planned and scheduled. These jobs, because they are better organized are often completed in less elapsed downtime reducing the overall cost because less labor is used. Generally, material costs are the same whether the job is planned or not. More planned work also reduces downtime.
- Maintenance supervisors make the greatest contribution to cost control through the productivity increases they bring about with quality supervision.

Based on this, the control of labor then becomes a vital maintenance function and indices that reflect how efficiently it is used should be included in the overall performance picture.

Overall Index of Maintenance Performance - Perhaps one of the most positive overall indices of maintenance performance is the cost of labor to install each dollar of material. If applied at the department level it provides a macro picture. But, it can be applied at the crew level as well revealing the relative productivity of the crew. See Figure 41-2.

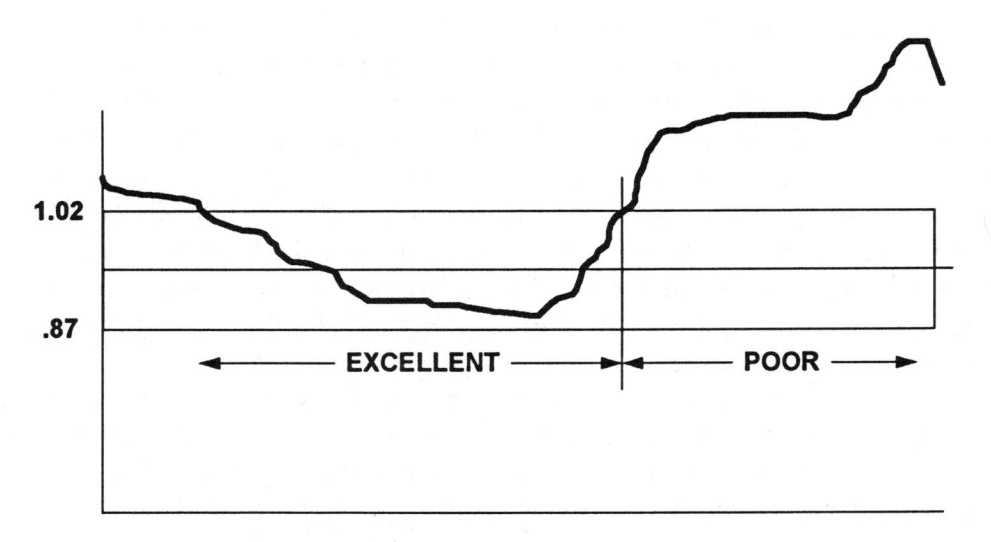

Figure 41-2. Based on the average 50 % split of maintenance labor and material costs, an efficient maintenance department should be able to install each dollar of material with between $.87 and $ 1.02 of labor. Labor costs in excess of that signal poor PM, little planning, ineffective work control and unsatisfactory control of labor.

Relationship Between Indices - A close relationship exists between operating costs, major repairs and total maintenance costs. As a result, one plant focused on the management of its labor. Their findings revealed that labor cost, as a percentage of total cost, was a good long term measure of labor efficiency. By upgrading PM and improving planning, the plant was able to reduce the ratio of labor use from 48 to 41 % of total maintenance costs. They realized a cost savings of over $ 95,000 per month. Thus, there is little question that labor data must be included in performance indices. However, a broader spectrum of indices is necessary to gain the overall picture of maintenance performance. In developing these indices, it is better to focus on the basics that assure better performance than to identify a poor performance but, not know how to improve it.

Most Useful Indices - For example, it is better to check the compliance with the PM program than to measure productivity. If PM is done on time, it will permit more planned work and assure better productivity. Similarly, indices must be meaningful. Measuring performance on single jobs is not particularly useful if the compliance with the total schedule is over 85 % and jobs were completed within $\pm$ 10 % estimated versus actual cost.

Downtime - Likewise, total maintenance downtime is rarely useful except to use maintenance as a 'whipping boy'. But, if downtime is divided into meaningful segments, it can bring about real improvements. For example, the downtime awaiting maintenance measures the effectiveness of internal maintenance communications. Downtime spent performing work reveals the efficiency of planning and control while the downtime awaiting materials states the quality of material control support.

Linkages of Indices - Many indices are interrelated. If, for instance, the maintenance supervisor spends little time on effective supervision, productivity will be poor. Generally, at least 60 % of the supervisors time must be spent on supervision to yield a minimum of 40 % productivity.

Different Indices for Different Objectives - As a result, the plant manager and the maintenance superintendent will utilize different indices with different objectives. The plant manager is interested in profitability and meeting production targets. See Table 41-1.

Table 41 - 1

Plant Manager

Total maintenance cost
Budget compliance
Capital expenditures new equipment
Percent of total operating cost
Cost per ton of product by plant
Cost per operating hour per fleet

The maintenance superintendent is interested in the actual performance of his department. Thus, his need for performance indices is more extensive and it covers areas that he must manage personally. Moreover, his indices must indicate exact problems, enabling him to identify the supervisor responsible for improving it.

In turn, that supervisor must be able to pin point the problem using available information. See Table 41-2.

Table 41 - 2

Maintenance Superintendent

Labor Control

- Productivity
- Percent of labor used on emergency repairs
- Percent of labor used on planned maintenance
- Backlog level
- Percent of overtime
- Compliance with vacation schedule
- Compliance with absentee policies

Control Measures

- PM compliance
- Compliance with weekly schedule
- Percent supervision time

Cost

- Labor cost to install each dollar of material
- Cost per productive hour
- Percent of maintenance to total operating cost
- Cost per ton in each plant
- Cost per operating hour by fleet
- Total maintenance cost
- Cost of overtime

Downtime

- Downtime to respond to operations requests
- Downtime to perform work
- Downtime awaiting materials

Summary - The proper performance indices can quickly summarize the overall performance of maintenance. Indices should be tied to available information so that maintenance can quickly identify and correct problems identified by exceptional trends in the indices.

VII

Evaluating to Sustain Effectiveness

42

Conducting Effective Evaluations

Evaluation is the First Step of Improvement

In this era of intense competition, only the profitable mines will survive. Maintenance, which often represents over 30 percent of operating costs, is one of the few major costs that can be controlled. If costs are not controlled and reduced, they deny profitability directly by costing too much. An ineffective maintenance program creates yet another unfavorable cost impact, equipment downtime. Downtime is three times as costly as the maintenance that could have avoided it. The lost product never reaches the marketplace and the downtime from poor maintenance diminishes profitability. [1]

Purpose of Evaluations - An evaluation establishes the current maintenance performance level by identifying those activities needing improvement as well as those being performed well. The evaluation not only confirms performance but, is the starting point for any improvement effort. Yet, the most important by products of a well conceived and effectively conducted evaluation are the education of personnel and their commitment to provide genuine support for improving maintenance. An evaluation that successfully points the way to improvement makes it more attainable and is a matter of pride to maintenance. Thus, the very people who can create the improvements are ready to help in a cause they believe in.

Establishing the Evaluation as a Management Technique

Resistance to Evaluations - Maintenance resists evaluations because there are too many factors over which they have no control that influence their performance. Typical:

- The warehouse may be poorly run because it is managed by a remote, disinterested accounting department.
- Equipment may be unreliable because it is being pushed beyond maintenance due dates by operators who are pursuing unrealistic production targets.

A primary reason for resisting evaluations is misunderstanding. This leads to suspicion and opposition. Typically, maintenance may hope to delay an evaluation by creating meaningless 'performance data' or warn of unpleasantness from unions. Under such circumstances, the plant may not go ahead with an evaluation. Thus, problems that should be corrected are never identified. 2.

Resistance must be eliminated by removing any threat to maintenance implied by an evaluation. Therefore, the evaluation must distinguish between those factors over which maintenance has responsibility and those over which it has no control, like the unresponsive warehouse run by an indifferent accounting department.

Responsibilities for performance over each group of factors must be clearly established. Thus, if poor maintenance performance is due to poor warehouse support, this must be brought out into the open and corrected. Maintenance can only be evaluated fairly if aspects such as operations cooperation, staff support and management sanction are assessed as well.

A Perspective of Evaluations - Maintenance, after all, is a service. They cannot compel operations to comply with their program. Yet, support and cooperation are required if maintenance is to be successful.

Conducting Effective Evaluations

Once it is clear that the factors over which maintenance has little control will also be evaluated, most resistance to evaluations will disappear.

Background of Maintenance Supervisors - Maintenance also resists evaluations because of the limited management background of many of its key people. Maintenance supervisors, superintendents and even managers rarely come from outside of maintenance. Rather, they are usually inbred and carry their management styles from the supervisor to superintendent level unchanged. Many are graduates of the 'school of hard knocks'. University degrees, not a requirement, are rare among maintenance leaders when compared with operations, for example. Few have had significant management training. They often assume their supervisory responsibilities with little preparation for the challenges they will face. Some may be unwilling supervisors and even uncomfortable in the role. The majority are ex craftsmen. Although they are competent in a single skill, they may be unfamiliar with the management techniques necessary to meet their obligations as supervisors. Further compounding the situation is the fact that a large number, as ex craftsmen, are former union employees who may harbor an 'us versus management' viewpoint.

A 10-Step Strategy for Successful Evaluations

To take advantage of the improvement that maintenance evaluations offer, an overall strategy is necessary. The strategy must identify actions to organize and conduct an evaluation and convert them into improvements. The 10 step strategy includes the following points:

1 - Develop a policy for evaluations.
2 - Provide advance notification.
3 - Educate personnel.
4 - Schedule the evaluation.
5 - Publicize the content of the evaluation.
6 - Use the most appropriate evaluation technique.
7 - Announce results.
8 - Take immediate action on results.
9 - Announce specific gains.
10 - Specify dates of the next evaluation.

This strategy recognizes that maintenance has the potential of contributing significantly to profitability. However, once the evaluation has taken place, the results must be converted into a plan of action and leadership must be exerted to develop and implement improvements.

A good evaluation will correctly identify improvements and provide the opportunity to initiate corrective actions. The evaluation, if conducted effectively, can be the first of step in achieving plant profitability through improved maintenance performance. 3.

1. Develop a Policy for Evaluations - There must be a strong policy requiring that maintenance be evaluated on a regular, continuing basis. Such a policy will preclude the resistance that many poor maintenance departments exhibit to discourage evaluations. This policy will help to redirect the energy of resistance into an effort to prepare for evaluations instead. A typical policy might be:

Maintenance makes a significant contribution to the overall profitability of the plant. To ensure that maintenance services are effective, we will evaluate maintenance on a regular, continuous basis. The evaluations will examine how effectively we perform our work, as well as how the plant supports and utilizes maintenance services. Evaluation results will be publicized so that we may know where we stand, what we have done well and where we need to improve. I look forward to your continued cooperation.

2. Provide Advance Notification of the Evaluation - Advise personnel about the evaluation and make a preliminary statement about its content, purpose and use of the results. Eliminate surprises with the announcement but, emphasize the policy of regular, continuing evaluations. The announcement should:

- Advise of evaluation dates.
- State the objectives.
- Summarize previous evaluation results.
- Restate improvement objectives.
- Thank everyone in advance.
- Advise that their help is critical to improvement.
- Offer to discuss announcement details.

3. Educate Personnel on the Purpose of the Evaluation - Explain that the evaluation is a checklist describing what should be done. Results describe how well they did and provide a basis for improvement. It identifies what is done well, not what is done poorly. It is an opportunity to help maintenance account for its contribution to plant objectives.

Conducting Effective Evaluations

Emphasize the positive aspects of the evaluation in the educational effort. Change unfavorable misconceptions of evaluations by telling personnel they are the means of finding out how maintenance can do better. When preparing personnel for the evaluation, acknowledge that there may be a genuine fear of audits and evaluations. This fear may be based on previous bad experiences, or it may have no basis at all. Many maintenance workers simply don't want anyone 'looking over their shoulders'. Other fears are fueled by managers who themselves may create uncertainty with remarks that mislead. One maintenance manager proclaimed that "auditors are those who come in after the battle is lost and bayonet the survivors!" Such attitudes contribute little to the mental preparation that must be made if a plant is struggling to improve. 4.

Some unions have been known to threaten strikes when an evaluation is announced. They fear that their 'sacred promise' of preserving jobs will be threatened by possible work force reductions as the result of an evaluation. The fact that the evaluation can also lead to improvements that will make their work easier may not occur to them. It follows that maintenance supervisors, caught up in the threat of their jobs being made harder by the union position, may resist as well.

Avoid Surprising Personnel with Short Notice - Avoid surprising personnel as this adds to resistance and creates distrust. People don't like surprises. Therefore, let personnel know what is coming and avoid resistance.

Handling the Evaluation of Dual Operations - Resistance to being evaluated also occurs when there are several operations that will be evaluated. The concern is that performance between several maintenance operations will be unfairly compared when, in fact, they are not similar and would be better compared on another basis. Regardless, comparisons will be made. The sponsor of dual evaluations, usually a general manager, must convey a supportive attitude. The sponsor should provide encouragement to conduct the evaluation and follow up to see that something constructive is done with the results. If help beyond the resources of a single plant is necessary for improvement, (like computer programming), the sponsor can help cement good relations by providing it.

Local Management Interest - In individual mines, local management will want to know how well their policies are understood and how effectively the procedures based on those policies are being carried out. Let them know. Although they are concerned with the quality of the maintenance program, they will be equally interested in learning how well, for example, production supports and cooperates with the program. Therefore, be less concerned with what others may think and get on with the evaluation. Assume that every other party is interested, concerned and ready to help.

4. Schedule the Maintenance Evaluation - Schedule the evaluation carefully, particularly if it will involve a physical audit lasting several weeks. In selecting the evaluation dates, be aware of potential conflicts that might distort evaluation results. For example:

- If a shutdown has just been completed, there may be distracting start up problems.
- Similarly, if a shutdown is coming up, preparation may compete for the attention of personnel.

Also, peak vacation periods may find key personnel away from the plant. Be aware of the expected absences of key people and weigh their non participation in deciding when the evaluation should be conducted. Similarly, recent personnel changes could limit knowledge of evaluation points, and personnel cutbacks or staff reductions might affect attitudes. In general, the evaluation should be carried out in a stabilized situation with as few distracting conditions as possible. With suitable advance notice, maintenance can prepare for the evaluation and look forward to learning how they are doing. 5.

Timing of evaluations is important. If they are conducted on a regular, continuing basis, people will look forward to them as an opportunity to demonstrate progress.

If maintenance personnel feel that the evaluation is constructive, they will prepare for it without hesitation. In subsequent evaluations, if they have accepted the evaluations and are convinced of their value, they will make a conscious effort to improve on previous results.

Conducting Effective Evaluations

5. Publicize the Content of the Maintenance Evaluation - Make sure the content of the evaluation is explained in advance. A maintenance evaluation covers such a variety of activities that it is unlikely any advance notice will constitute a dramatic shift in performance. By announcing what is to be evaluated, organizing for the evaluation is helped. Reports will be ready, personnel scheduled for interviews will be prepared, and the evaluation can be carried out more effectively.

Agreement on evaluation content also clarifies expectations of both the maintenance department and the evaluators. For example, if productivity is to be measured, hourly personnel should be advised so they do not misunderstand the intent.

Generally, the scope of most evaluations centers on three broad areas:

- The maintenance organization
- The maintenance program
- The environment in which maintenance operates

In assessing these areas, there are key points to be checked. For example, to evaluate the maintenance organization, examine:

- Efficiency in controlling personnel and carrying out work
- The effectiveness of supervision
- Work load measurement
- Labor utilization
- Productivity
- Motivation
- Effectiveness of craft training
- Quality of supervisor training

In evaluating the maintenance program, assess:

- Program definition
- Use of proper terminology
- The effectiveness of preventive maintenance
- The effectiveness of the work order system
- Whether necessary information exists
- The quality of planning
- Use of standards
- Whether scheduling results in good labor use

- If work control assures productivity
- How well maintenance engineering assures equipment reliability
- Effective use of technology

In evaluating the environment in which maintenance operates, examine:

- Whether management has assigned a clear, realistic objective
- If guidelines ensure common understanding and compliance
- Management support for maintenance
- Operations understanding and cooperation
- Staff department support and service quality
- Adequacy of material control
- Safety compliance
- Whether housekeeping preserves assets

The environment reflects the attitude of the plant toward maintenance. When poor, it is the most common reason for resistance to evaluations. Even a first rate maintenance organization, with an effective program, can fail if the environment does not provide receptivity and support.

Examining the environment includes answers to some difficult questions regarding:

- Management sponsorship
- Operations cooperation
- Level of service of staff departments such as purchasing

Not surprisingly, when maintenance personnel realize that the evaluation includes key factors over which they have little control, but which influence their performance, their resistance diminishes.

Be aware that an evaluation of the plant environment could reveal an unwilling operations department or an uncooperative purchasing agent. Therefore, it will be necessary to ask the support of the plant manager in conditioning these other departments for their participation as well as helping to improve their attitudes.

Examination of the environment can include steps that the plant manager must take to ensure the success of the program. Typically, these steps include a clear mission and realistic guidelines for other departments to follow in how they must support maintenance.

Conducting Effective Evaluations

6. Use the Most Appropriate Evaluation Technique - Evaluation techniques should be considered based on the plant situation. Some operations may require an evaluation in which every detail must be scrutinized. Other operations, having established the essential pattern of evaluations, may simply check progress by measuring only a few critical areas.

There are three techniques that can be used to evaluate maintenance performance:

- Physical audit
- Questionnaire
- Physical audit with questionnaire

Physical Audit - A physical audit is usually conducted by a team. Generally, the team is made up of consultants or company personnel (or both). They examine the maintenance organization and its program, as well as activities that affect maintenance (such as the quality of material support). The physical audit must examine the total maintenance activity first hand, by observing work, examining key activities (such as preventive maintenance, planning, or scheduling), reviewing costs, and even measuring productivity. It relies on interviews, direct observation of activities and examination of procedures, records, and costs. When done properly, the physical audit produces effective, objective and reliable information on the status of maintenance.

The physical audit by itself should be used when personnel cannot be frank, objective and constructive in completing a questionnaire. If the physical audit technique is used, be prepared to spend several weeks conducting the evaluation. The physical audit can be disruptive because of the time required to help explain procedures or participate in interviews. Therefore, it must be well organized in advance.

Questionnaire - A questionnaire gives a cross section of randomly selected personnel an opportunity to compare maintenance performance against specific standards. This cross section might include personnel from management, staff departments (such as purchasing), operations and maintenance. Appendix G - Maintenance Performance Evaluation, illustrates the type of standards against which maintenance should be evaluated. Although a questionnaire is subjective, the results are, nevertheless, an expression of the views of plant personnel. Therefore, participants will have committed themselves to identifying improvements they see as necessary. To most of them, this constitutes potential support for the improvement effort that must follow. See Figure 42-1.

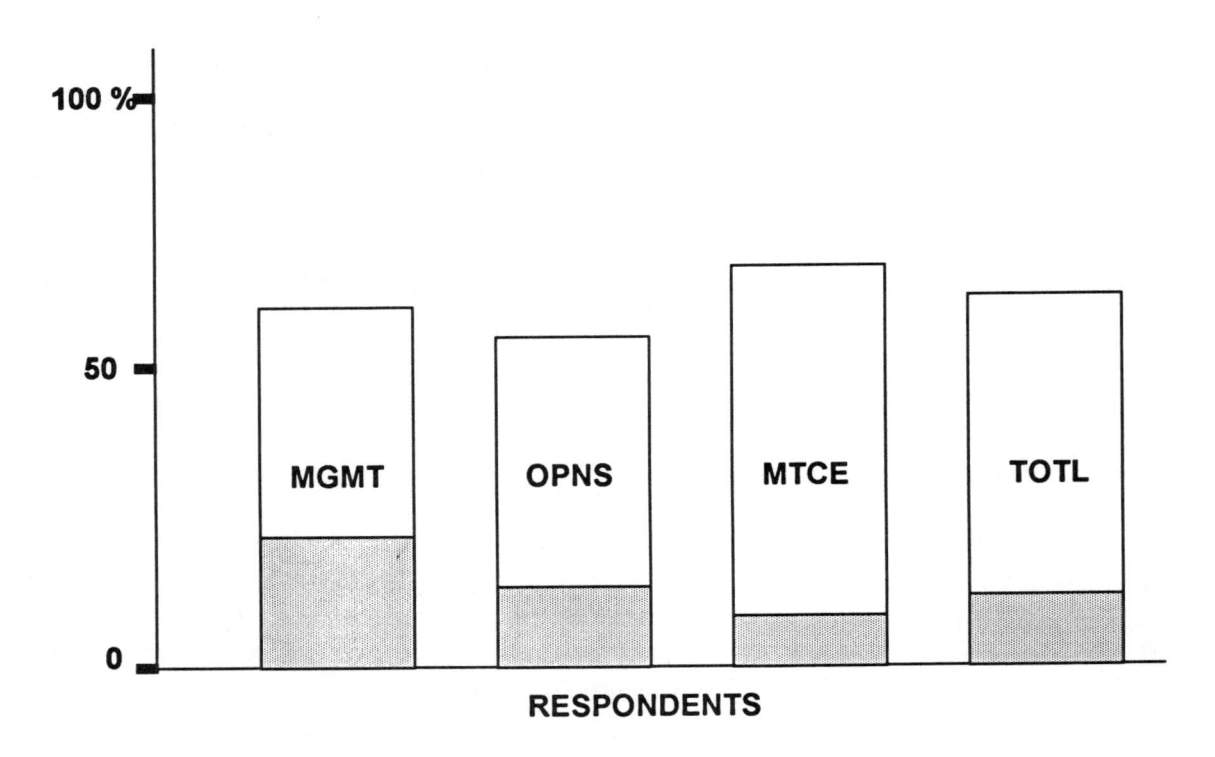

Figure 42-1. For all standards and each area, (like PM or planning), management, operations and maintenance all rate maintenance performance. In addition to the rating scores (average 61 % illustrated), they also indicate the percent of each group that did not know (shaded areas - 9 %). Thus, the questionnaire not only indicates performance but, it also provides a picture of possible training needed, procedures to be clarified or policies required.

Qualifications of Personnel Responding - When administering a questionnaire, be careful to ensure that personnel are qualified to respond. For example, participants outside of maintenance must respond only to those standards on which they have personal knowledge. The evaluation must be administered so that questions on evaluation points can be answered completely.

When carried out properly, the questionnaire can produce reliable results quickly while minimizing disruption to the operation. The questionnaire has the advantage of being administered often so that progress against a 'benchmark' can be measured. For example, one plant was able to establish areas in which improvement was still needed while setting aside areas in which good progress had been demonstrated by previous evaluations. The questionnaire is the best choice when a quick, non disruptive evaluation can serve as a reasonable guide in developing an improvement plan. It must be carefully crafted so that it embraces all of the elements of the maintenance program, like PM, as well as those activities that affect the program, such as purchasing.

A cross section of about 15 percent of the plant population, including operations and staff personnel (like accounting) is adequate to produce good results. Participants should have personal knowledge of maintenance performance for the standards against which they are comparing maintenance. There should be selectivity in who responds to what. For example, the accounting manager could evaluate the quality of labor data reported but, could not evaluate housekeeping.

Questionnaires are rarely of value if they are not administered in a controlled environment. Typically, if distributed for completion at the respondents' leisure, expect poor results because there are too many opportunities for misunderstanding. Questionnaires administered in a controlled environment, where participants can be oriented and their questions answered, are always best.

Physical Audit with Questionnaire - The combination of a physical audit and a questionnaire provides the most complete coverage. The techniques work together. The questionnaire, for example, provides confirmation of physical audit findings. This combined technique is preferred by consultants and corporate teams because it combines the objectivity of the outsider in the physical audit, while the questionnaire helps to educate personnel and gain their potential early commitment to improvement. As the insiders, their help is needed. Without it, little will happen. This combined technique is the best way of preparing for the improvement effort that must follow.

Handling Evaluation Results - By publicizing the evaluation results, there will be clear evidence that both the good and the bad are acknowledged. More importantly, there is a commitment to do something. By sharing the results of the evaluation with plant personnel, there is confirmation that their help is expected in attaining improvements. Conversely, keeping the results a secret will decrease credibility and make improvement actions more difficult. Evaluation results should be discussed openly and constructively so that the personnel who must later support improvements are being brought along.

Responsibility for Poor Performance - A maintenance superintendent should never rationalize that he was solely responsible when the evaluation reveals a poor performance. He probably got a lot of help from many people in the resulting poor performance. For example, plant management may not have provided the necessary guidelines to ensure the maintenance program could be carried out in a positive environment. Similarly, operations may have required so many equipment changes that maintenance was left with inadequate staff to maintain existing equipment.

Likewise, supervisors should not be blamed for disappointing results. Little is gained by trying to fix blame. Maintenance reaches into so many areas that few people would be without some degree of responsibility for the decline of maintenance performance. Always look ahead to determine the current performance level and move forward from there.

Maintenance can only offer their service and hope for cooperation and support. Always work toward creating an environment in which support and cooperation can be assured. By including operations and staff department activities in the evaluation, aspects over which maintenance has little control are assessed and corrections that help maintenance are identified for improvement.

Disclosing Evaluation Results - Mines performing poorly often do not show results to the personnel who participated. Better mines not only share results but seek help in interpreting the results and soliciting recommendations. For example, one successful maintenance manager observed that, "whatever the current performance is, it didn't get that way over night." He approached his task of improving maintenance performance by saying, "since this is what *we* think of *our* maintenance program then let *us* now consider what *we* must do about it." This plant demonstrated that it was already on its way to improving because the support and enthusiasm for doing better was successfully harnessed even before the evaluation was completed. There was sincere involvement.

Getting Support for Improvement - Most plant management personnel are uncertain of the roles they must play in creating an environment for a successful maintenance program. They acknowledge that stating a clear mission, providing guidelines and requiring a definition of the maintenance program are useful approaches to the creation of a positive environment for maintenance. Beyond that, uncertainty exists. Thus, a careful review of evaluation results with management will help considerably to support the improvement effort.

Production personnel generally do not realize the comprehensive nature of the maintenance program and how much its success depends on their support and cooperation. Therefore, the inclusion of operations personnel in the evaluation and a comprehensive discussion of their potential contribution to improvement is essential.

Staff in purchasing, warehousing, accounting and MIS often express a need to better understand the maintenance program so they can support it more effectively. Therefore, a discussion with them on the nature of their support will often assure better service.

Conducting Effective Evaluations

Many maintenance managers themselves acknowledge that their own maintenance programs are not adequately defined, much less properly explained to either maintenance personnel or the rest of the plant. Some admit a need to align basic program elements like PM, planning, scheduling, and maintenance engineering. Evaluations often reveal facts about the maintenance program that were concealed by defensive information reaching managers or the misleading attitudes taken by maintenance personnel.

8. Take Immediate Action on the Results - The most convincing way to demonstrate that the evaluation was a constructive step is to organize an improvement effort immediately. Obtain commitment to the constructive use of the results by converting them into an improvement plan and immediately organizing the improvement effort. This is the main objective of the evaluation. If the evaluation is one of a series, results should be compared with the previous evaluation. This demonstrates progress as well as the identification of areas that need more work. Separate the good from the bad. Offer congratulations on the good performances and organize the activities requiring improvement into priorities. Actively solicit help from anyone capable of providing it. Most will participate willingly. 6.

Thereafter, develop a plan for further improvement and implement corrective actions. If there are corrective actions beyond the capability of maintenance, don't hesitate to seek help.

Plant managers are usually pleased to be asked to help. It is also gratifying to learn that corporate managers, particularly those responsible for multiple plant operations, are eager to help as well.

Set up an advisory group and get underway. Let them first determine why certain ratings were poor. Then, ask for recommendations for improvement. Change the members of the advisory group frequently to encourage different views. As recommendations are made, try them in test areas before attempting plant wide implementation. See Chapter 43 - Converting Evaluation Results into Improvement Actions.

9. **Announce Specific Gains Resulting From the Maintenance Evaluation** - As soon as any gains that can be attributed to the evaluation can be identified, announce them and give credit to the appropriate personnel. People like to know how they did. Tell them. In the process, candor will invariably encourage a greater effort in future evaluations.

10. **Specify the Dates of the Next Maintenance Evaluation** - Announce the dates of the next evaluation immediately. As necessary, identify any additional activities that will be evaluated. Establish new, higher performance targets for the next evaluation. Reinforce the policy of continuing evaluations.

Evaluation Case Studies

There are valuable lessons in observing how several maintenance organizations successfully converted evaluation results into improved performance. The following case studies describe evaluation results and their conversion into corrective actions.

Improving Supervisor Training - One of the areas that evaluations most frequently find in need of improvement is supervisor training. Most plant personnel described a non existent supervisory development program. There were major problems. Newly appointed supervisors were expected to figure out their duties for themselves. Training on the maintenance program was often vague because the program itself was seldom well defined and documented. Supervisors identified for promotion were seldom trained on the broader duties they would be required to perform. There was little time allocated for supervisory training of any kind.

In the little training given, emphasis was given to technical skills, rather than 'foreman ship'. As a result, many supervisors were reluctant to move from 'one craft' crews to the 'multicraft crews' necessary for more flexible organizational arrangements like area maintenance.

There was an impression that management saw supervisors as those who 'fixed equipment' rather than those who managed the efforts of craftsmen in carrying out the maintenance program.

Generally, supervisors were criticized for their lack of supervisory skill. Few had ever received any supervisory training. They were rated well on the technical competence which they carried into their supervisory jobs as they moved from craftsmen to supervisor.

Conducting Effective Evaluations

Most supervisors displayed a tendency toward working directly with their crews rather than leading them. Generally, superintendents were criticized for acting like supervisors rather than managers. In turn, supervisors were criticized for acting like workers. Supervisors were often labeled as 'tire kickers' rather than managers. Poor marks were given for control of work and utilization of labor.

Solution:

- Training should emphasize foremanship.
- Make attendance possible.
- Require completion of training.
- Assess supervisory performance.
- Emphasize the training of new supervisors.
- Check progress.
- Reconsider supervisor selection criteria.
- Look for talent outside of maintenance.
- Weed out those with limited talent.

Monitoring and Improving Craft Training - Craft training programs were well rated and considered successful. Contractual requirements that training for hourly craft personnel be provided was often credited with their success. However, some craft training tended to emphasize what craftsmen felt they needed rather than what the maintenance situation required. One plant learned that crafts that would never use infra red testing techniques were receiving training on it instead of what their program required.

Solution:

- Conduct training needs analysis.
- Get reports on training progress.
- Review content to ensure it meets needs.
- Attend training to demonstrate interest.
- Talk with craftsmen to determine their satisfaction with training.

Defining the Maintenance Program - Few maintenance departments have a solid, well publicized definition of the way they carry out their program and what they expect of their own personnel as well as their customers. Program definition should explain how maintenance services are requested, organized, executed, controlled, assigned and performance measured. There was confusion on how to do these basic things.

Often, operating personnel acknowledged no role other than to submit work requests. There was little joint scheduling of major jobs, and confusion on what was to be done and when. Performance on major jobs was seldom questioned by operations nor were overall maintenance costs. Results usually indicated that since maintenance did not advise anyone how they operated, the misunderstanding was expected.

Solution:

- Verify that the program has been defined and explained.
- Follow up to determine it is understood and followed.

Developing a Maintenance Objective - Many maintenance departments had vague objectives like 'support operations' or 'get the coal out'. Those examined were compared with a reasonable standard:

The primary objective of maintenance is to maintain equipment, as designed, in a safe, effective operating condition to ensure that production targets are met economically and on time. Maintenance will also support non maintenance project work (like construction) as the maintenance work load permits. In addition, maintenance will maintain buildings and facilities and provide support services such as hoist operation or power generation.

Generally, the absence of a clear maintenance objective confused the maintenance program. Typically, maintenance, instead of focusing on preserving assets, ensuring quality product and providing dependable equipment was, too often, diverted into non maintenance activities like: performing 'process optimization' (usually improperly engineered modifications thought to make the process better) or moving, upgrading, modifying and installing equipment or performing construction work. There was also confusion in differentiating between maintenance and non maintenance work, not only within maintenance, but plant wide.

Maintenance is the repair and upkeep of existing equipment, buildings, and facilities to keep them in a safe, effective, as designed, condition so they can meet their intended purpose. Maintenance is an operating expense. Non maintenance, which includes actions such as construction and equipment installation is generally capitalized, depending on the cost. Maintenance departments without clear boundary lines and specific guidelines regarding the division of maintenance and non maintenance work usually failed to carry out the basic maintenance program adequately.

Maintenance managers with responsibility for non maintenance helped create the problem by giving undue emphasis to project work.

Some evaluation results suggested that several mines had what appeared to be low cost maintenance programs. However, in reality, maintenance resources were misused on projects, often over the objection of the maintenance. The maintenance budget appeared favorable but, only because little actual maintenance was accomplished. The rapid deterioration of poorly maintained equipment required its premature replacement. The net result was an unprofitable operation because capital costs were excessive.

Solution - Review the maintenance objective to ensure it clearly establishes the conduct of the maintenance program as the department's primary responsibility. If maintenance is required to perform non maintenance work, ensure there are safeguards to preclude improper utilization of maintenance labor. Make sure maintenance supervisors are not victimized by unrealistic emphasis of non maintenance work, such as modification. Verify that maintenance supervisors are not 'frustrated construction types' who would rather get involved in non maintenance activities like construction or modification while neglecting their primary task of maintenance, the repair and upkeep of existing equipment.

Establishing Maintenance Guidelines - Maintenance, as a service, cannot compel production to comply with its program. Consider therefore, an ambitious operating superintendent, who in a quest for quality and high output, pushes maintenance into 'process optimization' rather than allowing them to carry out the basic maintenance program. Certain 'process optimization' actions (usually modifications) may not be feasible or even necessary. Many should be approved by engineering but often aren't. Some that should be capitalized may be chopped up into incremental expensed actions to avoid the 'gauntlet' of getting capital funding approved. Should maintenance object, they are reminded that their indirect service role does not make a profit. Using this kind of leverage, the ambitious operating superintendent may get his way, right or wrong. In the process, he may, without malicious intent, diminish the maintenance program.

Abuses of the maintenance program also include ignoring published maintenance schedules in the interest of meeting production targets. Such an attitude can lead to equipment neglect. Supervisors know units require servicing or repair but, hope the next shift will have to cope with the breakdown. Downtime is then charged unfairly against maintenance. These scenarios are not uncommon and they undermine the maintenance program.

Evaluations should ask directly whether there are guidelines which could prevent this from happening. The results often show that such guidelines, which can only be issued by management, are weak or missing. As a result, maintenance may feel that operations views the maintenance program less seriously than they should. Very few managers clarified operations responsibilities for utilizing maintenance services effectively.

Solution - Seek management help in obtaining guidelines to preclude these abuses of the maintenance program. Ask management to consider making operations responsible for the cost of maintenance. They will become more demanding of quality work, completed on time and kept under budget. Expect a reduction in downtime as a result.

Standardizing Maintenance Terminology - Few maintenance departments adequately defined terms they used every day. Some vital terminology was often misleading, even within maintenance. The work load for maintenance often could not be identified, much less measured. Especially confusing was the meaning of preventive maintenance, emergency repairs and planned maintenance. Only a few workers could distinguish the backlog from the open work order file. Modification versus corrective maintenance and overhauls versus rebuilds were often confused. In one plant, there were seven different versions of the meaning of PM just within maintenance. The absence of adequate definitions emerges in other areas such as PM, where proper definitions are necessary to spell out effective procedures. Outside of maintenance there was confusion on basic terms.

Solution - Check basic terminology. If it does not exist, require that it be defined. If it does exist, make sure it has been published and verify its correct use. Typical of the type of terminology that must be defined is the basic work that maintenance performs. Appendix C - Maintenance Terminology provides a glossary of maintenance terms.

How to Establish the Maintenance Work load - Identification of the essential work to be performed by maintenance and the determination of a work force of the proper size and craft composition are often poorly carried out.

One maintenance superintendent determined work force size by adding staff until the overtime went down, hardly adequate but, typical of the lack of regard for this matter.

Conducting Effective Evaluations

Many maintenance departments did not know how to go about this task nor did they have elements in their information systems which permitted them to confirm or adjust work force levels.

Solution - Require maintenance supervisors to verify the size and craft composition of their work forces regularly. If they are hesitant, chances are good they have no idea how to do it. It is often necessary to require that they do it, then guide them through some reasonable procedure.

Measuring Productivity - Worker productivity is seldom measured in some organizations, nor are steps taken to explain its importance. Hourly personnel are usually fearful of productivity measurements. Most mines acknowledged that productivity was low, had not been measured and that the quality of labor control would not yield necessary improvements. Invariably, the maintenance supervisor could have improved productivity by giving better work assignments, ensuring work was pre organized and spending more time supervising his crew. Why he did not was seldom investigated.

Solution - Make productivity measurements mandatory. The most effective technique is random sampling, in which the pattern of what personnel are doing is very informative. For example, calculate what percentage of time is spent:

- Working
- Idle
- Traveling
- Waiting
- Performing clerical functions

Also critical is when personnel are doing these things. For example, if personnel are idle at the start of the shift, it usually means that work assignment procedures are inadequate. Similarly, if they are performing clerical tasks at the end of the shift, it often means that labor reporting techniques should be revised.

Productivity Verifies Labor Control - Productivity measurements are the most effective way of verifying the quality of labor control. However, it will be necessary to require maintenance to measure it. But, maintenance can ease the intimidation of the measurement process by measuring the factors that inhibit work. This more positive approach identifies the degree to which factors like not enough ladders rob personnel of productive time. The results provide specific improvement targets that respond to correction more readily because personnel see results when more ladders suddenly appear.

Good Supervision Aides Productivity - Beware of maintenance supervisors who consider anyone breathing to be productive. Educate them. Find out what these supervisors are doing and how well they actually control their crews. Their supervision time is critical to productivity, and they must spend at least 60 percent of their time on active supervision to assure at least 40 percent productivity (not very good). Therefore, make sure supervisors have enough time to supervise. If massive administrative tasks have been heaped on them, the resulting poor productivity of their crews may be not be their fault.

Conduct Effective Preventive Maintenance - The biggest problem with preventive maintenance (PM) is a misunderstanding of what it is. Views range from 'everything done before equipment failed' to 'the whole maintenance program'. This confusion can carry over into the effectiveness with which PM is scheduled, executed and benefits gained. Invariably, those who defined PM as routine, repetitive actions (i.e., inspection, lubrication, testing) had the best control over its effectiveness. Those who did poorly in PM also did poorly in defining terminology.

Many PM programs cannot be administered effectively because inspection and testing are mixed with repairs. Thus, the deficiencies resulting from inspection and testing, which should be separated and classified into emergency or unscheduled repairs or planned maintenance are not. As a result, the identification of work load elements is blurred and precludes determination of the proper work force size and craft composition.

Generally, good use is made of non destructive testing (like vibration analysis). However, such efforts are not always integrated into the overall PM program.

Solution - Require that maintenance personnel present their preventive maintenance program to operations and management. If they can't, they don't have one. If they hesitate, they need to have it challenged or evaluated. Most likely, they may not know what to do or how to organize PM.

PM services, like inspections for fixed equipment, must be organized into routes in which equipment in each is checked at specific intervals. Inspections should be 'detection oriented' and their completion verified. After inspection, deficiencies must be reported and decisions made on their importance before they are converted into new jobs. Although this concept is simple, it is often overlooked. PM is too important to ignore.

Conducting Effective Evaluations

Establishing Effective Planning - Planning is generally not a well executed function. Among the problems:

- There are no criteria describing work to be planned.
- Planning procedures are poorly described.
- Information for control of planned work is sparse.
- There are not enough planners.
- Planners are misused on other work or as relief supervisors.
- Few are adequately trained.

Many maintenance departments considered work to be planned if there was a pause between the work request and work execution. Most played with statistics to arrive at a plausible percent of work they claimed to be planned. Usually, there was a faulty classification procedure in which it was difficult for anyone to determine what work was planned. Often, the routine, repetitive PM services (such as weekly inspections or monthly lubrication services) were considered planned when, in fact, they are merely scheduled. Many large planning departments have failed to cut back their staffs when the use of the computer has displaced most administrative work formerly done by planners.

Solution - The most important ways to initiate planning improvements are to:

- Examine the planning procedure.
- Require a criteria for identifying work to be planned.

Establishing a Responsive Work Order System - Some organizations do not use a work order system. Usually, there is one piece of paper, inadequate for detailed planning but, overly complex for a simple request. No provision is made for the inevitable verbal orders. Efforts to suppress them result in no work order and no information about the job. Standing work orders become a burial ground for costs and repair history that should have been isolated unit by unit. Rarely is there an engineering work order to control non maintenance project work (like construction). Personnel try to use the maintenance work order to control this work when performed by a contractor with poor results. Typical shortcomings found in evaluating work order systems include:

- There is no real system.
- One piece of paper is used for all types of work.
- Big jobs can't be poorly controlled.
- Simple jobs are too hard to control.
- There is no linkage with accounting thus little information.

Solution - Organize a task force made up of accounting, operations, material control and maintenance personnel to evaluate and correct the inadequacies that often emerge under scrutiny. Maintenance will welcome such an evaluation because the work order system is an extension of the accounting system and their essential communication link to it. It is one of those items that affect maintenance performance over which they have no control.

Ensuring Accurate Field Labor Data - Labor use reported by hourly personnel is often carelessly recorded with more attention paid to 'filling up the 8 hours' than providing accurate, timely, and complete data. Some supervisors resort to filling out craftsmen's time cards themselves to get 'something better than what they were receiving'. Most supervisors admit that the labor data they record is largely conjecture when they have 10 - 12 people in the crew each performing 6 - 8 jobs per shift. Hourly personnel readily admit they are the best source of information on what they actually did. They should be given opportunity to report it directly. Often crew members do not report accurately because the importance of reporting is not explained. They are suspicious of the reports use. The end result is missing or misleading information because there is poor field labor data to support it. Most maintenance departments do poorly in reporting labor because of a lack of emphasis rather than a faulty reporting scheme.

Solution - The only way maintenance can reduce the cost of doing their work is to improve their effectiveness installing materials. This makes the control of labor important and it requires attention. Therefore:

- Ensure labor is reported initially by craftsmen.
- The supervisor should verify, not initiate, the information.
- Verify the accuracy, timeliness and completeness of labor data.
- Get accounting help to improve labor reporting.

Getting the Best Material Control Support - Material Control (i.e., inventory control and direct charge purchases) is usually well carried out when managed by material control professionals. When maintenance controls parts inventory, however, there is little continuity between inventory control and direct charge purchasing. Usually, it splits the material control function and creates confusion. Poorly administered maintenance programs usually produce poorly administered inventory control programs when maintenance is responsible for the function.

Identification of Parts a Problem - A major material control problem is the adequate identification of parts. Often supervisors are forced into this role because hourly personnel feel procedures are difficult or too time consuming. As a result, parts identification tasks seriously reduce the time supervisors should have been in the field controlling work. In turn, this situation produces lower worker productivity.

Solution - Check maintenance downtime against causes such as no materials available, wrong materials issued or waiting for materials. These factors are the best indicators of the state of material control. Make sure the material control function is not split, with purchasing under accounting and stores under maintenance, for example. It seldom works.

Watch several stock room transactions to learn how well maintenance craftsmen actually are able to identify materials. Check the supervisor's office. If it looks like a library of parts books, this clerical activity is probably his principal job. Check the 'bootleg' storage areas if you wish to learn the awful truth of why material costs are so high. Check the foreman's desk drawers and wall lockers if you want to find out where the real stockroom annexes are. Don't threaten a forced clean up, as too many usable items will end up in the waste dump rather than returned to the storeroom.

The resolution begins with a physical check such as the one outlined above. However, the solution lies in a serious, joint corrective effort between maintenance, accounting, purchasing and the stockroom staff. Within maintenance, instruct everyone on how the material control program works. Often they don't understand it.

Using Information Systems Effectively - Information systems seldom cover decision making information aspects adequately. This information includes: cost, repair history, backlog, labor utilization, the status of major jobs and performance indices such as cost per operating hour. There is often an over abundance of administrative information on items such as parts cross references, equipment lists or personnel who were absent.

Thus, while everyone can easily look up parts, few can adequately manage maintenance because information systems emphasize administrative rather than management information. Systems provided by the corporate level, in an effort to achieve uniformity between several mines, do not work well. Some maintenance organizations purchase stand alone software packages only to find that they need an integrated system.

But, they cannot obtain local support to create necessary communications software. Some package programs are incompatible due to both language and logic considerations. Often commercial packages fail to link adequately with field labor and material data from time cards and stock issues or purchasing documents. Most organizations are not satisfied with the timeliness, completeness and accuracy of information.

Solution - Specify the performance indices required by management such as cost to install each dollar of material. These will equate to specific information required to create the indices. Make certain this information exists. Ignore the clerical and administrative information such as error reports and parts lists. These are merely the means by which maintenance carries out its internal communications. Virtually every package program has these in great profusion. Instead, concentrate on whether maintenance actually has decision making information to manage the overall function. Mandatory are:

- Labor utilization
- Backlog
- Costs of units and components
- Status of major jobs
- Repair history

Once the right information has been obtained, ensure it reaches the people who need it to make their decisions. If required, instruct them how to use it.

Optimizing Maintenance Engineering - Maintenance engineers are often misused on non maintenance engineering projects like equipment installation or modification. Not enough attention is paid to functions that ensure equipment maintainability and reliability. A large percent of maintenance engineers are young and have degrees in a specific engineering discipline. They tend to gravitate toward what they understand best (project work) rather than maintenance engineering. Mines contribute to this problem by failing to adequately define maintenance engineering.

Solution - Require that maintenance engineering focus on equipment reliability and maintainability. Ensure that solutions like effective PM or repair standards are instituted to reinforce reliability objectives. Verify procedures to determine that newly installed equipment is able to be maintained. Does the equipment have maintenance instructions, spare parts lists, and wiring diagrams, for instance. Ensure that maintenance engineers are not misused on engineering projects.

Supporting Engineering Projects with Maintenance Resources - Invariably, major engineering projects are carried out with maintenance personnel rather than by a contractor. Too often, this work is not controlled and the maintenance program suffers from a shortage of available labor. Usually these projects are well planned and organized because they are in the spotlight. At the lower end of the cost scale, proper funding is questionable with frequent attempts to expense the work against the maintenance budget rather than seek capital funds through proper channels. Projects tend to be executed on time but, often at the cost of diverting maintenance labor from necessary maintenance work. Most projects have direct attention from management and are monitored by corporate personnel. These aspects, in part, explain their success.

Solution - Curb the tendency to misuse maintenance resources on non maintenance projects by developing guidelines that preclude the unfair diversion of maintenance labor from its own program. Check lower cost project funding practices to make sure maintenance is not unfairly paying the tab for a project that should have been capitalized.

Ensuring Support for Maintenance - Operations and staff departments show an appreciation for the value of a good maintenance program but, lack a specific means of providing support beyond saying "I'm for good maintenance."

Solution - Document specific instances in which operations have caused problems by not making equipment available when maintenance is due to be performed. Visit with the production superintendent and suggest how the situation may be improved through positive solutions such as a weekly, joint scheduling meeting. Measure how well staff services such as the warehouse are supporting maintenance and suggest how they might improve. If management needs to step in, suggest guidelines that could help classify the supporting role of other departments.

Attaining Better Housekeeping - There is a tendency in maintenance to allow discarded components to build up, creating junk piles. There is poor follow up in getting components into the rebuild pipeline, especially if done commercially.

Solution - Extend housekeeping inspections into the areas where maintenance tends to collect things they might be able to use it later. Encourage operations to clean areas before maintenance work begins. This not only helps work progress but, it builds good interdepartmental relationships.

Forming Maintenance Teams - Establishing maintenance teams from traditional organizations is proving difficult for many mines because they are not making the transition properly. Many are failing to ensure their maintenance programs will sustain the new organization. Others are trying to downsize concurrently with forming teams and are experiencing considerable difficulty gaining credibility from personnel involved.

Solution - A basic requirement for forming successful maintenance teams is to verify the maintenance program and ensure it will serve as a suitable framework for the transition and beyond. Downsizing and forming teams have mutual objectives of transforming less efficient larger organizations into smaller more effective ones. However, the massive behavioral changes necessary for either preclude trying them simultaneously.

Summary - Evaluations establish current maintenance performance level by identifying those activities needing improvement as well as those being performed well. The evaluation not only confirms performance but, is the starting point for any improvement effort. The most important result of an effective evaluation is the education of personnel and their commitment to providing support for improving maintenance.

1. - 6. - Images copyright, New Vision Technologies, Inc.

43

Converting Evaluation Results into Improvement Actions

Participation Yields Improvement Results

The successful implementation of improvements in maintenance will require changes in behavior of not only key personnel but, workers as well. Since change is difficult for many adults, expect resistance to change in implementing improvements.

Among the most effective means of reducing resistance to change is education and participation in the change process. Typically, if a person understands why a change is necessary, he will be more willing to consider how such a change can be brought about. Then with his direct participation, he can be in a position to influence exactly how the change is to be made. Often, if it affects him directly, his participation will remove any misunderstanding and potential threat to him or his feeling of security.

After a decision has been made on what the change should be, it is advisable to test the changes in the working environment. Thus, there is a further opportunity to modify the change to better suit operating circumstances. But, more importantly, the participants come to realize that these modifications, many of which they have suggested, increase the control they have over their own destiny.

Thus, with education and participation, resistance is soon displaced with acceptance of changes and positive, willing support.

The end result is improved performance through the successful implementation of changes suggested by an evaluation.

Technique

The use of two advisory groups has proven to be a very effective way to bring about the required changes while educating them and providing opportunity for participation. An action group and a decision group are typical of the way personnel can be organized to ensure active involvement in the implementation of changes in the maintenance organization and its program.

An action group might consist of middle managers and hourly personnel who must ensure that the maintenance program gets carried out properly. They would review improvement needs and recommend practical ways to implement them. Typically, an action group might consist of several craftsmen, a maintenance supervisor, a planner, an operator and an operating supervisor.

Working with the action group is a decision group consisting of management personnel who will approve or modify recommendations made by the action group prior to their implementation. Typically, the decision group would consist of the plant manager, the maintenance superintendent, the operating superintendent and staff personnel like the personnel manager or accounting manager.

Augmenting the action and decision groups is a coordinator who would schedule meetings, prepare agendas and document recommendations and actions approved by the decision group. The coordinator could be the maintenance superintendent, a planner or a supervisor not in the action group. Assisting the coordinator is a secretary who could be any person who can provide word processing support in the preparation of agendas or summaries and recommendations. Files would be created since the recommendations, as subsequently tested, would become the future maintenance manual.

Among the tasks that might performed by the action group are:

- Recommending training for craft personnel
- Helping to define maintenance terms
- Looking over the PM program for consistency
- Examining the method of identifying stock materials
- Reviewing the criteria for identifying work to be planned
- Review the method for recording equipment modifications

Converting Evaluation Results Into Improvement Actions

The action group would meet weekly (during lunch) for no more than 45 minutes. Each meeting would have a specific agenda dealing with a single topic. Written material corresponding to the topic would be provided. See Figure 43-1.

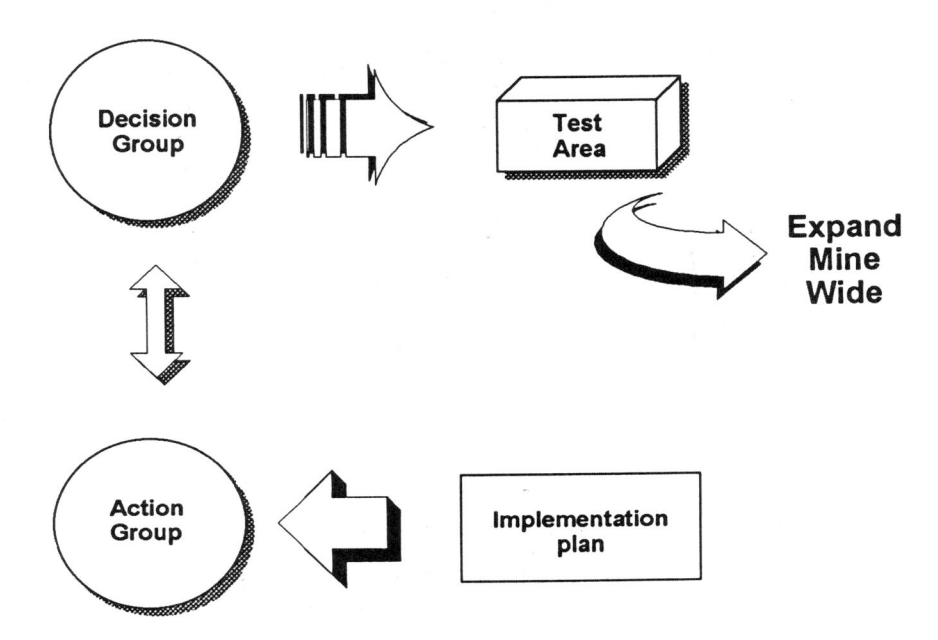

Figure 43-1. Using the implementation plan as a guide, the action group considers each improvement task and develops practical ways to bring them about. Recommendations are presented to the decision group who may approve, disapprove or modify them. Approved actions may be implemented directly or tested before plant wide implementation.

Each meeting on a new topic would define the problem clearly before the action group develops possible recommendations. Some topics might be under discussion for several meetings. Recommendations need not be written but, they must deal with the specific topic. At subsequent meetings, verbal recommendations will be made and recorded. Summaries of recommendations will be prepared for the record and to brief the decision group.

Every 4 weeks the action group would meet with the decision group to present its recommendations. Resulting decisions, if not provided during the meeting, would be communicated to the action group subsequent to the monthly meetings.

Each designated member of the action group would have a back up who would attend if the designated member cannot attend. Every 4 weeks at the meeting with the decision group, the designated members as well as the back up will attend.

At the conclusion of this meeting, the designated members will drop out of the group to be replaced by their back ups. As this happens, new back up members would be appointed. The same personnel may be in the action group but, membership for any two consecutive one month periods should be avoided if possible. A notebook is provided to each member of the action group to keep agendas, explanatory materials and summaries. The notebook will be brought to each meeting. It should be kept where either the designated or back up member can access it. Action group members are encouraged to discuss topics with anyone who can provide constructive comments.

Once a course of action has been agreed on, the activity should be tested in a pilot area. Action group members would observe results and, as necessary, modify the procedure being tested (with the advice of those operating the test area). Further testing would continue until the desired performance level is achieved. Thus, the final result represents the full participation of all personnel involved while providing them with an education on the changes themselves and why they are necessary.

The use of these dual level implementation groups offers some distinct advantages that are noteworthy. Action group members are deliberately rotated monthly because an underlying purpose is to educate as many as possible rather than just a few. Thus, the education process moves quickly and deliberately. The frequent turnover produces fresh solutions rather than the same solution to all problems. This avoids the loss of interest often experienced in permanent committees. Often, action group meetings are held at the test site rather than in a meeting room. This permits action group members to confer directly with personnel testing their recommendations. Testing is carried out in a controlled environment in which the before, during and after results are measured. Thus, results are confirmed. The monthly meeting with the decision group brings the action group face to face with the mines' decision makers. As a result, not only are discussions frank and honest but, agreement is reached quickly.

Summary - Involvement and participation are always the best ways to overcome the resistance to change which can imperil improvement efforts. Regardless of the technique used, the very personnel involved in the changes must be give sufficient opportunity to offer constructive recommendations and influence their own destiny. This assures successful implementation of necessary improvements.

Appendices

Appendix A

The Maintenance Objective

Establishing the Objective

The maintenance objective is the mission assigned by the plant manager to maintenance. It links maintenance with other key departments ensuring their mutual support of the mines' production strategy. Consider a typical objective:

The primary objective of maintenance is the repair and upkeep of production equipment to ensure that it is kept in a safe, effective, as designed, operating condition so that production targets can be met on time and at least cost. A secondary objective of maintenance is to perform approved, properly engineered and correctly funded non maintenance work (such as construction and equipment installation) only to the extent that such work does not reduce the capability for carrying out the maintenance program. In addition, maintenance will operate support facilities (such as hoist operation or power generation), but will ensure that necessary resources are allocated within its authorized work force and are properly budgeted. As appropriate, maintenance will also monitor the satisfactory performance of contractor services.

Key Points of Objective

There are several key phrases used in the illustrated maintenance objective, each with a specific intent. A primary as well as a secondary objective are provided to establish clear precedence. That is, production equipment maintenance is first while project work follows, resources permitting. Further, the use of phrases such as 'as designed' means that equipment modification is excluded from the primary maintenance task (it is not maintenance). This implies that all non maintenance work to be approved, properly engineered and correctly funded. Such work may only be done if it does not reduce the capability of maintenance to meet its primary function.

If maintenance is to perform operating functions such as power generation you should ensure that they are properly staffed and budgeted to do so.

With an objective such as that cited, there should be little question about maintenance priorities and limits as to what and how much it can do.

Based on a clear objective, maintenance knows its responsibilities exactly and may organize properly to carry them out. Its customers are also aware of maintenance limitations and will request support accordingly. The objective also helps to ensure that the intended maintenance program can be carried out effectively.

The absence of a clear objective can yield unfortunate consequences. For example:

In a large processing plant, the 300 person maintenance work force was having difficulty carrying out the maintenance program. During shutdowns it was necessary for them to be augmented by contractors to catch up. An investigation showed that up to 30% of work requests made by operations were not maintenance. Rather, they were equipment changes, modifications or installations. As a result, nearly 35% of maintenance manpower was used on this work detracting from manpower available for basic maintenance.

Once the plant manager saw these circumstances, he clarified the maintenance objective and educated personnel in the proper use of the maintenance work force. The objective helped to carry out this task successfully.

Appendix B

The Maintenance Work Load

Defining the Work Load

The maintenance work load is the amount of essential work performed by maintenance and the identification of the correct number of personnel by craft, working at a reasonable level of productivity, to carry it out. Any definition of the maintenance work load must begin with a definition of maintenance. Consider:

The repair and upkeep of existing equipment, facilities, buildings or areas in accordance with current design specifications to keep them in a safe, effective condition while meeting their intended purposes.

The work load for maintenance consists of the maintenance work they perform plus, in many instances, non maintenance work.

Maintenance

Scheduled Maintenance - Extensive major repairs such as: rebuilds, overhauls or major component replacements requiring advanced planning, lead time to assemble materials, scheduling of equipment shutdown to ensure availability of maintenance resources including: labor, materials, tools and shop facilities.

Preventive Maintenance - Any action which can avoid premature failure and extend the life of the equipment. It includes equipment inspection, testing and monitoring to avoid premature failures and lubrication, cleaning, adjusting and minor component replacement to extend equipment life.

Emergency Repairs - Immediate repairs needed as a result of failure or stoppage of critical equipment during a scheduled operating period. Imminent danger to personnel and extensive further equipment damage as well as substantial production loss will result if equipment is not repaired immediately. Scheduled work must be interrupted and overtime, if needed, would be authorized in order to perform emergency repairs.

Unscheduled Repairs - Unscheduled non emergency work of short duration often called 'running repairs'. Work that can be accomplished within approximately one week with little danger of equipment deterioration in the interim. Repairs are usually performed by one person, often in two hours or less, and in about 40% of instances parts are not required.

Routine Maintenance (repetitive work) - Janitorial work, building and grounds work, training, safety meetings or shop clean up and highly repetitive work such as tool sharpening.

Non maintenance

Non-maintenance (project work) - Construction, installation, relocation or modification of equipment, buildings, facilities or utilities. Usually capitalized.

Construction - The creation of a new facility or the changing of the configuration or capacity of a building, facility or utility.

Installation - The installation of new equipment.

Modification - The major changing of an existing unit of equipment or a facility from original design specifications.

Relocation - Repositioning major equipment to perform the same function in a new location.

Appendix C

Maintenance Terminology

The Language of Maintenance

Terminology is the language of maintenance. It includes definitions by which plant personnel can communicate clearly about maintenance. The most widely used terms are defined using common, practical definitions found in the work place.

Adjustments - Minor tune up actions requiring hand tools, no parts and less than one half hour.

Administrative Information - Information used to communicate within maintenance and operate the maintenance information system.

Area Maintenance - A type of maintenance in which one supervisor is responsible for all maintenance within a reasonably sized geographical area.

Autonomous Maintenance - The operators step by step activities when searching for the optimal conditions of operations and production while establishing basic conditions of cleaning, adjustment and lubrication in a self controlled environment. Often linked with TPM.

Backlog - The total number of estimated man hours, by craft, required to perform all identified, but incomplete planned and scheduled work.

Capital Funded - Non maintenance work authorized by a capital fund authorization.

Capitalized - Funding for work which expands the plant operating capacity, gains economic advantage, replaces worn, damaged or obsolete equipment, satisfies a safety requirement or meets a basic need.

Category - The types of work which make up the work load performed by maintenance: preventive maintenance, emergency repairs etc.

Component - A sub element of a unit of equipment: the belt of a conveyor, the motor of a crusher, the engine of a truck.

Concept - The means by which a major program, such as maintenance, is carried out in relationship to its objective and the other programs which it supports (operations) or on which it depends (purchasing).

Coordination - Daily adjustment of maintenance actions to achieve the best short term use of resources or to accommodate changes in operation needs.

Cost Center - A department or area in which equipment operates or in which functions are carried out.

Decision making Information - Information necessary to control day to day maintenance and determine current and long term cost and performance trends for management decisions.

Deferred Maintenance - Maintenance which can be postponed to some future date without further deterioration of equipment.

Downsizing - The conversion of a larger organization into a smaller one while taking necessary actions to preserve the efficiency and effectiveness of the organization.

Downtime - A time period during which equipment cannot be operated for its intended purpose.

Empowerment - To allow employees a definitive role in the control of their activities.

Engineering Work Order - A control document authorizing use of the maintenance work force or a contractor for engineering project work such as construction.

Equipment and Function File - A computer file which lists all equipment by type, with its components, plus functions carried out within a cost-center.

Expensed - Maintenance work charged to the operating budget.

Expert Level Maintenance - Maintenance based on long, proven experience in the correction of equipment problems.

Maintenance Terminology

Failure Analysis - Study of equipment failure data to determine and correct chronic, repetitive equipment problems and reduce or eliminate equipment failures.

Failure Coding - An indexing of the causes of equipment failure on which corrective actions can be based.

Forecasting - A projection of anticipated major tasks that are predictable based on historical data.

Function - An activity carried out on a unit or performed within a cost center: 250 hour service on haulage truck number 110, power sweeping in department 06.

Functional Maintenance - Maintenance in which the supervisor is responsible for conducting a specific function like pump maintenance for the entire plant.

Guidelines - General guidance provided by management as maintenance or other departments carry out their day to day functions. See policies.

Inspection - The checking of equipment to determine repair needs and their urgency.

ISO 9000 - A set of quality assurance standards that can be applied to any organization regardless of size or type. Used to develop a common approach to obtaining quality service or product.

Job Order System - See work order system.

Just in Time (JIT) - A state which assures that the necessary quantities of the right materials are available when needed.

Level of Service - The degree of maintenance performed to meet desired levels of equipment performance. A high level ensures little chance of failure while a low level meets minimum requirements risking breakdowns on less critical equipment.

Maintenance Engineering - The use of engineering techniques to ensure equipment reliability and maintainability.

Maintenance Information System - A means by which field data are converted into information so that maintenance can determine work needed, control the work and measure the effectiveness of the work done.

Maintenance Work Order (MWO) - A formal document for controlling planned and scheduled work.

Maintenance Work Order System - A means of requesting maintenance service, planning, scheduling, assigning, controlling work, and focusing field data to create information.

Maintenance Work Request (MWR) - An informal document for requesting unscheduled or emergency work.

Major Repairs - Extensive, non routine, scheduled repairs requiring deliberate shutdown of equipment, the use of a repair crew possible covering several elapsed shifts, significant materials, rigging and, if needed, the use of lifting equipment.

Minor Repairs - Repairs usually performed by one man using hand tools, few parts and usually completed in less than two hours.

Multicraft - Requires personnel to possess more than one trade.

Multiskill - Maintenance that blends necessary craft skills allowing personnel to do a job from start to finish. Also refers to flexible trades, cross trades, and multicraft maintenance.

Non Destructive Testing - Testing to determine equipment condition using methods like oil sampling or dye penetrant. See Predictive Maintenance.

Objective - The principal purpose for the existence of each line department (like maintenance) or staff department (like purchasing) and the roles they must play to assure that the plant production strategy is achieved.

Overhauls - The inspection, tear down and repair of a total unit of equipment to restore it to effective operating condition in accordance with current design specifications.

Performance Indices - Ratios which convey short term accomplishments and long term trends against desired standards.

Maintenance Terminology

Periodic Maintenance - Maintenance actions carried out at regular intervals. Intervals may be fixed (every six months) or variable (every 4500 operating hours).

Planning - Determination of resources needed and the development of anticipated actions necessary to perform a scheduled major job.

Policies - Management guidelines for the development of field procedures to assure achievement of plant profitability. See also guidelines.

Predictive Maintenance - Techniques to predict wear rate, determine state of deterioration, monitor condition or predict failure.

Principles - Logic, common sense, proven procedures or essential rules on which plant operation must be based.

Pro active Maintenance - The application of investigative and corrective technologies to extend equipment life

Procedure - The day to day method for carrying out elements of the maintenance program such as assignment and control of work. Field procedures should be built on policies.

Production Strategy - The plan for achieving plant profitability.

Productivity - The percentage of time that maintenance personnel are at the work site, with their tools, performing productive work during a scheduled working period.

Priority - The relative importance of a single job in relationship to other jobs, operational needs, safety, equipment condition, etc., and the time within which the job should be completed.

Project Work - Actions such as construction, equipment modification, installation or relocation to gain economic advantage, replace worn, damaged or obsolete equipment, satisfy a safety requirement, attain additional operating capacity or meet a basic need. Usually capital funded.

Purchase Order - The authorized document for obtaining direct charge materials or services from vendors or contractors.

Quality Standard - A standardized procedure for accomplishing a major maintenance task.

Quantity Standard - The standard resources required to meet the prescribed quality standard.

Rebuild - The repair of a component to restore it to serviceable condition in accordance with current design specifications.

Relocate - Move fixed equipment to a different location.

Reliability Engineering - Actions taken through the use of information, field experience and engineering techniques to design or redesign equipment to reduce or eliminate faults that imperil equipment reliability.

Reliability Based Maintenance - The combination of preventive maintenance, predictive maintenance and pro active maintenance to assure the reliability and effective life of equipment.

Repair History - A record of significant repairs made on key equipment used to spot chronic, repetitive problems, failure patterns and component life-span which, in turn, identifies corrective actions and helps forecast component replacements.

Repetitive Maintenance - Maintenance jobs which have a known labor and material content and occur at a regular interval.

Reposition - Move mobile equipment to a new working location.

Routine Maintenance - Maintenance or services performed consistently in the same manner.

Schedule Compliance - The effectiveness with which an approved schedule was carried out.

Scheduling - Determination of the best time to perform a planned maintenance job to appreciate operational needs for equipment or facilities and the best use of maintenance resources.

Maintenance Terminology

Self Directed Work Team - A team of maintenance or operations and maintenance personnel in which each member shares equally in decision making, control, conduct of work and accountability for results.

Specifications - Technical definition of equipment configuration or performance requirements to meet intended utilization of equipment or materials.

Standing Operating Procedures (SOP) - A written procedure used to ensure reasonable uniformity each time a significant task is performed.

Statistical Predictive Maintenance - Non destructive testing, predictive techniques combined with statistical techniques to help predict future maintenance needs.

Stock Issue Card - The authorized document for making stock material withdrawals.

Standing Work Order - A reference number used to identify a routine, repetitive action.

Team Coordinator - The focal point of work control within a team. Often rotated to ensure the team orientation is preserved.

Time Card - The authorized document for reporting the use of labor data.

Total Productive Maintenance (TPM) - Productive maintenance carried out by all employees through small group activities. TPM is equipment maintenance performed on a plant wide basis.

Type - All equipment of the same kind: conveyors, pumps, haulage trucks, loaders etc.

Unit - One unit of equipment of a specific type: conveyor 006.

Utilization - The percentage of time that a maintenance crew is available to perform productive work during a scheduled working period.

Verbal Orders - A means of assigning emergency work when reaction time does not permit preparation of a work order document.

Work force - The personnel who carry out the maintenance workload.

Work load - The essential work to be performed by maintenance and the conversion of this data into a work force of the proper size and craft composition, working at reasonable productivity, to ensure the maintenance program is carried out effectively.

Work Order System - A communications system by which maintenance work is requested, classified, planned, scheduled, assigned and controlled.

Work Sampling - The statistical measure of the utilization of labor to determine productivity.

Appendix D

Maintenance Policies

Guidelines for Maintenance

Maintenance policies are guidelines established by managers to preclude misunderstanding of the roles and responsibilities of key departments. Policies are the basis for developing day to day procedures. Typical policies are suggested below as they relate to specific maintenance activities.

Responsibilities

Each department manager will ensure compliance with the policies covering the conduct of maintenance.

Each department will develop and publish procedures by which other departments may obtain their services.

Operations will be responsible for the effective utilization of maintenance services.

Maintenance will be responsible for developing a pertinent maintenance program, educating personnel on its elements and carrying it out diligently. They will also make effective use of resources and ensure that quality work is performed.

Terminology

Maintenance will publish work load definitions and appropriate terminology to ensure their understanding and proper utilization.

Work Order System

The work order system will be used to request, plan, schedule, assign and control work.

Preventive Maintenance

Maintenance will conduct a detection oriented preventive maintenance (PM) program. The program will include equipment inspection, condition monitoring and testing to help uncover equipment deficiencies and avoid premature failure. The PM program will also provide lubrication services, cleaning, adjusting and minor component replacement to help extend equipment life.

PM will take precedence over every aspect of maintenance except bona fide emergency work.

No major repairs will be initiated until the PM program has established the exact condition of the equipment and elements of the repair have been correctly prioritized.

The overall preventive maintenance program will be assessed annually by maintenance engineering. They will ensure that it covers all equipment requiring services and that the most appropriate types of services are applied at the correct intervals. The performance of the PM program in reducing equipment failures and extending equipment life will be verified.

Equipment operators will perform appropriate preventive maintenance services to help ensure the reliable operation of equipment.

Planning and Scheduling

Planning and scheduling will be applied to comprehensive jobs (e.g. overhauls, major component replacements, etc.,) to ensure that work is well organized in advance, properly scheduled and completed productively and expeditiously.

A criteria will be provided to help determine which work will be planned, and all major repairs will be subjected to planning procedures unless an emergency repair is indicated.

Information

Maintenance will develop and use information concerning the utilization of labor, the status of work, backlog, cost and repair history to ensure effective control of its activities and related economic decisions such as equipment replacement.

Maintenance Policies

Minimum necessary administrative information will be developed and used.

Performance indices will be used to evaluate short term accomplishments and long term trends.

Organization

The maintenance work load will be measured on a regular basis to help determine the proper size and craft composition of the work force.

The productivity of maintenance will be measured on a regular, continuing basis to monitor progress in improving the control of labor.

Every effort will be made to implement and utilize organizational and management techniques like teams or Total Productive Maintenance (TPM) to ensure the most productive use of maintenance personnel.

Priority Setting

Maintenance will publish a priority setting procedure which allows other departments to communicate the importance of work and maintenance to effectively allocate its resources.

The procedure will facilitate the assignment of the relative importance of jobs and the time within which the jobs should be completed.

Maintenance engineering

Maintenance engineering will be emphasized to ensure the maintainability and reliability of equipment.

Current technology will be utilized to facilitate effective maintenance.

No equipment will be modified without the concurrence of maintenance engineering or the application of maintenance engineering techniques.

All new equipment installations will be reviewed by maintenance engineering to ensure their subsequent maintainability.

Material control

Established procedures for purchasing or the withdrawal of stock will be strictly adhered to by maintenance personnel.

Parts will not be removed from any unit of equipment and used to restore another unit to operating condition without explicit authorization from the maintenance superintendent.

Maintenance will return unused stock materials to the warehouse. Maintenance will not attempt to store such materials.

Non maintenance work

Engineering, operations and maintenance are jointly responsible for ensuring that all non maintenance projects (construction, modification, equipment installation etc.,) are necessary, feasible, properly engineered and correctly funded before work commences.

Maintenance is authorized to perform project work such as: construction, modification, equipment installation and relocation only when the maintenance workload permits. Otherwise, contractor support will be obtained subject to the current labor agreement.

Equipment modifications will be reviewed to determine their necessity, feasibility and correct funding prior to the work being assigned to maintenance. All such work will be reviewed by maintenance engineering before work commences.

Appendix E

Duties of Key Maintenance Personnel

Outline of Principal Duties

Maintenance Superintendents - The superintendent or manager is responsible for the entire maintenance function. He performs this through line supervisors such as general supervisor or first line maintenance supervisors. He uses maintenance engineers, in a staff capacity, to ensure the total maintenance program has continuity and that all new construction or installations can be supported by the maintenance program. He utilizes maintenance planners to develop plans for resource use, on major work that must be scheduled to ensure least interruption of operations and best use of maintenance resources. As necessary, he may use clerks to perform administrative tasks related to pay, vacations, absenteeism, etc. In small organizations, he may control purchasing and stockroom functions, if this is not done by accounting.

Supervisor - The principal duty of the maintenance supervisor is work control and the supervision of his crew. He is responsible for carrying out the workload assigned him. The supervisor generally carries out PM services, unscheduled and emergency work on his own initiative. Major planned and scheduled jobs appear on an approved schedule prepared by a supporting planner and jointly approved by maintenance and operations. The supervisor must instruct and train his crew members, arrange their vacation time and discipline them as necessary. He prescribes work methods and procedures. He also ensures the timely completion of work by adequately supervising it. He is responsible for the conduct of PM services, the control of labor (including overtime) and the procurement of materials for unscheduled and emergency work.

Maintenance Engineer - The maintenance engineer is a staff person responsible to the superintendent for ensuring that equipment and facilities are properly installed, modified correctly, if needed, correctly maintained and performing effectively. Specific responsibilities include: assessment of newly installed equipment to ensure maintainability, parts, maintenance, instructions, prints, adequacy of design and installation work. In addition, he monitors on going work to ensure that sound craftsmanship procedures are followed.

He also observes the adequacy of stockage of correct parts in proper quantity. Further, he monitors the quality of parts used to ensure good performance and recommends action to correct inadequacies. He reviews repair history and costs to determine repair, rebuild, overhaul or corrective maintenance needs. The maintenance engineer also reviews the conduct of PM services to ensure they are conducted on time, according to standard. He reviews equipment inspection deficiencies uncovered as a source of corrective maintenance. He helps to develop standards for individual major jobs, procedures for performing work, cost levels and the quantity of resources used. Periodically, he makes a cost and benefit review to determine make or buy actions. He also develops recommendations for training. As required, he prescribes methods for non-destructive testing and predictive maintenance. He reviews the information system to ensure its adequacy. He uses information to verify the work load and backlog with the view of making work force level change recommendations.

Maintenance Planner - The maintenance planner provides direct support in planning major actions such as overhauls, rebuilds or non maintenance work prior to execution by maintenance supervisors and crews. These planning actions include job organization, labor estimating, identifying and obtaining materials and coordinating work with operations or engineering. Once work is scheduled, the planner allocates labor according to priority and monitors the control of job execution. In addition, the planner monitors conduct of the preventive maintenance equipment inspections. He also analyses repair history to convert results into planned maintenance work. The planner should confine his activities to planned and scheduled maintenance and project work. Unscheduled and emergency work are controlled by maintenance supervisors.

Leadman - The leadman is regular maintenance craftsman who, in consideration of an hourly wage increment and any special talents, can be used temporarily to help a supervisor coordinate elements of major crew jobs. The leadman should not be considered as a supervisor since he cannot be expected to discipline his fellow crew members.

Team Member - Individual craftsmen who shares the responsibility for work control with other team members.

Team Coordinator - A rotating coordinator within the team who provides for the control of work.

Appendix F

Predictive Maintenance Techniques

Technique Applicability

The traditional preventive maintenance techniques of inspection, lubrication and testing along with cleaning, adjusting and minor component replacement are time based, occurring at prescribed intervals. Predictive techniques, on the other hand, are condition based. Equipment is monitored continuously and any abnormal condition signals the need for a corrective action.

Predictive techniques offer the advantage of not having to shut the equipment down to perform the task as with a physical inspection, for example. Thus, the cost of the surveillance done with equipment rather than personnel is less than the cost of labor required in applying preventive techniques like physical inspections.

The overall preventive maintenance program provides a framework for the control of both the traditional preventive maintenance tasks along with the newer, more sophisticated predictive techniques. Both preventive and predictive techniques have the same objectives of avoiding premature equipment failure and extending the life of the equipment. In turn, each contributes to the effectiveness of maintenance by finding problems before they result in equipment failure and finding them soon enough to allow work to be planned and organized. This contributes to better productivity and less down time to perform repairs.

While there is value in clearly identifying the various elements of predictive maintenance and applying them properly, they should not be considered as a separate program apart from preventive maintenance. They are a part of the overall preventive maintenance program with the same objectives, controls and purpose.

A far more important aspect of the maintenance program strategy is to ensure that the most effective combination of preventive and predictive techniques are being used to support the overall maintenance effort. Thus, this appendix has the objective of stating the purpose of each technique to allow the reader to establish the best combination within his maintenance program.

The most prominent of the predictive techniques are vibration analysis, infrared thermography, ultrasonic testing and oil analysis.

Vibration Analysis

Vibration analysis uses signals generated by a machine defect as the source to indicate a defect. Mechanical objects vibrate in response to the pulsating force of a machine defect. As defects grow in magnitude, the vibration increases in amplitude. Changes in the intensity of the signal indicate a change in the magnitude of the defect or increase in its deterioration. Interpretation of the signal permits diagnosis of the problem.

Electronic instrumentation is available today that can detect with accuracy extremely low amplitude vibration signals. They can isolate the frequency at which the vibration is occurring. When measurements of both amplitude and frequency are established, both the magnitude of a problem as well as its probable cause can be determined. Thus, the application of vibration analysis permits:

- Detection of impending problems
- Isolation of conditions causing excessive wear
- Confirmation of the nature of the problem
- Advance planning to correct the problem
- Performing the repair at the best time

Vibration analysis is probably the most widely used predictive maintenance technique. It can be used in any industrial environment where rotating or reciprocating equipment can cause financial penalty if it fails. Vibration analysis can reveal problems in machines involving mechanical or electrical imbalance, misalignment or looseness. It can uncover and track the development of defects in machine components such as bearings, gears, belts and drives.

Shock Pulse Diagnosis (Bearings)

A useful diagnostic tool for detecting failing bearings in operating machines is shock pulse analysis. It is practical for checks on the bearings of moving equipment, such as trucks and locomotives, and the bearings of stationary equipment. This technique is closely related to vibration analysis and is often applied when vibration analysis techniques are not able to be used in certain operating environments.

Predictive Maintenance Techniques

Shock pulse differs from conventional vibration monitoring in that it is insensitive to normal vibrations inherent in the structure of a working machine, and focuses strictly on the high frequency 'shock' signal characteristic of a damaged machine element. Like other predictive techniques, shock pulse gives ample warning of impending bearing failure, allowing maintenance to be done in a timely and orderly manner with the least disruption of machine and production time.

Instrumentation works on the principle that damaged bearings and other mechanical components generate abnormal amounts of high frequency energy from mechanical shock and friction. A special accelerometer attached to the structure of the machine transmits a signal containing the shock signatures. The analyzer unit filters out low frequency signal components. It integrates the resulting shock rate amplitude function, compares it with data stored in memory, assesses the mechanical condition of the bearing, and displays the result. Similar mechanisms can be inspected periodically and their shock functions compared with predetermined limits stored in memory. Out put can be a simple good or bad bearing indication.

Periodic inspection of a specific mechanism can also be made. A trend can be established by comparing current data with previous data. Thus, the rate of deterioration can be assessed and the expected life of the component predicted. On larger, more complex machines with numerous bearings, continuous diagnosis is used.

Infrared Thermography

Infrared (IR) thermography allows the detection of invisible thermal patterns that reveal potential equipment deficiencies. It identifies any problem that can be associated with excess temperature or lack of heat. The IR survey is conducted with equipment in normal operation and without equipment downtime.

While maintenance can apply a great variety of predictive techniques, most are restricted regarding the types problems they can detect. Thermography, however has a much wider scope of applications than most other predictive techniques. Electrical inspections, because they are so fast and cost effective are the principal utilization of thermography. Other maintenance problems, including mechanical problems where friction produces a thermal pattern, also reveal themselves thermally. Thermography is a useful supplement to vibration analysis.

Tank levels, under the right conditions, can be clearly seen because of differences in heat capacity. Valves and steam traps, if working normally exhibit a temperature difference that points to costly defects. As insulation or refractory breaks down or reduced thermal resistance produces a thermal pattern showing the extent of the damage. The insulation in flat roofs, when damaged by moisture or a leak, also can be located.

Don't always look for a 'hot spot'. A plugged transformer fin or line to an oil cooler, for instance, will be evident because it lacks heat when heat is the normal operation. Typical applications include:

Electrical

- Faulty connections on all types of equipment
- Defective potheads, bushings, surge arrestors and other sub station equipment
- Poor contact at fuse clips, disconnect switch blades, busway stabs and motor starter and relay movable contacts
- Defective circuit breakers and switch gear
- Phase unbalance on three phase circuits
- Overheating due to overload or harmonics
- Overheated motor or generator bearings or brush rigging
- Blocked transformer cooling fins and failed components such as capacitors and semiconductors (detected by absence of heat)

Process Equipment

Proper heat distribution or cooling in kilns, reactor vessels, coke drums, extruders and other process equipment

General Mechanical

- Defective bearings on all types of equipment
- Excessive friction on sliding surfaces
- Belt or pulley, clutch and brake slippage
- Improper gearing fit or lubrication

Steam and Other Fluid Systems

- Defective boiler refractory
- Faulty piping or insulation

- Valve leaks and improper, steam trap operation
- Air leaks (detected as excessive cooling)
- Blocked radiator fins
- Thermal distribution in cooling towers
- Sedimentation or improper liquid level in tanks

Buildings

- Identifying trapped moisture in roofs
- Determining heat and energy loss

IR imagers are basically infrared cameras. They provide a snap shot of how actual temperatures compare with the temperatures that are expected, based on temperatures surrounding the equipment area being scanned.

Output from a thermal imager can be recorded for later use or review. Typically, the image is recorded on conventional video tape that can be viewed or transferred into a computer. Several imaging systems now record directly to data disk, facilitating transfer into a computer database. The thermal image, once in the program, can be printed in a report format and stored along with pertinent temperature data for trending analysis.

Regardless of the type of imaging system used, data are interpreted from qualitative comparisons of the thermal patterns. Such comparisons form the basis for most of the analysis in predictive maintenance. Whether one phase is compared to another in an electrical system or the upstream and downstream sides of a valve or trap are compared, locating a problem by its relative temperature is usually the basis of analysis.

Thermography can define a problem better. The effectiveness of repairs also can be monitored. One of the best uses of infrared thermography is to show contractors how well their installations perform. It should be used to establish the baseline signature of new or modified equipment. That equipment should be inspected periodically, as it operates, to monitor deviations from the baseline. Thermal changes often precede problems. Solutions to problems that cannot be found with other test methods may be revealed with thermography.

Prices for complete imaging systems range from $12,000 to over $100,000. With such an investment, maintenance should clearly determine its requirements carefully.

Ultrasonic Testing

Ultrasonic inspection is used extensively on piping materials for the detection of surface and sub surface flaws and wall thickness measurement. Ultrasonic inspection methods utilize high frequency mechanical vibrations. The fundamental principle involves a controlled and uniform beam of ultrasonic energy which is directed by a transducer into a test material. The energy will be transmitted with little loss (or attenuation) through a homogeneous material. It will be either attenuated, or reflected, by discontinuities or defects in the physical structure existing in a second location in the same material or in another material. The measurements of the energy change, either by attenuation or reflection, are the criteria employed.

To illustrate, steam and air leaks can be quite costly if steam is wasted through faulty valves, malfunctioning steam traps, or poorly designed systems. In a steam system with 150 LB of pressure and a production cost of $ 6 per thousand pounds, a leak 1/32 " in diameter, no larger than the tip of a ball point pen, can cost $ 300 a year. Such leaks in any plant system can be detected easily with the help of ultrasonic equipment.

Sound is usually caused by a vibrating body. The sound source produces a vibration that causes surrounding molecules to vibrate back and forth a small distance. As a result, each particle transmits motion to its adjoining particle away from the sound source. A sound wave has two main characteristics: tone (frequency) and loudness (amplitude). Tone is defined by the number of vibrations or cycles per second. Loudness depends on the force of the vibrating object and is expressed in decibels.

An ultrasonic contact module is placed against bearing to detect early wear of failure. As the bearing goes from good to bad, the amplitude of the signal increases from 12 to 50 times. Ultrasonic sounds are sound vibrations above the range of human hearing. For most humans, 16 to 18 kHz is the upper limit, although a few people can hear sounds up to 21 kHz. Most operating equipment in an industrial plant emits ultrasonic signals that can warn of impending failure.

The basic ultrasonic detection unit contains the electronic circuitry to translate the ultrasound into the audio range. The short waves of ultrasound have difficulty penetrating more than two media (for example, air and metal) therefore, it is best to use the contact method to listen for leaks in by pass steam traps, valves, or behind walls.

Steam Traps - One of the sources of the greatest steam losses is faulty steam traps. Using ultrasonic tools, maintenance personnel can isolate the area being tested by eliminating confusing background noises. Once accustomed to the various sounds, personnel can quickly adjust to recognizing differences among the three types of steam traps: mechanical, thermostatic, and thermodynamic.

Valves and Piping - It is often extremely difficult to determine if a valve by pass or blockage is occurring within a piping system while it is on line. An ultrasonic contact module can easily monitor valves. If a valve is leaking internally, ultrasonic emissions are generated at the orifice. As a liquid or gas flows through a pipe, little or no turbulence is generated except at bends or obstacles. In the case of a by pass valve, the escaping liquid or gas goes from a high to a low pressure area, creating turbulence on the low pressure or down stream side. The ultrasonic component of this noise is much stronger than the audible component, especially in minute leaks. When partial blockage exists, a condition similar to that of a by pass valve is produced. The blockage also generates ultrasonic signals. If a partial blockage is suspected, a section of piping should be inspected at 2 to 3 foot intervals. The ultrasound generated within the piping will be greatest at the site of a partial blockage. Trying to detect leaks in piping behind walls, under floors, and underground can be frustrating. Because ultrasound is transmitted at the source of the leak, and is directional in nature, leaks can be located by probing the outer surface (wall, floor, ground) with a contact module. In underground leaks, when loosely packed dirt or sand do not transmit ultrasound well, a rod or wave guide should be used.

Pressure Leaks - Ultrasonic sensors can simplify locating pressure and vacuum leaks in most systems and reduce the search time. An ultrasonic emission is created by turbulence at the orifice site of both pressure and vacuum leaks, and is heard as a hissing sound through the audio channel of an ultrasonic probe. Internal leakage can be determined using a stethoscopic detection head. When placed in contact with exterior wall, the unit detects sound emissions at the orifice site. The detection head pinpoints sound emissions that may occur in pressure of vacuum leakage situations.

Ultrasound is directional. As a result, the location of a leak can be pinpointed readily. Because of the small radiating surface of a leak, the ultrasonic emission signal is strongest when the sensor is in direct line with the leak.

Vacuum Leaks - Vacuum leaks are located in the same manner as pressure leaks except the ultrasonic emission is drawn toward rather than driven into the atmosphere. Therefore, the signal will not be as intense as a pressure leak with the same flow rate and orifice size. Ultrasonic sensors also are capable of monitoring and detecting electrical discharges. An arc or a corona discharge emits ultrasound at the site of emission. An electrical discharge can be located quickly by screening. As the point of discharge is approached, the signal in the ultrasonic range intensifies. It may be heard as a frying or buzzing sound in a headset. Often, the intensity of the ultrasound is relative to the severity of the problem. High voltage lines can be scanned easily for corona, wires in a junction box can be checked for poor contact and arcing in circuit breakers can be pin pointed without removing cover plates.

Tone Test - Some ultrasonic sensors, using a transmitter, are capable of testing non pressurized items for potential leakage as well as outright leakage. The ultrasonic transmitter is small, compact, and battery operated. The signal it transmits penetrates a void but not a solid. If the sensor is placed in a pipe, the sound cannot penetrate a solid wall but, it can penetrate a void in the pipe. When an exterior wall is scanned, the signal is detected at the site of penetration (the void). This method can be used to check empty pipes and tanks before they are put on line. It is also useful in spotting faulty welds, seals and gaskets. Air infiltration problems around doors and windows can also be checked.

Heat exchangers and condensers can be tested for weakened and leaking tubes. The tube bundles are surrounded with intense ultrasonic transmitters. The ultrasound penetrates leaks in the tubes and actually vibrates thin spots in the tubes. When the tube sheets are scanned with an ultrasonic sensor, the tubes with thin walls and leaks carry the tone. The good tubes do not. In tone test, tone generators are placed around tubes, surrounding bundles with ultrasound. Scanning the open end of the tube sheet locates leaking or thin walled tubes. A thin walled tube does not register as intensely on a meter as a leaking tube.

Bearings - Ultrasonic testing of bearings can help avert costly failures by detecting faults, and can permit scheduling of bearing changes at convenient times. A contact module is used to detect the ultrasonic signals. The modules contain a wave guide (probe) connected to a transducer. The vibrations excite the transducer, and the signal is converted into an audible output suitable for human hearing. Some ultrasonic instruments are provided with a meter display in addition to the audible output. The meter display is very important when readings are taken on a periodic basis by different personnel.

Predictive Maintenance Techniques

When the audible sound translated by the ultrasonic instrument is heard as a soft rushing noise, it indicates a good bearing. A much louder grinding or crackling noise indicates a bearing that is beginning to fail. Because the ultrasonic signals do not carry through more than one interface, the signal does not have interference from background noise. This feature is very helpful in high noise areas. The audible signal also helps determine what the bearing problem might be. For example, a severely loaded bearing with a signal that increases sharply, accompanied by a sharp cracking or grinding noise, should be replaced at the first opportunity. A gradual rise in the meter reading, accompanied by a smooth rushing noise, indicates possible lubrication failure. If cleaning and re lubrication return the meter reading to the original level, the problem is solved. Otherwise, the bearing is worn and should be replaced.

Other Malfunctions - Operating machinery can malfunction for reasons other than bearing failure. Among these are valve disorders, gear problems, and internal hydraulic leaks. When a valve malfunctions, a change in ultrasonic emission occurs. If a valve is stuck open, passage of a gas or fluid creates ultrasonic turbulence. When it is closed, no sound is emitted. When a valve is leaking, for example, on a compressor, an ultrasonic contact module can be employed. A quick determination is possible using the translated audio signal of the ultrasonic sensor. By touching the surface near the valve seat with the sensor, the worker can listen to the distinct sound of the valve opening and closing. He can hear the valve by pass as a muffled clicking sound. This sound can be compared to the definite clicking sound of a properly functioning valve. While a gear is in operation, it emits ultrasound. Changes in the regular movement of a gear create changes in the ultrasonic pattern. Worn gears or gears missing teeth generate a shock wave at the site of the fault. This shock wave is easily detected through ultrasound because of the short wave nature of the transmitted signal. It may be a cyclic rubbing or clicking sound. Internal hydraulic leaks are difficult to isolate. Often many man-hours are lost trying to trace the telltale dripping of fluid to its origin. Touching the exterior surface of the suspected areas with a contact ultrasonic module makes it possible to determine the leaking site. The ultrasonic turbulence, a rushing sound, is greater at the point closest to the leak source.

Oil Analysis

It is estimated that about 30 % of all engine failures are caused by metal particles contaminating the lubrication systems. Oil degradation causes residues and sludge to form, which decreases equipment efficiencies and can often destroy machinery components.

As one of the most prominent non destructive testing techniques that make up the preventive - predictive arsenal, oil analysis identifies specific wear mechanisms and contamination levels. Generally, there are two types of testing for:

- Metals to detect wear rates, inorganic contaminants and inorganic additive levels
- Physical properties to detect contamination from glycol, water, fuel, combustion by products, particulates, acidity, basicity and viscosity

Wear particle analysis is an excellent method for examining the lubricated components of a machine without requiring down time for analysis of wear particles. Lubricating oil carries within it particles generated by a machine during operation. These particles can be analyzed for relative size and concentration electronically. When examined, the size, shape, and make up of these particles will give valuable information on the wear mode and wear rate of a machine. This analysis can make early detection of developing problems possible, allowing maintenance to determine when the machine should be repaired, before unscheduled failure.

Often overlooked in the preventive maintenance program is the additional factor of contaminant inducted failure. Contaminants may be solid particles, air, moisture and chemicals. Therefore, it is important to consider actions such as monitoring fluid cleanliness and ensuring adequate filtration to round out the oil analysis effort.

Magnetic Particle Testing

Magnetic particle inspection is a non destructive method for detecting cracks, seams, porosity, lack of fusion or penetration, or laminations at or near the surface in ferro magnetic materials. Since the process depends on the magnetic properties of the materials to be tested, this method of examination cannot be used on aluminum, brass, copper, bronze, magnesium, stainless steel or other non ferromagnetic materials. The magnetic particle inspection method utilizes magnetic fields to reveal discontinuities in materials. A magnet will attract magnetic materials only when it has ends or poles. If there are no external poles, magnetic materials will not be attracted to a magnetized material even though there are magnetic lines of force flowing through it. When a discontinuity is created in the circular ring, the effect is to establish minute magnetic poles at the edge of the discontinuities.

This pushes some of the magnetic lines of force out of the metal path and creates an attraction for the formation of magnetic particles at the discontinuity. Satisfactory results may generally be obtained when the surfaces are in the as welded, as rolled, as cast, or as forged condition. However, surface preparation by grinding or machining may be necessary in some cases where surface irregularities would otherwise mask the indication of discontinuities. Prior to magnetic particle examination, the surface should be dry and free of dirt, grease, lint, scale, welding flux or other extraneous matter.

The surface to be examined is magnetized by means of high amperage current, either by placing prods on both sides of the area to be inspected, or by wrapping the surface with a coil. The prod method is the most widely used for the inspection of piping materials and weld. With the current on, the area under investigation is dusted with finely divided iron particles to provide a light, uniform coating. A stream of air is generally used to remove particles not affected by the magnetic field. Regardless of the manner of producing the magnetic flux, the greatest sensitivity will be to linear discontinuities lying perpendicular to the lines of flux. To ensure the most effective detection of discontinuities, each area to be examined should be tested twice with the lines of flux in one test approximately perpendicular to the line of flux in the other. If there is a crack or defect in the part either at or near the surface, it will cause a discontinuity in the magnetic flux lines. The iron particles will collect at this point, outlining the defect. Surface defects will produce sharp, distinct patterns which are tightly held to the surface of the part. The patterns become broader and fuzzier as the defect causing indication is progressively deeper beneath the surface. There are certain metallurgical discontinuities and magnetic permeability variations which may produce magnetic particle indications similar to an indication from a mechanical discontinuity. Although these types of indications are termed non relevant, they cannot be accepted as such when they exceed the criteria of acceptance unless they are proven to be non relevant by additional testing.

Electrical Predictive Maintenance

Less commonly know among non electricians are a multitude of test instruments that are used quite routinely in electrical predictive maintenance.

Many of them are used in conjunction with the more traditional preventive maintenance techniques while others have a singular application.

A Guide to Test Instruments for Electrical Preventive Maintenance

- Multi tester, ac and dc clamp on ammeters, industrial analyzer, watt meter, power factor meter, high-voltage 'hot-stick' volt meter, magnetic voltage indicator, digital meters - Measure power and control circuit voltage, current, resistance, and power; useful for trouble shooting and obtaining values for preventive maintenance comparison records.
- Insulation-resistance tester, thermometer, psychrometer - Test and monitor insulation conditions of conductors, motors and transformers. Use thermometer and psychrometer for temperature and humidity corrections.
- High-potential dc tester - Proof-testing of insulation of conductors and motors.
- Recording meters - For permanent record of voltage, current, power and temperature for analytical study.
- Osciliographic recorder - For analysis of wave forms. Provides permanent record.
- Circuit breaker and motor overload relay tester - Checks trip point and elapsed time of molded case CBs and over load relays.
- Tachometer - Determines rpm of rotating machines
- Transistorized stethoscope - Detects faulty rotating machine bearings and leaking valves.
- Vibration analyzer - Detects excessive vibration and locates imbalance on rotating components; for dynamic balancing and alignment.
- Temperature indicating meter - Determine operating temperature of equipment.
- Infrared sensor - Useful in detecting failing bearings and connections
- Phase sequence indicator - Determines direction of motor rotation prior to startup.
- Low-resistance tester - Checks low resistance paths of switches and contacts.
- Fault locator - Detects opens, shorts and grounds in cable and determines location of fault.
- GFCI tester - Determines condition of ground fault circuit interrupters.
- Oil dielectric tester - Tests transformer, OCB and other insulating oils.
- Vacuum tube multimeter, tube and transistor checker, oscilloscope, capacitance and resistance bridge, expanded scale meters - Used for electronic and solid state circuit testing.
- Protective relay tester - Used to set, calibrate and test switch gear protective relays.
- Maximum load indicator - Useful for testing for overloads and balancing load on phases.
- Noise level meter - Determines sound levels of equipment.

Current Monitoring Techniques

Typical of the techniques used in industry to monitor equipment performance are on board computers used to monitor the status of important haul truck, loader or shovel operating functions from systems like drives or engines. Monitored data are stored at intervals along with any faults as they occur. Information such as payload, cycle times and number of loads hauled are monitored and stored during each haul cycle or working period.

Data are accessible from the operators instrument panel and can be down loaded for subsequent analysis. Typical of the functions monitored are:

Lighting

- Clearance lights
- Dynamic retarding lights
- Headlights, fog lights, stop and tail lights
- Ladder lights
- Payload lights
- Rotating beacon
- Turn and clearance lights

Safety Equipment

- Brakes, parking, brake lock, low brake pressure
- Differential pressure
- Horns
- Steering bleed down solenoid
- Steering pressure
- Warning lights and alarms
- Windshield Washer

General Equipment

- Air system after cooler heater
- Air tank drain valve heater
- Body up signal
- Empty loaded body signal
- Engine starting
- Ether start control
- Hoist limit solenoid

- Override switch, electric drive
- Payload sensors and inclinometer
- Radiator fan clutch
- Radiator pressure
- Radiator shutters
- Steering system filter

Cab

- Accumulator precharge
- Air pressure
- Alternator tertiary current
- Battery boost current
- Battery equalizer monitor signal
- Battery voltage
- Display dimming
- Engine air filter vacuum
- Engine coolant temperature
- Engine oil pressure
- Engine speed (tachometer)
- Fail diode signal
- Fuel level
- Gage display selection
- Blower pressure
- Hoist pump filter switch
- Hoist up limit override control
- Hour meter
- Hydraulic oil filter
- Hydraulic oil level
- Hydraulic oil temperature
- Ignition system relay
- Machine speed (speedometer)
- Motor armature current
- Motor field current
- Payload
- Retard setup signal
- System fault signal
- Throttle solenoid signal
- Volts alternator, field
- Wheel motor over temperature
- Wheel motor speeds

Predictive Maintenance Techniques

One of the obvious problems with the level and sophistication of the data that can be generated is the problem of what to do with it.

A well organized maintenance organization with an effective program and a competent information system will benefit tremendously from such data. Conversely, a poor maintenance organization will be buried in data and get little value out of their investment in such monitoring devices.

This underscores the value of arranging the various predictive techniques into a well ordered maintenance program in which the potential of each can be realized.

INEFFECTIVE	BASIC	EFFECTIVE
Reaction to problems as they occur: • Emergencies • Downtime • High cost • Low productivity	Application of: • Inspection • Lubrication • Testing Cost improvement Less downtime Better productivity	Addition of : • Predictive techniques • Reliability Engineering Dramatic Improvements
Progressive Phases of Program Development		

As predictive techniques are added to the traditional preventive maintenance actions, a maintenance organization can affect dramatic improvements in cost, downtime and productivity.

As the overall maintenance program is being developed, organized or improved, there is a need to focus the preventive actions and predictive techniques so they will yield their greatest potential.

Maintenance should avoid a piece meal assignment of predictive techniques throughout the organization, with separate groups responsible for different functions. While a degree of specialization is necessary, the results of the complete application of all the predictive techniques should be brought together organizationally so that their impact can be realized.

Therefore, serious consideration should be given to a formal maintenance engineering organization in larger mines and plants so that the overall predictive program can function effectively. In addition, the performance of the predictive program in identifying equipment problems and correcting them should be presented regularly to key managers to make them aware of the potential of these techniques in avoiding unnecessary down time and operating costs.

Familiarity with the techniques of predictive maintenance is only the starting point of realizing their potential. They must be placed within the context of an effective maintenance organization with an efficient program to assure their value.

Appendix G

Maintenance Performance Evaluation

About the Maintenance Performance Evaluation (MPE)

The MPE allows a plant to determine the performance of its maintenance organization by having a cross section of personnel rate maintenance against a series of performance standards. These standards are contained within the 34 sections listed below.

01 - Objective	18 - Planning
02 - Program definition	19 - Scheduling
03 - Policies	20 - Work control
04 - Terminology	21 - Engineering
05 - Organization	22 - Standards
06 - Teams	23 - Material control
07 - Supervision	24 - Safety
08 - Labor Control	25 - Housekeeping
09 - Workload	26 - Operations
10 - Productivity	27 - Staff
11 - Backlog	28 - Management
12 - Motivation	29 - Shop operation
13 - Craft training	30 - Support services
14 - Supervisor training	31 - Mobile equipment
15 - Preventive maintenance	32 - Non-maintenance
16 - Work orders	33 - Cost control
17 - Information	34 - General

Ratings are converted into percentile scores to permit the organization to identify sections and standards within each section that are in need of improvement. Thereby, a plan of action can be developed. Based on the scores, improvement priorities can be set within the improvement plan. In addition to scores for each standard, there are separate scores for management, operations and maintenance. Thus, if there are widely differing scores for the same standard, the group associated with each score can be identified. Each standard and each group within the standard also shows a score indicating the percent of the group that did not know the answer.

Thus, widely differing scores between two groups may be due not to a poor performance but, rather to an need to better educate personnel.

Just as sections and standards in need of improvement are identified, so also are sections and standards in which a good, solid performance is indicated. In either instance, MPE results should be reported promptly to plant personnel, especially those who participated in the evaluation. This step gains credibility for the evaluation and support needed to implement necessary improvements. Similarly, plant personnel are always pleased to learn they are doing well.

Administration of the Maintenance Performance Evaluation - The plant should randomly select personnel from management and staff, operations and maintenance to participate in the evaluation. Management and staff include personnel such as the plant manager, the purchasing agent, the warehouse supervisor or an accounting clerk. Personnel from operations would include anyone in operations from the superintendent to an operator. Maintenance includes anyone in maintenance. Regardless of plant size, no more than 25 should participate in the ratio: management and staff - 20%, operations - 32% and maintenance - 48%. In the case of 25 total participants, 5 would represent management and staff while 8 would be from operations and 12 from maintenance. There should be a vertical slice of each organization in each group. The maintenance group, for example, might include the superintendent, a general supervisor, the maintenance engineer, a planner, four supervisors and four craftsmen of different crafts. Each participant would receive a booklet and an answer sheet. The answer sheets are coded indicating the group. A is management and staff, B is operations and C is maintenance. Each group answer sheet is further divided into levels, as follows: 1 - manager, 2 - superintendent, 3 - supervisor or team leader, 4 - an engineer, 5 - a planner or scheduler and 6 - an hourly employee. Thus, the answer sheet coded C3 would be filled out by a maintenance supervisor or a team leader. In the event of a self directed team, all team members would fill out the C3 answer sheet rather than C6. Individual answer sheets have a 0 printed in selected spaces indicating that the individual should not rate the standard. **After distributing the booklet and answer sheets to each participant, the administrator should read the instructions below with the participants.** You are to rate each standard that is not marked with a 0 from 1 to 10 (highest). If you do not know the answer mark an X indicating you do not know. The X's are important because they signal a need for education or a policy or procedure. Read every standard and in the event that you can answer a standard marked with a 0, strike through the 0 and rate the standard from 1 to 10. A rating of 1 to 3 indicates a poor performance, while 4 - 6 indicates a fair performance. A rating from 7 - 8 is a good performance while a 9 is excellent and a 10 leaves no room for further improvement.

Maintenance Performance Evaluation

Be as frank and objective in your ratings as possible. Take special care to identify standards in which you simply do not know the answer for these are activities in which improvement can be easily achieved simply through clarification. The selection of different levels of personnel within a group also helps in the validity of the evaluation. A superintendent, for example, may have issued what he feels is a procedure worthy of an 8. However, personnel trying to make the procedure work may be having difficulty. Thus, a lower rating might signal that the procedure is difficult to implement or in the case of the X rating, some personnel may not be aware of it. The evaluation requires about 3 hours to complete. Answer sheets should be checked to ensure that every space has been completed or has a 0 already there. In a multi department maintenance organization separate MPE's should be administered to each department. In a multi-phase processing plant, for example, each plant would perform a separate MPE. The rated answer sheets are scored by computer. If the MPE is administered on site by us, the report will be ready the day following MPE administration. If the plant administers the MPE and mails the answer sheets, the report will be returned within one week.

The MPE Report - The report is divided into four parts:

Part 1 describes the technique for presenting the results to plant personnel, developing an improvement plan and organizing personnel to help in the improvement effort.

Part 2 lists the percentile scores of all 34 sections and the overall MPE score.

Part 3 is a narrative of the scores for each section. It explains what each score means, suggests the standards that require improvement and identifies the activities in which a better explanation or education may be necessary.

Part 4 provides the detailed scores for each standard in every section in the following format:

		MGMT	OPNS	MTCE	TOTL
The standard being rated is printed	R	53	56	70	57
in this location	X	20	33	00	21

In the case of the scores illustrated, management and staff rated the standard at 53% with 20% who did not know the answer. Operations rated the standard at 56% with 33% not knowing the answer. Maintenance rated itself at 70% and everyone could rate the standard (X =00). The weighted average score was 57 % with 21 % not knowing the answer.

Maintenance Performance Evaluation Performance Standards

01 Objective - The maintenance objective is a clear statement of the essential activities to be performed and their priorities. A typical maintenance might be:

The primary objective of maintenance is to maintain equipment, as designed, in a safe, effective operating condition to ensure that production targets are met economically and on time. Maintenance will also support construction as its resources permit.

1. There is a maintenance objective and it is stated clearly.

2. The objective describes the important functions to be performed to support the plant's production plan.

3. The maintenance objective has been effectively communicated to the organization.

4. The objective has eliminated confusion about what maintenance should be doing and its priorities.

02 Program Definition - The program by which maintenance services are requested, organized and carried out should be clearly defined. The definition should ensure that procedures for obtaining, controlling and executing services are described. As a result, plant personnel are able to carry out or support the program effectively. Program definition may take the form of diagrams or written procedures however, the simpler, the better.

1. The maintenance program has been well defined.

2. It describes how maintenance services are requested, planned, scheduled, assigned, controlled, carried out and their cost and performance measured.

3. Program definition has been communicated to all plant personnel.

4. As a result of program definition, plant personnel carry out or support the program effectively.

03 Policies - Because maintenance is a service organization, it cannot compel operations to comply with their program. Therefore, management must provide policies to guide departments as they provide, receive or support maintenance. Policies ensure that important procedures are followed plant-wide. Typical:

Maintenance will develop and carry out a preventive maintenance program to inspect, lubricate, test, adjust and clean equipment to extend equipment life and avoid premature failures.

Operations will cooperate and ensure that all prescribed services were carried out effectively as scheduled.

1. Management has provided policies to guide maintenance in the development of its program.

2. Operations has policies to follow in utilizing maintenance services.

3. As a result of clear policies, the maintenance program is made easier to carry out.

04 Terminology - Terminology includes definitions by which plant and maintenance personnel can communicate effectively. Definitions should be communicated to plant personnel and used correctly.

1. Terminology has been defined.

2. It has been communicated to plant personnel.

3. Terminology is complete. It includes, for example, definitions of workload elements like preventive maintenance or emergency repairs.

4. Terminology is used correctly. Personnel, for example, can distinguish an overhaul from a rebuild.

05 Organization - The maintenance organization should be responsive to the needs of the plant while using the talents of its personnel most effectively.

1. Organizational charts are complete and correct.

2. Organizational charts cover all shifts.

3. The organizational structure facilitates communications.

4. Job descriptions describe actual duties.

5. The organization encourages working efficiency.

6. Criteria exist for the selection of key personnel.

7. People respond quickly to changes and new procedures.

8. Procedures on planning, work control etc., are complied with.

9. Maintenance controls functions like preventive effectively.

10. Maintenance responds promptly to work requests.

11. Personnel make decisions automatically on actions over which they have control.

12. During peak vacation periods, maintenance still gets its work done.

13. Crews are assigned to shops or production areas based on actual work required.

14. Craft jurisdiction is not a problem.

15. Different crafts work together well.

16. Personnel are not overloaded with extra duties which keep them from carrying out their principal duties.

17. Maintenance performs activities such as boiler operation or construction well.

18. The maintenance organization is checked periodically to ensure it can meet its objective.

06 - Teams - Team members should be empowered to make decisions regarding control and conduct of work.

1. Team members are well informed on the maintenance program.

2. Team procedures for the receipt and classification of work are effective.

Maintenance Performance Evaluation

3. Procedures for assigning work to team members function well.

4. Assigned work is performed cheerfully.

5. Team members have a criteria for determining which work requires planning.

6. Team members have worked out an effective procedure for planning work.

7. The most important work is identified and given proper priority.

8. Team members report work completion correctly.

9. Required field data such as labor or parts used are correctly reported.

10. Team members are able to identify and obtain materials used in their work properly.

11. Each team member can use the computer effectively to perform functions required in his work area.

12. Necessary administration such as updating schematic wiring diagrams or equipment specification files is done correctly

13. A cooperative working relationship has been established with operations personnel.

14. Non maintenance work is coordinated with engineering before it is attempted by team members.

15. Team members work together harmoniously.

16. Skill levels among team members are clearly established.

17. Training to improve team skill levels is provided.

18. Morale is excellent among team members.

19. Team productivity is outstanding.

20. Overall team performance is excellent.

21. Team achievement has inspired other groups to follow the implementation pattern used by the team.

22. Management has openly complimented team members for their accomplishments.

23. Numerous beneficial recommendations for further improvement of performance have been offered by team members.

24. Team members are confident of their accomplishments.

25. As a result of demonstrated performance, other plant organizations are anxious to follow the example of the team performance.

07 Supervision - General supervision provides overall direction for the maintenance work force. First line supervisors control the personnel who perform the work.

1. The best candidates are chosen as supervisors.

2. Selection of senior maintenance supervisors is not restricted to maintenance personnel. (Talent from production or engineering is used to enrich maintenance).

3. A clear line of promotion exists for supervisors.

4. Promotion criteria are prescribed to reach each supervisory level.

5. The position of supervisor is eagerly sought after by qualified craft personnel.

6. Supervisors have specific responsibilities and sufficient authority to carry them out.

7. Supervisors have adequate time and opportunity to carry out their key duties.

8. Supervisors are rated on their performance and advised of the results.

9. Supervisors who perform poorly are counseled to help them develop more effectively.

10. Incompetent supervisors are promptly replaced.

Maintenance Performance Evaluation

11. Supervisors promoted from craft ranks seldom request return to their tools.

12. Maintenance supervisors understand the maintenance program.

13. The maintenance superintendent delegates effectively.

14. Supervisors carry out the maintenance program effectively.

15. Supervisors counsel their crew members effectively.

16. Supervisors are effective in leading and motivating their crews.

17. Maintenance supervisors are technically competent.

18. Supervisors are effective in their administrative tasks (such as control of absenteeism and overtime, follow up on status of jobs, preparing reports etc.).

19. Cooperation between maintenance supervisors is effective.

20. Cooperation between maintenance and production supervisors is first rate.

08 Labor Control - Maintenance controls cost by the efficiency with which they install materials. This makes the control of labor an important function. They must also control absenteeism and minimize overtime. They should utilize work control measures to maximize productivity.

1. Policies emphasize the control of labor.

2. Absenteeism is well controlled.

3. There is an effective vacation policy.

4. Overtime is properly controlled.

5. Labor use is checked to ensure good control.

6. Labor is allocated for scheduled work.

7. Labor used on planned jobs is estimated and actual use measured.

8. Personnel can be shifted between production areas to meet peak workloads.

9 On the completion of major jobs, actual labor use is analyzed to check labor control effectiveness.

10. Supervisors are accountable for the effective use of crews.

11. Supervisors assign work effectively.

12. Supervisors control work efficiently.

13. The maintenance work force is well controlled.

09 Workload Measurement - This is the identification of the essential work to be performed and its conversion into a work force of the proper size and craft composition to carry out the required work.

1. The proper size and craft composition of the maintenance work force has been determined.

2. Non-maintenance work (like construction) does not interfere with performing maintenance.

3. The work force size is changed if there is a significant increase or decrease in work.

4. Types of maintenance work are carefully identified.

5. Management requires solid justification before increasing the size of the work force.

10 Productivity - This is the measure of the effectiveness of the control of labor. It is the percentage of time that a maintenance crew or an individual craftsman is at the work site, with tools, performing productive work.

1. Management encourages productivity checks.

2. Productivity improvement targets have been established.

Maintenance Performance Evaluation

3. Productivity is measured regularly.

4. The work force understands the need for periodic checks.

5. Management has explained why productivity checks are necessary and, as a result, they are accepted.

6. Productivity is measured using indices such as percentage of hours worked versus traveling, idle etc.

7. Causes of delays are analyzed and reduced to help improve productivity.

11 Backlog - The backlog measures the degree to which maintenance keeps up with the generation of new work. Backlog data is also used to help adjust the size and craft composition of the work force as the work load changes.

1. The backlog is measured in estimated man-hours by craft.

2. The backlog is not measured by the number of jobs remaining to be done.

3. The backlog does not include emergency repairs because such work must be done immediately.

4. The backlog does not include unscheduled repairs because each job requires so few hours.

5. Preventive maintenance is not included in the backlog because the man-hours required each week are approximately the same.

6. Management accepts maintenance backlog data as sufficient reason to increase or decrease the work force size or its composition.

12 Motivation - This is a stimulus to action. When applied properly, it results in improved productivity, better quality work and greater job satisfaction. An individual is motivated when he feels that his job is important, his role in carrying out that job is recognized and his reward for performing that job is regarded at the same level at which he sees it. Motivation applies to all individuals, from manager to laborer. While the form of reward at different organizational levels may differ, the essentials of motivation at each level must be met.

An otherwise excellent maintenance program will not be successful unless its work force is motivated to make it successful.

1. The maintenance superintendent motivates the maintenance work force.

2. Maintenance supervisors motivate their crews.

3. Supervisors, themselves, are motivated.

4. Staff personnel, like planners, are motivated.

5. Craft personnel are motivated.

13 Craft Training - There should be an overall craft training program with strong management backing. Craft skills must keep pace with current technology. Basic skills must be kept up to date. Training materials and methods must be effective. Those giving the training must be well qualified. Training should challenge and motivate those being trained. Personnel who are to be trained should meet certain minimum standards to qualify for training. Testing should be carried out to ensure that training is effective and those trained did learn. Attendance and progress records should be kept. Progression from one level to the next should be conditional upon demonstrating necessary skills. Higher skill levels should never be awarded on the basis of seniority. When greater skill levels are achieved, personnel who qualify should be acknowledged.

1. A training program provides for the selection and training of personnel to fill craft needs.

2. Technical training for craftsmen is provided.

3. Craftsmen are trained in the production process and its hazards.

4. Safety training is given to all personnel, including contractors.

5. Craft personnel are skilled in diagnosis and repair.

6. Specialized training is conducted on specific problems.

7. Employees' suggestions for training are acted on.

Maintenance Performance Evaluation

8. Craft skills are updated regularly.

9. Training by equipment manufacturers is provided.

10. Training techniques, such as video, are matched with training needs.

11. New employees are properly indoctrinated.

12. Records are kept on training.

13. Higher skill levels are only awarded on the successful completion of training.

14. Training is provided on the maintenance program.

15. The training program has resulted in tangible benefits such as less rework and fewer overtime hours.

14 Supervisor Training - The supervisor training program should be encouraged by management. Supervisors must be taught how to control work properly within the framework of the maintenance program. They must also keep their supervisory skills up to date and be able to train crew members, other supervisors, planners or craftsmen. Thus, communication skills must included in training as well as maintenance management techniques.

1. There is an effective training program for maintenance supervisors and selected staff personnel such as planners.

2. Management supports the program.

3. Supervisors have opportunity for training.

4. New supervisors are trained before they assume their new duties.

5. Supervisors receive training in the production processes and its hazards.

6. Supervisors train their crew members effectively.

7. Supervisors conduct realistic, practical safety training.

8. Supervisors receive training on work related matters such as labor control.

9. Supervisors receive training in human relations.

10. Supervisors have been trained in the maintenance management program.

11. Supervisors' capabilities are enhanced by job rotation and seminars.

12. A procedure exists for the determination of supervisory training needs.

13. Supervisors are trained in the duties of maintenance planners.

14. When supervisors are identified for promotion, they receive appropriate training.

15. Training records are up to date.

15 Preventive Maintenance - The PM program should successfully extend equipment life and avoid premature failures through timely inspection, monitoring, testing, lubrication, cleaning, adjustment and minor component replacements. As a result, there should be fewer emergency repairs and more work should be planned. As the planned work is scheduled and performed, maintenance personnel will work more productively and their work will be of higher quality.

1. There is an effective overall PM program.

2. Management understands and strongly supports PM.

3. The PM program successfully uncovers problems before equipment failure.

4. The PM program emphasizes safety.

5. The PM program emphasizes the preserving assets.

6. The PM program has reduced emergency repairs.

7. The PM program has resulted in more planned work.

8. Manpower required for PM services is known.

9. Completion of PM services is verified.

10. New equipment is added to the program promptly.

Maintenance Performance Evaluation

11. The PM program is updated periodically.

12. PM services are carried out effectively.

13. Supervisors ensure PM services are done on time.

14. Operating personnel cooperate fully with the PM program.

15. Where appropriate, operating personnel perform PM tasks.

16. Predictive techniques such as vibration analysis are integrated into the PM program.

17. Each PM service has a standardized checklist.

18. PM services are identified by a standing work order number to help control the work.

19. Extensive repairs are not carried out until the PM service is completed.

20. PM services are scheduled at the correct intervals.

21. PM services for fixed equipment are linked together in routes to avoid unnecessary travel.

22. PM services for mobile equipment are scheduled to avoid unnecessary interruption of operations.

23. Operators and craftsmen cooperate fully in the conduct of PM services.

16 Work Orders - The work order system provides a means of requesting work. Once received, work order documents, standing work orders or verbal orders are used to plan, schedule, assign and control work. The work order system then focuses labor and material data from the time card, stock issue card or purchase order into the information system where it is converted to decision making information.

1. The work order system covers all types of work.

2. The work order system is used correctly.

3. The work order system is tied to the information system to produce decision making information.

4. Work orders link with time cards, stock issue and purchasing documents to obtain data on labor and material used.

5. The work order system permits the cost and performance of major jobs to be measured.

6. Work order documents are simple and easy to use.

7. Small jobs can be handled in the simplest way possible.

8. The work order system can control routine, repetitive activities like shop clean up.

9. Instructions on how to use the work order system are complete and clear.

10. The work order system can control non maintenance jobs like construction.

11. Provision is made for controlling contractor work.

12. Provision is made for job planning and estimating.

13. A practical priority setting scheme exists.

14. Work order approval is required for selected major work.

15. Verbal orders are used primarily for emergency work.

16. Verbal orders get work done with no loss of control or job information.

17. Standing work orders are used only for routine, repetitive functions like shop clean up or to represent groups of low maintenance cost equipment.

18. All personnel understand the work order system.

19. Compliance with the work order system is good.

Maintenance Performance Evaluation

17 Information - Decision making information is used to manage maintenance and is the basis for decisions on the control of work, use of resources and cost. It must be complete, accurate and timely and made available to the right personnel and presented in a useful format. Administrative information provides clerical data on such matters as absenteeism.

1. Information to manage maintenance is available.

2. Information content and use are explained.

3. Performance indices are provided.

4. Operations reviews maintenance cost and performance.

5. Management reviews cost and performance.

6. All who need information can get it.

7. Recipients know how to use information.

8. Actual and estimated man-hours are compared on major jobs.

9. Information shows labor use on PM versus emergency repairs or scheduled maintenance.

10. Information exists on absenteeism.

11. Overtime information is available.

12. The backlog shows whether maintenance is keeping up new work.

13. Backlog information is of sufficient reliability that management will change the work force size if the data so indicates.

14. An open work order list shows all work orders open or being worked on.

15. Cost and performance on major jobs are available.

16. At the completion of a major job, cost and performance are summarized.

17. Equipment costs are summarized on a month and year to date basis.

18. Costs are available on components (like drive motors).

19. Costs are available on functions like grass cutting.

20. Cost information is summarized by cost centers.

21. Actual costs are compared with budgeted costs month and year to date by cost center.

22. Repair history traces significant equipment repairs and failure patterns.

23. Repair history includes the life span of critical components.

24. Report or display formats are clear and easy to read.

25. The information system satisfies genuine needs.

26. The information system is the plants primary communications system for control of work and cost.

18 Planning - Planning ensures that jobs can be completed more efficiently. It determines labor, materials, tools and supporting equipment needs in advance to assure their availability for specific jobs. Planned jobs use less manpower, are completed with less downtime and the work is done more effectively. Thus, repairs are made more deliberately resulting in better work quality. There should be a criteria for determining which jobs are to be planned. There must be an effective procedure for planning. Planners should be used primarily for planning work.

1. Important work is planned.

2. Criteria specify work to be planned.

3. Procedures are complete and clear.

4. Work order documents support planning.

5. Information is available on planned jobs.

6. A forecast identifies most major future jobs to be planned and scheduled. Each task is accompanied by a standard task list, bill of materials and tool list.

Maintenance Performance Evaluation

7. Repair history provides data on the life span of major components for the forecast.

8. PM inspections and equipment monitoring are a major sources of planned work.

9. Planned work is developed from failure analysis.

10. Planned work is derived from testing such as oil sampling.

11. Crews prefer work that is planned.

12. Planned work includes construction work.

13. Major jobs must be approved before work is done.

14. Job approval is based on cost and need.

15. Planned work is prioritized.

16. All planned work is scheduled.

17. Labor is allocated to planned work by priority.

18. Planning has reduced downtime required.

19. Operations prefers maximum planned work.

20. Criteria exist for the selection of planners.

21. Planners have ample time to plan.

22. Planners are well trained.

23. There are enough planners.

24. The percent of planned work is increasing.

25. Maximum planned work is possible in the operating environment.

19 Scheduling - Maintenance and operations schedule major work jointly to establish the best timing for a job and to avoid interruptions to production while utilizing maintenance resources most efficiently. Repetitive tasks like PM services area scheduled routinely.

1. Policies require effective scheduling.

2. Procedures are well explained and effective.

3. Scheduling has reduced downtime.

4. A weekly schedule is negotiated by operations and maintenance.

5. Operations approves the weekly schedule.

6. Manpower for scheduled work is allocated by priority.

7. Operations makes its equipment available according to the approved schedule.

8. Non maintenance work, like construction, is included in the weekly schedule.

9. Material shortages rarely interfere with the schedule.

10. Management seldom overrides an approved schedule.

11. Man-hours spent on scheduled work are measured.

12. Scheduled compliance is measured.

13. Major maintenance is planned, scheduled and completed in a reasonable time.

14. Daily work plans are prepared by each crew based on the weekly schedule.

15. Operations and maintenance meet daily to coordinate work to be accomplished in the next 24 hour period.

20 Work Control - Work control ensures quality work, done on time. Procedures begin with direction from the superintendent and continue with effective planning and scheduling of major jobs. The most critical application of work control is exercised by the supervisors or team leaders.

Maintenance Performance Evaluation

They assign work, ensure timely job completion and quality work. Work control procedures guide the efforts of teams making them more effective.

1. Policies require effective work control.

2. Supervisors and team leaders assign work effectively.

3. Major jobs are planned with efficient activity sequences.

4. Major jobs start on time, remain on schedule and are completed on time.

5. Standard task lists, bills of materials and tool lists are applied for most repetitive major jobs like major component replacements.

6. Cost and performance are assessed to check job progress.

7. Variances, such as excessive cost, must be explained.

8. Supervisors or team leaders have adequate time to control work.

9. Craftsmen have adequate time to do work.

10. Multiple major jobs are easily controlled during the shift.

11. Several crafts working together are always well controlled.

12. Job support like transportation is carried out effectively.

13. Operations is advised of work completion and invited to inspect work.

14. Repetitive work is regularly assessed to improve procedures.

15. Supervisors and team leaders evaluate work control performance.

16. Verbal instructions can be used without loss of control or information.

17. Quality control is emphasized.

18. There is seldom any loss of control of work.

19. Supervisors and team leaders control work effectively.

21 Maintenance Engineering - Maintenance engineering assures the maintainability and reliability of increasing complex production equipment. Maintenance engineering tasks include guidance for: developing the PM program, organizing periodic maintenance tasks, establishing commissioning procedures for new equipment, assessing equipment modification requests and verifying the adequacy of work performed by contractors. Maintenance engineering tasks may be carried out by a dedicated staff or by regular supervisors and craft personnel.

1. Equipment maintainability, reliability and performance are assessed.

2. Repair history is analyzed to determine corrective actions.

3. Field investigations are used to confirm equipment problems.

4. New equipment is added to the PM program promptly.

5. PM services frequencies are reviewed regularly.

6. Predictive techniques like vibration analysis are included in the PM program.

7. Equipment performance is monitored.

8. Repair techniques are prescribed and training provided.

9. Data identifies equipment with the greatest repair cost.

10. Newly installed equipment is fully maintainable before put into service.

11. Job scopes are carefully developed.

12. The most effective, up to date repair methods are used.

13. A modern technical library exists.

22 Utilization of Standards - Standards provide targets against which maintenance performance can be assessed. Quality standards prescribe how the job should be done and what the final product should be. They often include task lists, bills of materials and tool lists. Quantity standards prescribe the amount of manpower by craft, the job duration etc. Historical data such as repair history is used to develop standards for repetitive jobs.

Maintenance Performance Evaluation

1. Standards are used for jobs on which work quality, cost and performance are important.

2. The work order system supports the use of standards by comparing actual performance with standards.

3. The information system provides feedback on job performance which can be used to develop historical standards.

4. A procedure exists for the preparation, use and updating of standards.

23 Material Control - Material control includes procedures to identify, stock, purchase or manufacture materials in proper quantities to make them available to maintenance as requested. To obtain materials, maintenance must determine what it needs and specify the quantity and the time when they are required. The work order system, material control and information systems permit maintenance to obtain information on material costs and usage.

1. Material control procedures are well explained and effective.

2. Critical spare parts lists exist for equipment.

3. Maintenance specifies stock items and quantities.

4. Out of stock materials are promptly replenished.

5. The stock room is properly staffed.

6. The stock withdrawal procedure is well organized.

7. A procedure exists for returning unused stock parts.

8. Accountability for stocked parts is excellent.

9. Maintenance personnel follow material issue procedures.

10. Stock materials are easily identified.

11. Parts interchangeability is well documented.

12. Craftsmen can identify needed parts properly.

13. A procedure exists for reserving stock parts in advance.

14. There are standard bills of materials for major repetitive jobs.

15. Stock issues are made on presentation of approved work orders.

16. Stock can be issued against either work orders or equipment numbers.

17. Material costs by equipment are summarized.

18. Material costs can be obtained job by job.

19. Purchasing has enough lead time to obtain materials.

20. Purchase order status can be tracked.

21. Drawings for manufacture of spare parts are available.

22. An on site material delivery system exists.

23. There are few unauthorized material storage areas.

24. Overall material control is excellent.

24 Safety - Safety training and adherence to proper practices are essential in creating the environment in which maintenance can carry out its work with confidence that accidents and injuries can be prevented.

1. Personnel attend safety meetings regularly.

2. The safety record is constantly being improved.

3. Personnel use prescribed safety equipment.

4. Personnel obtain safety clearances for jobs.

5. Personnel are well trained in the safety considerations of their jobs.

Maintenance Performance Evaluation

6. Fire protection is effective.

7. Vessel entry procedures are followed.

8. Welding and burning permits are obtained.

9. All personnel are involved and committed to safety.

25 Housekeeping - Good housekeeping practices have the objective of keeping the plant environment in an orderly, attractive condition while helping to preserve assets. Good housekeeping results in the creation of a working environment in which personnel are motivated to work more effectively.

1. Housekeeping inspections are conducted.

2. Tools, materials and debris are removed promptly from job sites on work completion.

3. Operations personnel carry out housekeeping tasks effectively.

4. There is a neat, clean, and tidy appearance everywhere.

26 Operations - Operations personnel need reliable equipment to meet the production targets on time and at least cost with quality product. Therefore, they must understand, support and actively participate in maintenance activities.

1. Operations personnel have good knowledge of maintenance requirements of the equipment they operate.

2. Operations releases equipment promptly for service or repair.

3. Equipment is operated carefully and correctly.

4. Operations is familiar with all aspects of the maintenance.

5. Operations helps to identify the timing and cost of future work like overhauls.

6. Operations checks the timely completion of PM services.

7. Operators report problems promptly.

8. Operators clean, adjust and service equipment.

9. Operations is concerned with job cost and quality of work.

10. Operations does not overload maintenance with unexpected work or unreasonable demands.

11. Work requested by operations is necessary, feasible and correctly funded.

12. Maintenance and operations cooperation is evident at every level of their respective organizations.

27 Staff - Maintenance depends on reliable service by staff departments such as warehousing, purchasing, data processing etc. To be able to provide reliable service, staff personnel must understand the maintenance program and provide procedures that are compatible with the maintenance program.

1. Staff personnel understand the maintenance program.

2. Staff personnel are service oriented.

3. Recommendations for better services are acted on.

4. Staff personnel confer with maintenance to improve service.

28 Management - Management creates an environment for successful maintenance by assigning clear department objectives, providing guidelines for the conduct of maintenance and ensuring that the maintenance organization is responsive within the framework of a well defined, appropriate program.

1. Management has assigned an objective to maintenance.

2. Policies prescribe the conduct of maintenance.

3. Management has approved the maintenance program.

4. Management verifies the responsiveness of maintenance.

Maintenance Performance Evaluation

5. Management measures maintenance performance.

6. Management limits construction work to ensure maintenance is done first.

7. Management does not add unexpected or unreasonable work without first determining if maintenance can do it.

8. Management has created a positive working environment for maintenance.

29 Shop Operation - Shop facilities such as a machine shop, fabrication or welding shop, carpentry shop, paint shop, plumbing shop etc., provide support for field maintenance work as well as for installation or construction.

1. Shop activities like welding are arranged to ensure an efficient flow of work.

2. Planning is related to work stations to ensure effective work control.

3. The work order system is used to control work.

4. Information is provided on the status of jobs.

5. Emergency work does not hinder shop work progress.

6. Shop drawings are complete and well organized.

7. Procedures for requesting work are explained

8. Work requesting procedures are followed.

9. Shop performance and productivity are measured.

10. A workable priority system exists.

11. When required, shop personnel are able to be assigned to field work.

12. Shop materials are properly stocked and organized.

13. Job costs are available.

14. Tool control is effective.

15. Shop personnel are skilled.

16. Shop work is well controlled.

17. Shop planners are effective.

30 Support Services - Transportation to move people and materials, supplies, rigging, mobile cranes and operation of heating or compressed air systems are types of support services provided by maintenance. In addition, buildings and grounds work or custodial services are commonly carried out by maintenance.

1. Transportation and equipment is easily obtained.

2. Rigging is well organized.

3. Personnel operate support equipment carefully.

4. Operating costs are charged against the job.

5. There is a sufficient amount of mobile support equipment.

6. Buildings and grounds maintenance is carried out well.

7. Debris is collected and disposed of promptly.

8. Roads are well maintained.

31 Mobile Equipment - Mobile equipment maintenance can represent a total program, as in an open pit mine. However, in a plant, the use of mobile equipment may be limited to a support role. *Therefore, rate all 20 standards only if your total operation is carried out with mobile equipment.* <u>Otherwise, rate only standard 20.</u>

1. There is a well organized area for units awaiting work.

2. Units are washed before work is performed.

3. PM services are carried out in designated areas.

4. Major repairs are not done until PM services are completed.

Maintenance Performance Evaluation

5. PM inspection deficiencies are reviewed to determine further repair actions.

6. The garage is divided up into efficient working areas.

7. Areas are designated for long term repair activities.

8. The garage is organized to maintain all types of equipment.

9. Major maintenance is well planned.

10. PM services and major repairs are scheduled conveniently for operations.

11. Operations complies with the approved schedule.

12. Warehouse facilities are properly located and well run.

13. Garage utilities like compressed air lines are convenient.

14. Areas for welding or tire repair are properly located.

15. Garage configuration enhances maintenance activities.

16. Internal communications operate effectively.

17. Completed work is picked up promptly by operations.

18. Quality control like road testing is carried out.

19. Field repair is done efficiently.

20. Overall maintenance of mobile equipment is well done.

32 Non maintenance - Maintenance departments are often called upon to perform construction, equipment installation or equipment modification. While the same craft skills may be used, limits should be observed to ensure this work does not displace regular maintenance work.

1. Non maintenance work does not interfere with the maintenance program.

2. A priority scheme successfully allocates labor for major maintenance versus non maintenance work.

3. Project engineers plan and control major projects.

4. The work order system controls non maintenance work effectively.

5. Information is provided on the cost and performance of non-maintenance jobs.

6. Modifications are reviewed and approved before being made.

7. Non maintenance projects comply with approval and funding procedures.

8. Non maintenance projects must have adequate instructions before they are carried out by maintenance.

9. Contractor installations are checked to ensure their maintainability before they are put into service.

10. Maintenance crews are given adequate engineering instructions and on-site support for project work.

33 Cost Control - Maintenance costs are forecasted against production targets relative to time periods: monthly, annually etc. This budgeting process establishes expectations against which actual expenditures are compared. With such a yardstick, performance is measured by comparing man-hours used, labor cost accrued, material cost accumulated and total costs committed. The focus of these costs is equipment, buildings or facilities maintained, activities performed or jobs completed. Maintenance controls costs primarily by the effective use of labor. But, they can also influence costs by encouraging production to operate equipment correctly. The overall cost control effort should include adequate preventive maintenance, emphasis on planned work and the use of cost related information to help anticipate the need to prepare for and plan major work.

1. The maintenance budget is based on production targets and their relationship to maintenance costs.

2. Capital expenditures are budget separately.

3. Overhauls are budgeted separately.

Maintenance Performance Evaluation

4. A standardized budgeting technique is used.

5. A cost summary compares actual and budgeted costs, on a month and year to date basis.

6. Maintenance costs can be isolated for equipment and components.

7. Overhauls are subjected to cost and performance evaluation.

8. Supervisors and team leaders explain excessive costs.

9. Supervisors and team leaders have cost related performance targets such tool loss reduction, accident prevention or productivity improvements.

10. There are regular cost review meetings held by maintenance and production.

11. Maintenance has improved productivity through effective labor control.

12. Maintenance reviews warehouse stock to check consumption rates and identify obsolete parts.

13. Maintenance anticipates purchasing needs to avoid extra costs like air freight.

14. There is an effective cost control effort.

34 General - Rate the general conditions described below, considering all aspects of maintenance that you are *personally* familiar with.

1. The responsiveness of maintenance.

2. The quality of its work.

3. Cooperation with production.

4. Cooperation with staff departments.

5. Cooperation with management.

6. The competency of managers or superintendents.

7. The competency of supervisors or team leaders.

8. The competency of planners.

9. The competency of craftsmen.

10. The quality of the PM program.

11. The adequacy of the work order system.

12. The quality of information.

13. The timeliness and completeness of cost information.

14. The adequacy of repair history.

15. The quality of work on major jobs.

16. The quality of internal maintenance cooperation.

17. The effectiveness of planning and scheduling.

18. The productivity of workers.

19. Cooperation by operations in carrying out maintenance.

20. Level of service provided by staff personnel.

21. Interest by management in a quality maintenance program.

22. Success in implementing the principles of programs like Total Productive Maintenance (TPM) or Just in Time.

23. Success in implementing maintenance teams.

24. Success in implementing interdepartmental teams.

25. Success in empowering employees to make practical decisions on matters like work control.

26. Success in the involvement of personnel in beneficial change.

27. Success in correcting long standing problems like wasteful labor practices.

The **Maintenance Performance Evaluation** is available as a service. It may be self administered, or you may have it administered on site.

For more information contact:

Paul D. Tomlingson Associates, Inc.
Management Consultants
1905 Glencoe Street
Denver, Colorado 80220
Tel: (303) 377 5585
Fax: (303) 377 5569

List of Illustrations

Action - decision group participation, diagram, 373
Action verbs for to short title job descriptions, 272
Adequacy of predictive maintenance, 179
Adherence to management principles, 88
Advantages of forecasting for warehouse, 168
Advantages of maintenance program definition, 123
Advantages of package systems, 243
Advising operations of PM deficiencies, diagram, 129
Analysis of current organization, 88
Area Organization, 64
Assembling elements of engineered standards, diagram, 191
Assembling weekly schedule elements, 277
Assigning PM services, diagram, 127
Assigning work developed from PM inspections, diagram, 151
Backlog as a method of work force adjustment, diagram, 229
Backlog definition, 229
Backlog interpretation, diagram, 298
Backlog Summary Report, 299
Barriers to multiskill implementation, 110
Causes of work order in backlog too long, 277
Characteristics of supervisor in trouble, 98
Checklist for establishing an information system, 215-216
Classification of PM deficiencies, diagram, 150
Communications function of work order system, 260
Comparing resources of planned versus unplanned work, diagram, 155
Component replacement signals manpower requirements, diagram, 166
Computer graphics for parts identification, diagram, 201
Computer graphics used in parts identification, 307
Computing the backlog, diagram, 302 and 322
Conduct of planning meeting, diagram, 133
Contents of traditional craft training program, 106
Control of shop work and support services, diagram, 194
Controlling non maintenance work, 317
Coordination of component replacement work, diagram, 132
Coordinator controlling operations and maintenance, 69
Coordinator using craft pool, 70
Coordinators in future plant organization, 68
Cost of Performing Work, 45
Cost to install each dollar of material, desirable ratios, diagram, 342
Craft organization, 62

Craft pool supporting mobile operations, 71
Craftsman's information requirements, 281
Crew member training on system use, 248
Criteria for planning work, 157
Cycle of arbitrary decisions on work force reduction, 91
Data job preparation and craft delays for standards, diagram, 189
Days elapsed between reporting deficiencies and repair, diagram, 146
Decision making information, 227
Decision making reports, 239
Decisions on handling PM deficiencies, diagram, 130
Deficiencies reported as PM program is implemented, diagram, 145
Definition of the maintenance program, schematic, 135
Determining supervisory coverage, three formats, 295
Determining the backlog level, diagram, 301
Determining timing of planned work, diagram, 274
Development of a plan for information system use, 252
Disadvantages, fully integrated system, 236
Disadvantages, partially integrated system, 236
Disadvantages, stand alone system, 237
Division of work, area crew and craft pool, 65
Downtime information interpreted, 312
Duties at Various Organizational Levels, 24
Effect of PM on manpower use, diagram, 147
Elements of standards, 190
Engine module removal, task list, 167
Engineering work order for conveyor installation, format, 319
Engineering work order in overall work order system, diagram, 318
Enhancements in work completion due to planning, 156
Equipment cost by type and component, format, 316
Equipment deterioration versus PM detection orientation, diagram, 143
Equipment management program, 329
Equipment numbering schemes, table, 270
Equipment specification program, two diagrams, 306
Estimating the work load, six diagrams and formats, 293 - 294
Facts about mining maintenance, 60
Failure codes, table, 271
Failure trend. illustration, diagram, 271
Field data and accounting documents merge into information, diagram, 234
Field data reported into information system, 249
First line Supervisory Activities, 23
Fleet cost and performance report, 316

List of Illustrations

Flow of information between operations and maintenance, 328
Focus of Activity versus Time, 25
Focus of information versus time, schematic, 279
Forecast aids material control, diagram, 199
Forecasting aids to purchasing agent, 168
Format for measuring work delays, 338
Four phases of establishing an information system, 210
Fully integrated system, advantages, 236
Function of work order in isolating job cost and performance, 269
Gains of defining maintenance program, 219
General supervisors information interests, 280
Goals for schedule compliance, 288
Guidelines for handling emergency repair data, 289
Identification of parts via computer, diagram, 200
Identifying PM services due, diagram, 128
Impact of PM on labor control, diagram,, 228
Impact of PM on Manpower Utilization, 39
Improvement in job execution due to planning, 156
In house system development assurances and drawbacks, 242
Information of interest to maintenance superintendent, 280
Information provided by inventory control program, 305
Information pyramid, 232
Information required for effective maintenance engineering, 181
Information system development options, 241
Information system implementation steps, schematic, 253
Information versus management level, diagram, 340
Initiating new work, diagram, 126
Interface of work order and accounting system, schematic, 266 - 267
Job duration versus deviation in standards, diagram, 190
Job duration versus type of work, diagram,, 264
Job plan, diagram, 159
Job status classifications, 288
Labor control ensures, 227
Labor information objectives, 292
Labor reporting data, 291
Levels of stock information, 306
Line staff organization, 62
Linkages of engineering work order, diagram, 263
Macro productivity trends, 297
Maintenance absentee summary, 297
Maintenance labor utilization report, 296

Maintenance overtime summary, 296
Maintenance program definition requirements, 124
Maintenance superintendents performance indices, table, 344
Maintenance work control overview, schematic, 212
Maintenance work order, format, 261
Maintenance work order -EWO, 76% complete, format, 320
Maintenance work order -EWO, just started, format, 320
Maintenance work order -EWO, materials only, format, 321
Maintenance work order status, closing details added, 286
Maintenance work order status report, open, format, 285
Maintenance work order status report, work in progress, format, 286
Maintenance work order status report, completed, format, 287
Maintenance work request, format, 262
Making effective organization changes, how to:, 73
Manager: Adequate maintenance program ?, 324
Managers performance indices, table, 343
Manpower balance for PM program, diagram, 152
Manpower requirements for component replacement program, diagram, 166
Material disposition for overdue jobs, 303
Mean time before failure (MTBF), diagram, 164
Measuring successful PM program, 153
Plant services manager in future organization, 69
Mobile equipment PM services scheduled on operating hours, format, 282
Mutliskill implementation gains, 110
New organization benefits to operations and maintenance, 72
Opening work orders, 268
Operations guidelines in maintenance partnership, 328
Operators assisting crews with PM services, diagram, 129
Operators helping crew members and vice versa, diagram, 126
Organization performance indices, 46
Organization training for supervisors, 118
Organizing and operating the PM program, schematic, 148
Overhauling front end loader, quality standard, 184 - 185
Package system disadvantages, 243
Partially integrated system, advantages, 236
Periodic performance measurements, 231
Planners information needs, 280
Planners training on system use, 248
Planning gains, before work begins, 156
Planning steps, schematic, 158
Planning unique major jobs, diagram, 133

List of Illustrations

Plant areas (as in area organization), 64
PM control network for identifying services due, diagram, 149
PM inspection results signal component condition, diagram, 165
Predictive techniques, 144
Preparing for and conducting scheduling meeting, schematic, 172
Priority range versus time to complete work, 300
Priority scheme, maintenance, table, 275
Productivity versus Manpower Savings, 45
Program Definition Presentation Technique, 31
Program execution indices, 46
Project bar chart, diagram, 264
Project plan for engineering work order, 319
Project priority rating, 276
Purchase order versus work order linkages, 268
Quality of interaction indices, 46
Quantity standard, diagram, 185
Questions personnel have on maintenance shortcomings, 118
Reaction to work load changes during downsizing, 95
Reasons for not measuring productivity, 336
Reasons for purging backlog, 302
Reasons for work delays, 336
Reducing the backlog 299
Repair history, format, 315
Reporting work completion, diagram, 134
Responsibilities for component rebuilding, diagram, 201
Responsibilities of supervisors and planners (training), 117
Review of PM program by maintenance engineer, diagram, 130
Rewarding what is important in craft training, 113
Routine PM, 144
Scheduling major component replacements, diagram, 127
Scheduling PM dynamic versus static services, diagram, 128
Scheduling PM services on fixed intervals, format, 282
Scheduling procedure for PM, diagram, 152
Section manager for mobile operations, 71
Section managers in future plant organization, 68
Selected overall performance indices, 231
Selection of work orders from backlog by priority, diagram, 171
Setting standards, diagram, 184
Shop control guidelines, 195
Skills at Various Organizational Levels, 24
Slotting standards, diagram, 187

Specifying PM services according to equipment need, diagram, 149
Stand alone system, advantages, 237
Standing work order use, table, 262
Steps to facilitate work force reductions, 96
Successful transition of operator maintenance, 331
Supervisor training on system utilization, 247
Supervisors information requirements, 281
Supervisory training on production strategy, 117
Table Resistance to change factors, 84
Task areas for engineered standards, table, 189
Ten step strategy for successful evaluations, 347
Ten step work force reduction strategy, 94
The weekly scheduling cycle, diagram, 283
Three objectives of maintenance information, 211
Time card labor data, 235
Time card linkage with work order, 269
Time standards for work elements, diagram, 188
Time to complete work by priority rating, 276
Total plant organization (future), 67
Total Productive Maintenance, division of responsibilities, table, 330
Traditional organization, 61
Training administration objectives, 106
Training results for craftsmen, 106
Two levels of cost information, 231
Types of administrative reports, 233
Types of non maintenance work, 273
Typical maintenance management system, schematic, 238
Updating operating hours for mobile forecasting, diagram, 131
Use of standards, 185
Verification of PM program, 179
Verifying material and labor for scheduling, diagram, 171
Volume of jobs versus cost of work, diagram, 265
Weeks of backlog,, 302
What maintenance expects of material control, 202
What the maintenance engineer wants to know, 280
What's not in the backlog, 229
Window of opportunity for planners, diagram, 170
Work completion times by type of work, 275
Work order status, diagram, 230
Work order system coverage, 259
Work order system elements, 260

Index

A

ability to reserve parts, 203
abnormal condition, 144
absenteeism, 231, 235
    trends, 94
abusing equipment, 96
acceptance, 84
accountability, 16, 197
accounting, 101
    department, 197
    documents, 234
    system, 13, 181, 211
action
    group, 373
    group is a decision group, 372
    verb, 272
active resistance, 84
actual data is used, 251
adequacy work order system, 207
adequate program, 324
adherence to principles, 11
adjust inventory levels, 136
adjusting, 144
administrative
    duties, 101
    information, 8, 232, 227
    jobs, 93
    procedures, 199
    reports, 233
advisory panel of craftsmen, 113
allocate labor, 171
allocation of manpower, 161
allowances for job preparation, 190
analysis current organization, 88
apprenticeship programs, 108
approved
    schedule, 138, 161
    weekly schedule, 275
arbitrary work force reductions, 91

area
    crew, 93
    managers, 68
    organization, 63, 64
    organizations, 74
    supervisor, 65
assemble materials, 137
assembly requirements, 160
assign emergency jobs, 150
assignment of work, 22
attaining better housekeeping, 369
attrition, 95
availability of materials, 173, 174
awarding pool personnel, 66

B

backlog, 15, 102, 120, 229, 170,171
    computation, 299, 303 - 304
    control, 299
    data, 218
    of pending jobs, 139
    summary report, 299
bar chart, 159
bar coding wands, 225
basic
    maintenance, 92
    craftsman qualifications, 105
    requirements of operations, 328
    rules of accountability, 203
behavioral changes, 79
behaviorally, 83
benchmark, 189
    jobs, 187
    tasks, 189
best evaluation technique, 53
better productivity of crew, 156
bill of materials, 14, 185, 200

buildings and grounds, 92, 196

C

capital
   funded project work, 263
   funds, 1
   intensive non maintenance work, 206
   projects, 92, 93
capitalized, 41, 205
carpentry support, 196
categories of work, 38, 272
causes of poor productivity, 44
central control, 22, 72
chain of command, 34, 35, 90, 124
checklists, 151
clarification, operations role, 121
clarify the program, 140
clarify working relationships, 157
classification procedure, 201
classifying deficiencies, 151
cleaning, 144
clearer job instructions, 156
CMMS, 151, 198
commitment of resources, 137
common understanding, 38
communications network, 13, 210
company welfare, 92
competitiveness, 108
completeness of planning, 173
completion time, 159
complex activities, 11
compliance with PM program, 147
compliance with policies, 7
compliance with the program, 153
component (cp), 200
   life span, 230
   rebuild vendors, 180
   rebuilding, 180
   replacement, 127, 132, 164, 166
   replacement intervals, 165
   replacement program, 39, 103, 163, 166
components, 201
computer, 136, 144
   graphics, 201
   graphics display, 224
   integrated technologies, 111
   literacy, 248
conceptualization, 210, 217
concerns of employees, 111
condition based system, 144
condition monitoring, 77, 127, 141, 214
conduct
   adequate training, 256
   effective preventive maintenance, 364
   maintenance, 11
   scheduling, 172
   PM services, 150
   scheduling meeting, 286
   training, 253
   meeting, 172
confer with operations, 161
confirm
   job scope, 159
   maintenance program, 253
   effectiveness, 211
   priorities, 175
   roles of key personnel, 253
construction, 273
construction craftsman, 205
consulting
   assistance, 241, 245
   organization, 245
consumer of materials, 16
continuous analysis, 144, 223
contractor, 10, 157, 263
contractors, 180, 181
control

Index

document, 235
  labor, 16
  network, 149
  labor, 9, 227, 291
  preventive maintenance, 281
  repetitive major jobs, 163
  work, 39
  information, 211
controlling, 23
  absenteeism, 297
  cost and performance, 309
  emergency repairs, 289
  labor, 290
  materials, 305
  non maintenance work, 317
  overtime, 296
  planned and scheduled work, 283
  training, 112
converted into labor costs, 136
coordinate
  on going jobs, 162
  shutdown, 157
  meetings, 170
coordinator, 68, 69, 99, 119
correct funding, 10
correct installation, 102
corrective maintenance, 42
cost, 120, 231
  benefit reviews, 102
  and performance, 162
  centers, units, functions, 235
  control, 252, 268
  detail, 231
  information, 230
  of maintenance, 139
  of performing work, 45
  per operating hour, 181
  reduction, 76, 162, 225
  report, 316
  summary, 231
  versus performance, 341

craft
  allowance, 191
  code and employee number, 235
  composition, 9, 13
  organization, 62, 63, 222
  organizations, 63
  organizations, 96
  oriented, 220
  pool, 64, 65, 66, 67, 70, 71
  ranks, 115
  skills, 66, 80, 105
  training, 105
  training program, 105, 106, 107, 108
  training strategy, 111
crafts
  combining, 93
  involved, 159
craftsman, 97, 134, 281
  check productivity, 336
  technical skills, 215
crew
  assignment sheet, 151
  groups, 90
  members, 35, 136, 137, 139
  productivity, 94
criteria
  administrative reports, 232
  planned work, 150, 157
critical components, 144
critical utilities, 196
custodial services, 92
cycle (or frequency), 14

D

daily
  coordination meetings, 138, 161
  schedule, 128, 223
  work plan, 15
data

processing department, 241
   processing services, 242
date and shift, 235
day to day procedures, 12, 239
decision
   group, 373
   makers information, 279
   making, 87
   making information, 278
   making reports, 239
deficiencies, 145
   found, 129
   reported, 142, 145
   list, 150
degree of control, 264
delay status, 288
department relationships, 7
departmental objectives, 239
detection of problems, 87
detection orientation of PM, 13, 143
determine resources, 160
developing
   maintenance objective, 360
   maintenance standards, 185, 187
   communications software, 243
   team, 125
diagnostic skills, 205
direct charge purchases, 243
direct field supervision, 101
disclosing evaluation results, 356
displacing outside contractors, 76
distribution of craft man-hours, 294
division of work, 22, 273
document program, 2
documentation, 251
downsizing, 16, 108, 225
   warehouse staff, 203
   maintenance, 91, 182
downtime, 181, 213, 220, 311
downtime data, 139
drawings, 157

dual reporting, 243
duration of the job, 157
duties at various organizational levels, 24
duties of key maintenance personnel, 30, 118
dynamic PM services, 128, 141, 277

E

early detection of problems, 142
ease of reporting labor, 291
economic decisions (overhaul), 231
educate personnel, 7, 35, 52
education, 33, 81, 112, 124, 125
effective
   maintenance organization, 19
   operating condition, 87
   operation of equipment, 11
   organizational changes, 73
   priority setting, 300
   tool control procedure, 203
   training program, 116
   work order system, 259
   bay utilization, 231
electrical and instrument supervisors, 222, 248
elements of the work order system, 265
emergency
   repairs, 41, 39, 273
   work, 8, 229
employee
   empowerment, 20, 59
   productivity, 76
engineered
   performance standards, 188
   time standards, 186
ensuring
   accurate field labor data, 366
   support for maintenance, 369

Index

enter
    new deficiencies, 152
    new work, 249
environment, 27, 28, 54
equipment
    availability, 108
    chronic problems, 120, 178
    commissioning standards, 180
    cost by type/component, 316
    deficiencies, 136, 249
    deterioration, 143
    downtime, 110
    failures, 141
    maintainability, 180
    maintainability, 178
    management program, 329
    management programs, 111
    modification, 6, 92, 93, 180, 205
    numbering, 243
    numbering procedure, 262, 270
    reliability, 107, 111, 149, 178, 179
    repair patterns, 120
    reporting, 250
    specification program, 306
    specifications, 251
    included in program, 149
    type (tp), 200
    user, 138
establish
    core training group, 253
    manpower by craft, 160
    policies, 22
    schedule and objectives, 253
    responsive WO system, 365
    maintenance guidelines, 361
estimate cost, set priority, 160
evaluate maintenance, 96
evaluation, 10, 49, 122
    advance notice, 52, 348
    avoid surprises, 349
    best technique, 53

    checklist, 122
    dates of next, 52
    favorable environment, 51
    management technique, 346
    policy, 52
    technique, 353
EWO, 207, 241, 263, 267, 317 - 319
ex craftsmen, 119
exception
    information, 232
    reporting, 126
excessive administrative, 233
execution of the program, 46
existing equipment, 206
expensed, 41
exploded view, 224
extending the life of equipment, 141

F

fabrication of assemblies, 137
facsimile of the reports, 250
failure
    coding, 251, 271
    trends, 120
fear of loss of power, 84
feasibility, 10
feasibility of the work, 206
feed back on job progress, 252
feedback from PM inspections, 165
field
    check, 252
    data, 209, 231, 234, ,239, 286
    investigation, 138, 315
    labor data, 228, 234
    planner, 194
    repair, 225
fill in supervisors, 119
first line supervision, 15, 22
first step of improvement, 345
fiscal requirements, 239

fixed
  assignments, 92
  equipment environment, 169
  interval, 142, 151
fleet cost/performance report, 316
flexible WO system, 290
flow charts, 125
flow of information, 328
focus of activity versus time, 25
focus of information versus time, 279
forecast, 137, 164, 165, 166, 176, 199
  data, 167
  event, 165
  intervals, 164
  jobs, 137
forecasting, 39, 163
  procedure, 165, 168
forming maintenance teams, 370
free issue items, 202
fully integrated system, 236
function within the cost center, 270
future craftsmen (profile), 109
future maintenance organization, 219, 220

G

general
  backlog, 171
  data, 189
  manager, 69
  supervisor, 101, 138, 280
getting material support, 366
goals, 14
good planning, 156
graphics, 307
ground rules, 7, 28, 206
guidelines, 1, 2, 12

H

handling
  dual plant evaluations, 349
  emergency and unscheduled work, 288
  evaluation results, 355
  results, 52
  unscheduled repairs, 289
hands on simulation, 112
historical standards, 186
hourly worker level, 115
human element, 20

I

identification of parts, 198, 306
identify essential work, 21
idle time, 110
impact
  successful PM, 154
  computers, 111
  information, 213
  productivity, 45
implement a team, 77
implementation, 210, 374
  groups, 374
  process, 90
  steps, 252
  of operator maintenance, 331
importance, reporting problems, 147
improved performance, 168
improving, 93
  equipment performance, 180
  productivity, 297
  supervisor training, 358
  work quality, 224
in house system, 241, 242
inbreeding, 3
incompetent worker, 80
industrial engineer, 336

Index

information, 119, 181, 209
   checklist, 215
   needs, 47, 87, 343 - 344
   pyramid, 232
   revolution, 219
   system, 13, 57, 139, 210, 241
   versus management levels, 340
infrared scanning, 144
initiate new work, 124
insecurity, 84
inspection (equipment), 127, 144
installation, 273
installation of materials, 197
instructors, 124
interaction, 30
interface of systems, 265
interface with accounting, 237
internal administration, 232
interpreting the objective, 6
inventory control, 165, 220, 224, 243, 244, 305
investigate job in field, 158
invoice, 207
involve plant management, 11
involvement, 112
ISO 9000, 180
isolate the cost and performance, 268

J

janitorial and custodial work, 93
Japanese, 75, 76
Japanese businessmen, 76
job
   approval, 138, 160
   descriptions, 115
   duration, 190, 230
   plan, 138, 156, 159
   performance, 112
   preparation time, 191
   priority, 160
   scope, 138, 156
   site, 159
   status, 195, 285
   steps, 159
justify actions, 211

K

key activities, 87
key maintenance personnel, 97
keyboard entries, 225

L

labor and material data, 269
labor
   control, 228, 325
   data, 235
   information objectives, 292
   productivity, 95
   reporting document, 291
   standards, 315
   turnover, 94
   utilization, 120, 295
   utilization report, 15
large crew and power tools, 203
largest controllable cost, 27
layoff, 94
lead
   men, 98, 103, 119
   time, 142
   time for planning, 146
leadership during change, 88
leadership skills, 100
least disruption, 14, 93
legend, 134, 212
level of deterioration, 143
level of service, 196
level with employees, 76
leveled time, 188
levels of management, 22

life span of components, 131, 181, 315
limits for non maintenance, 15, 16
line
   of progression, 99, 100
   staff organization, 62
linkages with other systems, 30
list deficiencies, 150
listing
   codes, 250
   of significant repairs, 120
load files (computer), 253 - 254
local vendor shops, 203
local management interest (evaluations), 350
logical sequence to planning, 158
long range planning, 101
long term
   cost trends, 227
   performance, 112
   employees, 92
   repairs, 224
look up part numbers, 252
loss of power, 84
lubrication, 144

M

Machiavelli, the prince, 1532., 90
machining, 160
maintainability, 177
maintenance
   operations cooperation, 81
   and operations superintendents, 100
   communication network, 249
   contract services, 6
   departments, 205
   downtime, 343
   engineer, 34, 97, 102, 167, 220, 280
   engineering, 9, 70, 80, 177, 178, 182, 314, 315, 325
   engineering group, 220
   evaluation, 33, 42
   field data, 234
   functions, 270
   information, 234, 325
   information system, 232
   labor reporting, 291
   management information system, 209
   management system, 237
   manager, 280
   objective, 5, 325
   organization, 21
   performance, 13, 33
   personnel, 136, 205
   planner, 97, 101, 102, 170, 220
   planning, 72, 96
   policies, 7
   preparation, 170
   productivity, 45
   program, 7, 8, 29, 32, 77, 101, 116, 119, 123
   resources, 14, 161
   superintendent, 3, 34, 37, 101, 344
   supervisor, 98, 99, 100, 136, 137
   support group, 70
   system conceptualization, 236
   team, 79
   work control overview, 212
   work force, 98
   work order, 261, 318
   work order status, 320
   work request, 261
   work load, 9
major
   components, 131, 163
   repairs, 8
   work, 101
make or buy actions, 102
management, 33, 100
management principles, 88

Index

managerial
   functions, 26
   processes, 88
   skills, 100
managers guidelines, 239
managing unscheduled work, 290
manpower, 147
   availability, 174
   for the total program, 153
   needs, 293
   required for all PM, 152
   used, 147
manufacturer's warranty, 157
master mechanic, 220
material, 197
   control, 9, 57, 197, 201, 202, 203, 204, 326, 334
   department, 197
   function, 197
   information, 200
   personnel, 204
   coordinators, 72
   cost data, 235
   data, 235
   identification, 200
   procurement, 203
mathematics of the backlog, 303
Maynard operation sequence, 191
mean time before failure (MTBF), 164
measure
   performance, 17
   productivity, 37
   program effectiveness, 149
   maintenance work loads, 94
mechanical craftsmen, 248
medical problems, 94
meeting chairman, 169
merging systems, 309
Methods time measurement, 191
mid level supervision, 22

minor repairs, 90
MIS, 251
mission, 4, 20
mobile
   cranes, 196, 221
   equipment, 217
   equipment components, 163
   equipment maintenance, 218
   equipment maintenance organization, 219
   equipment maintenance program, 217
   PM, 282
modem, 225
modification, 6, 42, 273
monitor
   accomplishments, 253
   development progress, 245
   equipment condition, 136
   job execution, 162
   services, 149
   system use, 257
   training, 359
monitoring maintenance performance, 3
Motion time analysis, 191
multiskill, 109, 110, 111, 112
mutually supporting objectives, 239
MWO, 264, 266, 267, 277, 269, 278
MWR, 262, 267, 277, 278

N

negotiate with operations, 175
network, 223, 224
new
   installations, 180
   maintenance mission, 111
   organizations, 107
   technology, 98, 109
   testing techniques, 214

work, 126, 158
equipment, 107
nomograph, 190
non destructive testing, 102, 141
non maintenance projects, 10, 207
non maintenance work, 10, 13, 38, 96, 206, 207, 273, 276
non repetitive, 187
normal versus an abnormal condition, 144
number of supervisors, 295
numbering equipment, 270

O

object, 272
objective, 2, 6, 12, 21, 55, 86, 169
objectives of preventive maintenance, 154
obtain schedule approval, 175
obtain stock materials, 136
older craftsmen, 108
older supervisors, 116
on board computers, 108, 111, 219, 220, 223
on screen
    diagram, 222
    reporting, 243
on site material delivery, 138
on the job training, 112
open MWO, 268
open week, 274
operating
    costs, 231
    delays, 138
    hours, 131
    the program, 148
operations, 33
    approval, 138
    delays, 161
    personnel, 205

supervisors, 100, 129, 139
maintenance partnership, 327
maintenance scheduling, 96
operations maintenance team, 225
operators, 139
opportunity for planning more work, 143
optimizing maintenance engineering, 368
organization, 25, 54, 61, 118, 325
    alignments, 13
    change, 11, 83, 90, 25, 73, 88, 89
    chart, 21
    concepts, 86
    goals, 22
    performance, 46
organizing, 23, 148
organizing the program, 148
overall maintenance program, 80, 131, 134, 163
overall performance, 252
overdue WO, 302
overhaul, 14, 42, 181, 218
overhead conveying systems, 225
overhead cranes, 196
overtime, 47, 110
ownership, 134

P

package programs, 241, 242, 243, 246
paper work control system, 84, 249
partially integrated system, 236
participation, 85, 86, 371
parts identification, 306
parts manufacturing, 308
passive resistance, 84
payroll, 244
peer coaching, 112
percent of failures, 271
performance, 46, 122, 170, 326
    benchmark, 96

Index

evaluations, 122
goals, 122
indices, 1, 8, 47, 121, 231, 139, 251
information, 311, 313
measurements, 231
picture, 339
periodic maintenance, 137, 284
performance standards, 113
personal
   computer, 248
   pool, 94
   qualifications, 22
   react to changes, 83
persuasion, 85
phase out all procedures, 254
philosophy of the organization, 87
physical audit, 53, 353
physical audit with questionnaire, 53, 355
plan
   achieving profitability, 239
   information use, 252
   major jobs, 14
planned
   maintenance, 155
   reductions, 95
   work, 146
planner, 101, 168, 280
planning, 23, 56, 137, 155
   activity, 101
   and scheduling, 8, 132
   criteria, 221
   performance, 286, 288
   procedures, 8
   steps, 133, 158
plant
   configurations, 13
   environment, 163
   shutdown, 117
PM
   compliance reporting, 153
   inspection, 165
   module, 243
   program, 130
   services, 80, 141, 282
   work load, 292
policies, 10, 56
policy for evaluations, 348
pool craftsmen, 222
power
   structure, 21
   tools, 203
practice reporting, 250
predetermined time standards, 187, 190
pre ordered, 156
predicting wear out, 215
predictive maintenance, 102, 141, 136, 144, 177, 179, 210
preliminary
   form, 134
   plan, 138, 284
   program, 125
   schedule, 161, 171, 172, 173
preparation time, 189
prepare for the meeting, 172
presentation techniques, 31
preventive and predictive maintenance, 141
preventive maintenance, 13, 56, 127, 144, 151, 177, 179, 273, 325
preventive maintenance program, 9, 148
preventive maintenance services, 78
primary
   information system, 96
   maintenance, 6
   source of field labor data, 235
principal duties, 100
principles
   of maintenance management, 11
   of organization, 19, 20

priorities applied, 276
priority, 160, 171
priority rating scheme, 272, 273
    maintenance, 275
    non maintenance, 276
pro active maintenance, 177, 178
production
    data, 220
    managers, 251
    personnel, 251
    standards, 190
    strategy, 1, 4, 5, 12, 28, 78, 100, 239
    supervisors, 251
    targets, 11, 239, 274
productivity, 20, 43, 56
    and performance, 155
    versus manpower savings, 45
professional consultants, 113
professional consulting assistance, 245, 246
profitability, 1, 46, 112
profits, 92
program definition, 12, 30, 31, 36, 54, 124, 125, 133, 134
project
    control, 207, 208
    engineer, 320
    engineering work order, 264
    manager, 264
    plan, 319
    status, 208
    work, 205, 276
    work load reductions, 93
promotion, 99
proper
    backlog level, 300
    operating condition, 143
    size (work force), 9
publicize evaluation, 52, 351
publish procedures, 7
purchase
    order, 207, 264, 266, 268,
    order status, 220
    order tracking system, 160
    requisitions, 203
purchasing, 2, 16, 101, 168, 203, 244
    agents, 249
    information, 307
purpose of evaluation, 345

Q

qualification to respond to questionnaire, 354
quality
    assurance, 183
    control, 225
    material support, 16
    of information, 250
    of maintenance information, 234
    of parts, 102
    planning, 156
    standards, 184
    training, 247
    interest (in), 312
    standard, 184, 185
query repair history, 252
questionnaire, 53, 353
quick fixes, 224

R

random sampling, 336
range of time estimates, 187
re organization, 89
ready status, 223
real time recording of events, 225
realistic plan of jobs, 161
rebuild, 42
rebuilt correctly, 178
receipt of deficiencies, 150
recommended job plans, 252
recruit the new team, 78

Index

reduce resistance to change, 85
reduce the warehouse staff, 204
reducing
    cost, 154
    work force, 93
    work load, 96
reference lists, 232
refresher training, 180
regular work force, 95
reliability, 177
    based maintenance, 154
    engineering, 141, 220
    of equipment, 214
    technician, 112
relocation, 273
reorganization, 59, 88
repair history, 102, 120, 139, 230, 164, 230, 315
    data, 164
    information, 181
repair techniques, 214
repetitive
    functions, 262
    tasks, 103, 186
replace equipment, 231
replacement minor components (PM), 144
replacing major components, 126
reporting and failure, 146
reporting field data, 249
reports
    Absentee Summary, 297
    Backlog Summary, 299
    Cost by type/component, 316
    EWO, 319
    Fleet cost, 316
    Labor Utilization, 296
    Mobile PM, 282
    MWO/EWO, 320 - 321
    Overtime Summary, 296
    Repair History, 315

    WO Status, 285 - 287
requesting work, 195
require no formal planning, 137
reserve stock parts, 203
resistance
    testing, 144
    to change, 85
resource use, 14
response to emergencies, 99
responsibility, 86
    for poor performance, 355
    of tool repair, 203
responsive organization, 13
restructured, 89
review deficiencies, 130
rigging, 138, 157, 196, 203, 221
road test, 225
ROI - return on investment, 276
role
    of operations, 121
    of key personnel, 253
routine maintenance, 273
    preventive maintenance, 144
    services, 14
    repetitive activities, 13, 141
running condition, 144

S

salaried
    employees, 79
    positions, 115
    status, 79
salary level, 115
schedule compliance, 170, 288
schedule
    major jobs, 14
    the evaluation, 52
    the maintenance evaluation, 350
scheduled
    event, 165
    maintenance, 272

maintenance work load, 293
scheduling
   work, 169
   meeting, 138, 174
   procedure, 151, 152
schematic diagram, 31, 134, 140, 239
scope
   of services, 14
   of job, 133
section manager, 68, 71
security, 85
seek responsibility, 120
selection from an table file, 250
self
   directed team, 61, 77, 79, 82
   study courses, 112
senior managers, 125
sensors, 144
serial number, 286
service
   activities, 196
   due, 127, 128
   organization, 27
sharing information, 311
shift coverage, 295
shock pulse, 144
shop
   activities, 195
   clean up, 273
   commercial, 308
   control, 194
   drawings, 195
   materials, 195
   operations, 193
   performance, 195
   personnel, 137
   support, 138
   work, 160, 173, 194
   work completion, 160
shops, 16
shutdown schedule, 169
shutdown times, 173
simple work order element, 264
skill training, 80
skilled and unskilled personnel, 94
skills
   certification, 113
   of shop personnel, 195
solve material control problems, 199
sonic testing, 144
soundness of program, 253
source of information, 87
sources of new work, 124
special equipment, 157
spread sheet, 188, 189
staff
   departments, 27, 33, 35, 101, 119, 121, 333
   personnel, 30
   staff specialists, 88
   the organization, 22
   training, 115
staffing, 23
stand alone system, 237
standard
   bill of materials, 103, 131, 137, 163, 165, 217, 272, 308
   of quality, 157
standard
   construct, 272
   setting techniques, 186
   task list, 131, 163
standardize equipment, 231
standardizing
   jobs, 272
   maintenance terminology, 362
standards, 14, 102, 165, 168, 217, 179, 183, 188
standards program, 186
standing work order, 261, 262
static services, 128
status of cost and performance, 139
status of major jobs, 120, 252

Index

stock, 160
   issue, 266
   issue card, 134
   issue documents, 133
stocked parts, 200
strategies of multiskill use, 110
strategy for evaluations, 51, 347
structure of people, 20
successful PM program, 143, 153
successfully implement a team, 77
superintendent, 97, 101, 125
supervision, 55, 100, 231
supervisor, 97, 168, 281
   selection, 100
   supervisors, 97, 116
   trainer, 250
supervisors material needs, 203
supplemental objectives, 6
support services, 193, 194
support maintenance program, 136
supporting engineering projects, 369
SWO, 267
system
   acceptance, 247
   compatibility, 239
   designer, 228, 237
   developer, 232, 233
   integration, 236
   programming, 234
   testing, 251

T

task area:, 189
task list, 14, 103, 167, 200, 217
task time, 189
task time standard, 188
team, 36, 75, 116, 125, 165
   approach hastened by downsize, 331
   coordinator, 103, 152
   decision making, 79
   development, 110
   environment, 105
   firm work control, 80
   members, 77, 79, 82, 97, 103, 152
   operation, 107
   organization, 15, 63, 75, 100, 116, 119, 129, 152
   probably won't succeed, 76
teams, 20, 80, 84, 202
technical
   complexity of equipment, 214
   repair skills, 214
technically-oriented supervisor, 100
technological advances, 116
technology advances, 108
terminology, 8, 12, 41, 141
test areas, 90
test data, 251
testing and calibration, 144
time card, 133, 134
time to completion, 276
time values for job preparation, 189
timing of future replacements, 120
tool list, 14, 103, 131, 137, 163, 185, 200, 217
top level supervision, 22
total
   communications and management system, 216
   plant organization, 67
   mining operation, 117
   total productive maintenance, 9, 16, 20, 40, 121, 329
TPM, 60, 84
trade groups, 251
traditional, 25, 61
   craft training, 105
   maintenance job roles, 111
   maintenance organization, 97
   managerial functions:, 23
   organization, 22, 61, 78

repair skills, 114
supervisor, 78
training, 117, 180
　administration, 106
　and recruiting, 88
　procedure, 250
　results, 106
　sessions, 134
　supervisors, staff personnel, 17
travel time, 191
travel zones in the plant, 190
troubleshooting, 112
type of work, 264
types of standards, 184

U

ultrasonic testing, 144
uncover deficiencies, 136
union
　business, 101
　officials, 89
unique major jobs, 133, 138
Universal maintenance standards, 191
unscheduled, emergency repairs, 137
unscheduled jobs, 229
unscheduled repairs, 272
unused stock materials, 8
updating
　basic skills, 180
　of existing skills, 221
use
　information, 8
　labor to install materials, 16
　of labor, 138
　of manpower, 145
　of resources, 169
　of schematic diagrams, 125
　of standard titles, 272
　of standards, 183, 185
　of time, 25
using
　information systems effectively, 367
　the backlog effectively, 298
　the computer, 139
　the planner, 251
　verbal orders, 289
utility operation, 196
utilization, 210
　of labor, 139
　of maintenance, 11

V

vacation planning, 101
variable intervals, 142, 151
vendor training, 251
verbal orders, 263, 266
verify
　field data sources, 253
　manpower, 174
　most data, 249
　network and hardware, 253
　hardware and networking, 255
vibration analysis, 136, 144
video based programs, 112
visual inspections, 144
vital
　maintenance function, 16
　maintenance information, 227
volume of jobs versus cost, 265

W

warehouse, 8, 16, 35, 203
warehouse tool room, 203
warehouseman, 249, 250
warehousing, 168
warehousing, 2, 111
weekly
　accomplishments, 252
　cycle, 158

Index

    schedule, 121, 128, 169, 277
    scheduling meeting, 161, 170, 176, 274
window of opportunity, 274
work
    attitude, 66
    completion, 134, 136
    completion time, 273 - 274
    control, 20, 80, 124
    control procedures, 222
    cultures, 110
    description, 272
    factor, 191
work force, 9
    balance, 95
    level, 95, 102
    reduction, 92, 94, 96, 108
    size and composition, 15, 229
    work load, 37, 38, 92, 102
work methods and procedures, 100
work order
    administration, 287
    element, 264, 277
    numbers, 235
    status, 162, 230, 285
    status reports, 160
    system, 8, 13, 124, 218
    system elements, 260
    systems, 57, 111
work requests, 266
work sampling, 186, 231
worker centered learning , 113
working environment, 4
working relationship operations, 81
work load, 15, 108, 292
    computation, 292 - 294
    variations, 95
world class maintenance, 59

Y

younger workers, 108
younger supervisors, 116